Plumbing 1

Plumbing 1 and *2* have been published simultaneously to meet the needs of the revised syllabus of City and Guilds subject 617 (Plumbers' Work) and the changeover to metric units.

The author's well-known previous books, *Plumbing First Year* and *Plumbing Second Year*, were intended to be part of a four volume series to cover each year of the course. Although *Plumbing 1* and *2* are based substantially on these books, the subject matter has been thoroughly revised and extended to employ the metric system and to cover the total area of the craft certificate course.

A. L. Townsend M.I.P., M.R.S.H.
Formerly Senior Lecturer in Building Services at the School of Architecture, Oxford

Plumbing I

Stanley Thornes (Publishers) Ltd

Originally published in 1962 by Hutchinson Education
First published as *Plumbing — First Year* 1962
Reprinted 1966
Second edition 1969
Reprinted 1970, 1974, 1976, 1977, 1979, 1980, 1981, 1982, 1984, 1987, 1988

Reprinted 1990 by
Stanley Thornes (Publishers) Ltd
Old Station Drive
Leckhampton
CHELTENHAM GL53 0DN

ISBN 0 7487 0315 2

Printed and bound in Great Britain
by Courier International Ltd, Tiptree, Essex

Contents

APPLICATION

The author and publisher are grateful to the Nuralite Company Limited, Whitehalf Place, Gravesend, Kent, for the generous provision of materials and advice for Chapter 18, 'Non-metallic roof weathering.'

City and Guilds Craft Theory Syllabus

1 Drawings and calculations

The following items should be introduced at appropriate points throughout the syllabus: they should not be dealt with in isolation. All examples should relate to craft practice.

The use of decimals and fractions. Percentages. The use of simple formulae. Mensuration of quadrilaterals, triangles, and circles. The properties of the right-angled triangle. Development of surfaces of pipe intersections. The setting out and lettering of drawings. Use of scales. Rule assisted proportionate sketching. Simple projections: plans, elevations and sections of craft details.

2 Materials

Application to and suitability for use in plumbers' work of the metal and plastics in common use. Their simple physical properties. Materials used as protective coatings to pipes, fittings and sanitary appliances.

3 Alloys, solders and fluxes

The composition of alloys such as brass and gunmetal. The composition and properties of lead—tin solders used in plumber's work. Working temperature range of plumbers' solder. Purification of plumbers' solder. Bronze welding, silver soldering and low temperature brazing alloys. The application of fluxes and reasons for their use.

4 Pipes, tubes and rainwater goods

Selection of pipes and tubes used in plumbers' work for hot and cold water supplies, town gas, flush, overflow, waste, soil and ventilating pipes, drainage and domestic central heating—lead, copper, steel, cast iron, asbestos cement, pitch fibre, plastics. Their commercial sizes, weights and gauges. Methods of jointing, manipulation and fixing. Capillarity as applied to capillary joints. Methods of jointing, supporting and fixing cast iron, steel, asbestos cement and p.v.c. rainwater goods.

5 *Sheet weatherings*

Roof weatherings and coverings in sheet lead, copper, aluminium and non-metallic flexible sheets (other than bituminous felt). Coverings to small flats, canopies, gutter and valleys, soakers, flushings, weatherings to chimneys, mansard roofs, dormers, skylights, lantern lights and roof glazing. Condensation and secret gutters. Methods of jointing and fabrication including solid rolls, drips, standing seams, single and double lock welts and laminated joints. Finish at eaves, verges, ridges. Methods of fixing. Capillary and its prevention.

The preparation of building surfaces to receive sheet weatherings. The weights and gauges, grades and temper of sheet materials used for various positions in roof work. Methods of allowing for thermal movement; dimensional limitations of materials. Deterioration and corrosion of sheet weatherings. Electrolytic action. The use of copper and lead as damp-proof courses to solid and hollow walls.

6 *Cold water supply*

Properties of water from common sources of supply; their effect on various materials with which water comes in contact. Hard and soft waters: permanent and temporary hardness. Plumbosolvency and cuprosolvency. Effects on pipes of corrosive soils and building materials. Connections to water mains; laying of and protection to service pipes. Domestic systems of cold water supply. Connections to showers and mixer fittings. Cold water storage; cisterns of metal and plastics; correct methods of fixing. Siting and protection of cisterns. Capacities of storage cisterns for domestic supplies.

Principles of working of bib and pillar taps, stopvalves, gate valves, plug cocks and drainage taps. Portsmouth and diaphragm ballvalves. High and low pressure ballvalves. The principles of the lever.

The expansion of water on freezing. Maximum density of water. Frost bursts. Protection of service pipes and water installations against frost. Suitable insulating materials. Causes and remedies of noises in cold water systems; water hammer. Liquid pressure; pressure per unit area due to head of water.

7 Domestic hot water supply and heating

The effect of heat; expansion of solids, liquids and gases due to changes of temperature. Transmission of heat by conduction convection and radiation. The expansion of water on heating. Circulation of water in pipes and boilers.

The cylinder system of domestic hot water supply. Use of secondary circulation. Indirect systems of domestic hot water supply (excluding the use of steam supplies). Reasons for use of indirect systems. Use of and connection to horizontal cylinders. Connections to towel rails, drying coils, shower baths and mixer fittings. Fixing of back boilers and independent boilers for hot water supplies and domestic heating. Types and positioning of immersion heaters and associated thermostats. Installation of gas-fired auxiliary circulators. Precautionary measures for the prevention of explosion of boilers, collapse of cylinders and loss of heat. Safety valves. Air locks in pipes. Air release pipes. Causes of noises in hot water systems.

The measurement of temperature; thermometers, Centrigrade and Fahrenheit scales, including conversions. Quantity of heat; The British thermal unit, calorie and therm. The calculation of capacities of pipes rectangular tanks and cisterns, and cylinders with flat ends. Electrolytic action.

8 Sanitation

Description of domestic baths, wash basins, sinks, wash-down and siphonic water closets, bidets, W.C. cisterns, urinals and automatic flushing cisterns and their fixing. Principal dimensions of common sanitary appliances and capacities of flushing cisterns.

Atmospheric pressure; the action of the siphon. Traps to sanitary appliances; depth of seal; the unsealing of traps by evaporation, induced siphonage, self-siphonage and momentum. Methods of prevention of loss of seal. The arrangement of soil, waste and ventilating pipes for two-pipe, fully ventilated one-pipe and single-stack system of sanitation in buildings, not exceeding five storeys. Sizes of soil, waste, ventilating and trap-ventilating pipes. Connections to drains. Testing of soil and waste pipe systems; function and use of the manometer. Natural lighting and ventilation of W.C. apartments.

9 Drainage

The layout of domestic drainage systems. Separate and combined systems. Support and protection of drains; ventilation; access.

Fittings for domestic drainage including gullies and intercepting traps. Testing of drains.

10 Gas supply and appliances

Gas service pipes; use of siphons; connections to meters. The measurement of gas pressures. The principles of the luminous flame and Bunsen burners. The common products of combustion. The fixing and ventilation of gas fires, domestic water heaters and boilers. Thermostatic devices. Construction of flues to gas appliances with special reference to condensation and back draught. Room-sealed gas appliances; balanced flues.

11 Lead and hard metal welding

Working knowledge of apparatus for oxy-coal gas and high pressure oxy-acetylene lead and hard metal welding. Safety precautions to be observed in assembly and use of equipment. The preparation of sheet lead and pipes for lead welding. Simple forms of bronze welded, silver soldered and low temperature braze joints in copper tubes. Preparation and set up of simple welded joints in mild steel sheet. Work hardening of metals. Hardening, tempering and annealing, as applied to plumbing processes.

Introduction

Plumbing services are of vital interest to everyone. Imagine your own home without its water supply, W.C., basin, bath, sink and drains. Then just try to visualise the dismal discomfort that would be your lot.

By law, a habitable building must be weather-proof, have a pure water supply, correctly fixed sanitary appliances, and drains. The plumber is concerned with all these important requirements. He also deals with hot-water supplies, and the warming of buildings, particularly domestic dwellings, by central heating.

The history of the trade is a long one. It began in this country about two thousand years ago, with the Roman invasion. The Romans arrived to find unfriendly tribes and none of the living comforts—villas with central heating, water supplies and drainage—to which they were accustomed. Even in these early times the Roman plumber—a skilled worker in *plumbum*, or lead—had an important place in the community. He was quickly called upon to provide these health- and comfort-giving services, without which the Romans found life unthinkable. Your technical college or municipal library will have books which show how well these Roman plumbers did this work during the four hundred years of occupation. Their work was simple, even crude by modern standards, but it was effective. Many examples of it are still to be seen in museums, and are well worth looking at.

Eventually the Roman legions were withdrawn from Britain and the skill of the plumber was almost forgotten for many centuries. By about 1400, however, the population had grown to some three million, transport had improved a little, trading began to thrive, and people came to live in organised groups or townships. Skilled lead-workers again found their work in demand, for weathering buildings in the new towns, for making water-cisterns and for various other purposes, such as making weights for the merchants' scales.

Plumbing craft guilds were set up to protect the plumber's interests and to govern his trade. Among other things, the Guilds ruled that no plumber would be allowed to work in the trade unless certified 'by the best men in the trade that he knows how well and lawfully to do his work'. Thus only trained plumbers were employed, and severe penalties were imposed on any who did work of a poor quality. In this way the good name of the trade was upheld, and those who used the plumber's services were protected from the evil results of bad work by unqualified men.

The Worshipful Company of Plumbers is one of the many ancient Craft Guilds of the City of London. Its right to make rules for the government of the trade was approved by Edward III in 1365. Further rules, or ordinances, were approved in 1488 and 1520, and in 1588 the Company was honoured with a Grant of Arms by Elizabeth I. Its crest, still in use today, depicts St Michael the Archangel, the patron saint of plumbing. In 1611 the Company was granted a Royal Charter by James I.

These marks of royal approval show the regard in which the plumber and his work were then held. Today, with the health of over fifty million people to be safeguarded in Britain, the importance of the plumber is greater than ever. The Worshipful Company continues to influence the right and proper conduct of the modern trade.

The Institute of Plumbing (acting with the Worshipful Company) is a modern outcome of the Company's work. The British Medical Congress of 1883 expressed grave concern at the inferior plumbing which resulted from the work being done by men with little or no training. This state of affairs had been brought about by social and industrial changes, over which the Guilds had no control particularly at the time of the

Industrial Revolution. In the following year the subject was again discussed at the International Health Congress in London, which recommended that, in the public interest, some sort of register of really competent plumbers should be made. The recommendation was put into effect in 1886, when the Worshipful Company set up a voluntary scheme of registration for plumbers able to show that they were well trained and competent. The scheme has continued ever since, and in 1969 its administration was handed over to the Registration Council of the Institute of Plumbing.

One purpose of the Institute may be summarised as follows: 'To maintain a high standard of plumbing work, and thereby to improve the quality and efficiency of that work'. In this way it aims to raise the status of the competent plumber, and to ensure that only well-trained men are employed in the work, thus safeguarding the future of the trade. You will notice how close these aims are to those of the ancient Craft Guild.

The work of the Institute and its large and growing membership has been increasingly recognised by architects, doctors and the ordinary householder. Its general office is at Scottish Mutual House, North Street, Hornchurch, Essex, from which further information about it can be obtained.

The City and Guilds of London Institute is another example of the influence of the Crafts Guilds, this time aimed at the provision of better technical training for those working in the trades. The poor standard of work in the 1880's troubled them as much as it did doctors of the time. Together with the Mayor and Corporation of the City of London, they recommended the setting up of a body to be responsible for the development of technical training courses for the trades. In 1887 the Institute was established, and since then the training of countless plumbers has been guided by its syllabuses and courses. The offices of the Institute are in Portland Place, London W.1, where it constantly reviews all technical changes in the trades, and revises training schemes and syllabuses accordingly. It also prepares and conducts examinations for those who have reached a satisfactory standard.

A major revision of the plumbing craft syllabus has recently been completed and is now in operation. A Craft Certificate, issued to all successful candidates in the Institute's examination at the end of the three-year course, indicates that the holder is well trained in the principles and practice of the trade and fitted to work as a plumber. The Advanced Craft Certificate, for which an examination is held after two more years of technical training, marks its holder as a top-grade plumber, and every student should try to attain this standard.

An examination syllabus published by the C.G.L.I. appears on pages ix–xii. It is included to give you some idea of the plumbing topics on which examinations might be set at the end of the Craft Certificate course.

The Institute of Plumbing also aims to promote good standards of plumbing design, administration and practice. It publishes technical booklets, which deal with drainage; soil, waste and ventilating pipework; hot- and cold-water installations; and so on. These and other information about the work of the Institute may be obtained from its office at Scottish Mutual House, North Street, Hornchurch, Essex.

The National Association of Plumbing Teachers devotes much time to seeing that student·plumbers can make the best use of the many facilities provided by the technical colleges.

The Plumbing Trades Union is actively concerned with conditions of employment. The P.T.U. looks after the interests of the apprentice and journeyman plumber in many ways, concerning itself, for example, with wage rates, hours of work and technical training. Further information may be obtained from the General Secretary, 15 Abbeville Road, Clapton, London, S.W.4.

The British Standards Institution has contributed greatly to the efficiency of the plumbing trade by standardising the quality of materials and the dimensions of appliances. The Institution also publishes *British Standards* and *British Standard Codes of Practice*, which are compiled by experts and explain the recommended ways of working. A list of publications can be obtained free from the Institution's office at 2 Park Street, London, W.1, and, of course, they will be in the college library.

The Building Research Station at Watford, Herts, does valuable research into modern plumbing problems. A notable example of its work is the *single-stock system*, which, if fixed in the way the Station advises, is cheaper but no less efficient way of installing soil and waste pipework. The development of a new ball-valve for cisterns, and special types of spray taps which have been found more efficient and economical, are other results of the application of scientific thought to plumbing work.

The Material Development Associations, particularly those for lead sheet and pipe, copper, zinc, and aluminium, all work to improve the quality of materials and their application to plumbing work. The addresses of these bodies may be obtained from the trade journals, which again are a valuable source of plumbing information. These may be seen in your technical college or municipal library, if indeed you do not get them for yourself.

Bye-laws are enforceable by law, and anyone proved guilty of breaking one is liable to a fine. It is a bye-law that lights should be carried on vehicles after lighting-up time. This is clearly sensible, and necessary for the safety of all road users. In the same way Building, Sanitary and Water Undertakings' bye-laws protect both plumber and householder, and it is only sensible to know and abide by them.

Bye-laws prescribe the materials to be used in plumbing work, and the way in which they should be fixed and tested. More will be said about them later in the book.

Plumbing is a progressive trade. It is proud of its fine traditions, but it is fully alive to the needs of the present, and quickly absorbs the new ideas in materials and methods which modern science can offer. Up-to-date ideas in plumbing design and installation help to make our work better and more efficient. Of course, new ideas can be irksome. Sometimes it is necessary for one to re-learn an old skill in order to make use of helpful new methods; and the fact that some new materials need different techniques means that one may have to learn a brand-new skill. More and more, however, the application of scientific thought to plumbing adds interest to the work.

All these new ideas tend to complicate the technical as well as the practical parts of the plumber's training and work. Nevertheless, you will find the technical side just as interesting, and no more difficult to learn, than the practical side, particularly if a little time is spent in discovering the simple principles on which all plumbing is based.

Principles

1

The international metric system (S.I.)

The United Kingdom is committed to a change from the Imperial to the metric system of mensuration and the S.I. system is to be adopted.

A phased programme has been prepared and it is expected that by 1972 all measurements and calculations will be worked in metric.

Meanwhile, a vast amount of work has to be undertaken in consultation with manufacturers and other interested bodies before all relevant British Standards and Codes of Practice are revised to new metric dimensions. This involves more than simple conversion of existing dimensions to metric equivalents because in many cases the existing dimensions will have to be altered to comply with commonly used sizes in continental countries already established in metric usage.

Eventually, the Building Regulations and Water Bye-laws will have to be metricated. Until they are, and this cannot be until all British Standards and Codes of Practice have been metricated, Imperial dimensions contained therein will have to be converted to exact metric equivalents.

The examinations of the City and Guilds of London Institute will be set in metric terms by 1970 and the aim of this chapter and the metrication of this book is to offer some explanation of the implication of going metric, in so far as information is available at the time of going to press, and give practice in metric usage. (See also worked examples in Chapter 30).

With so many unresolved factors, some of which are outlined below, the ultimate metrication of all plumbing materials as far as design, dimensions and calculations are concerned, must be left until a firm lead is given by those responsible for the manufacture of materials and equipment, by those who compile the British Standards and Codes of Practice, and eventually by the legislators who prepare Regulations and Byelaws.

Pipe sizes

1 *Lead* So far no decision has been made and consultation with representatives of the Water Board will have to take place before metrication of lead pipes can be dealt with.

2 *Copper* Consultation is proceeding at international level (the International Standards Organisation). This may lead to the early metrication of the present BS 659, 1967, BS 1386 and BS 3931. When these new standards become available they should be consulted. It is likely that metric-sized tubes will be specified by their outside diameters (o.d.). In this case present sizes, for example $\frac{1}{2}$ in, $\frac{3}{4}$ in, 1 in, $1\frac{1}{4}$ in, $1\frac{1}{2}$ in and 2 in, would become 15 mm, 22 mm, 28 mm, 35 mm, 42 mm and 54 mm, all o.d., respectively.

However, this is not finalised and so, in this book, nominal bore sizes are shown in Imperial nominal bore sizes together with nominal metric equivalents. Thus, $\frac{1}{2}$ in, $\frac{3}{4}$ in, 1 in, $1\frac{1}{4}$ in, $1\frac{1}{2}$ in, and 2 in are shown as 13 mm, 19 mm, 25 mm, 32 mm, 38 mm, and 50 mm. In one or two cases where o.d. sizes are quoted, these are clearly indicated as o.d. dimensions and will be found as outlined in the previous paragraph.

3 *Mild Steel Pipes* have been metricated and are completely interchangeable with those described in BS 1387 and are in fact the exact metric equivalent of them. They are specified by nominal internal pipe diameters, or bores, as follows: $\frac{1}{4}$ in–8 mm, $\frac{3}{8}$ in–10 mm, $\frac{1}{2}$ in–15 mm, $\frac{3}{4}$ in–20 mm, 1 in–25 mm, $1\frac{1}{4}$ in–32 mm, $1\frac{1}{2}$ in–40 mm, 2 in–50 mm, and, to give an example of one larger size, 4 in–100 mm.

4 *Stainless steel* tubes which use either solder capillary or Type A compression fittings, will have to follow dimensional tolerances as described for copper tube, when these are finally resolved. Incidentally, the 'Kingley' fitting which is a Type B manipulative fitting, can be used to advantage on S.S. tube.

5 *Plastic tubes* in polythene will be designated by o.d. dimensions when BS 1972 is revised. Wall thickness will vary according to the working pressures to which the pipe will be subject. Since the outside diameters of the tube will be the same throughout the pressure ranges indicated by Classes A, B, C, and D, the bores will diminish slightly as the wall thicknesses increase. It is likely that nominal bore size will also be given in *Imperial* dimensions, although o.d.s and working pressures will be in metric.

A few typical examples: Nominal sizes, given in imperial dimensions, ½ in, ¾ in, 1 in, have o.d.s of 21·5 mm, 26·9 mm, 33·7 mm, respectively.

Pressures will be indicated in *bars* and in metres head. One *bar* = 100,000 N/m^2 or 100 kN/m^2, which is very nearly the same as atmospheric pressure commonly stated at 14·7 lbf/in^2 or, in metric, 101·3 kN/m^2. (See later note on *Pressures.*)

6 *P.V.C. tubes* will probably follow the pattern outlined above for polythene.

7 *P.V.C. soil-waste pipe systems* have long been available in metric sizes, the nominal 4 in being 110 mm o.d.

Now, Building Regulation N 4, among other things, states that 'any soil pipe, waste pipe or ventilating pipe shall be of adequate size for its purpose, but in no case shall the internal diameter of a soil pipe or waste pipe be less than the internal diameter of the outlet of any appliance which discharges into it . . . .' The internal diameter of soil pipes shall not be less than

(a) 2 inches (50·8 mm), if it exclusively serves one or more urinals, or
(b) 3 inches (76·2 mm) in any other case.

Taking note of this (in particular the prescription (b) above, which allows for 3 in diameter soil pipes *if* 3 in W.C. outgo-diameters are used to satisfy the earlier prescription—soil pipe diameter to be as large as largest connection to it) one enterprising p.v.c. soil pipe manufacturer in conjunction with a pottery firm which is producing washdown pans with 3 in diameter bore outgoes, has marketed an 82·5 mm soil pipe which satisfies the Regulations minimum 3 in soil pipe bore diameter. Pipes of this size are shown capable of taking discharges likely to occur in a two-storey house.

3 in to 4 in adaptors are available to connect this neater and cheaper soil-waste system to the 4 in (101·6 mm or 100 mm nominal) soil drain which is the least diameter allowed by the Regulations.

Unfortunately, the possible use of 3 in soil discharge pipes has not met with universal approval by all who administer the Building Regulations and this will delay this innovation coming into common use.

Sheet lead will be specified in mm thickness instead of lb/ft² as hitherto. Code numbers for lead thickness will tally with previous lbs/ft². *See footnote.*

Sheet copper thickness will be specified in mm. Present thicknesses will be rounded off to an exact mm size, *not* converted from imperial thicknesses to metric. Consult the new BS 1569 as soon as this becomes available.

Sheet aluminium thickness will be specified in mm, as described for copper sheet. However, for a time *both* the present imperial standards and the new metric one for aluminium will operate and the industry will offer materials with both ranges of standards.

Sheet zinc thickness will be expressed in mm. 12 and 14 zinc gauge materials will become 0·6 mm and 0·8 mm thick materials respectively.

For a long time, tables have been published giving *conversion factors* to enable changes from imperial to metric and, whilst these may be helpful in the early stages there should be no

B.S. 1178 Code N°	SI Unit Kg/m²	Imperial Unit
4	20·41	4lb/ft²
5	25·36	5lb/ft²
6	28·36	6lb/ft²

undue reliance on such conversions. The sooner you can become accustomed to thinking in metric terms the better you will adapt to the new system. An abridged conversion table is included in this chapter so that readers may make conversions or verify a metric calculation example by conversion back to imperial values—that is, until such time as confidence in metric working is obtained.

CONVERSION FACTORS

	Units			
Item	*Metric*	*Imperial*	*Metric to imperial*	*Imperial to metric*
Length	metre (m)	foot (ft)	3·28 ft	0·305 m
	millimetre (mm)	inch (in)	0·039 in	25·4 m
Mass	kilogram (kg)	pound (lb)	2·2 lb	0·454 kg
Area	square metre (m^2)	square foot (ft^2)	10·76 ft^2	0·093 m^2
	square millimetre (mm^2)	square inch (in^2)	0·015 in^2	645·1 mm^2
Volume	cubic metre (m^3)	cubic foot (ft^3)	35·3 ft^3	0·028 m^3
Capacity	litre (l)	gallon (gal)	0·22 gal	4·54 l
Pressure	newton/square metre (N/m^2)	lbf/square inch (lbf/in^2)	0·000145	6894 N/m^2 or 6·894 kN/m^2
Density $\left(\frac{\text{Mass}}{\text{Volume}}\right)$	kilogram/cubic metre (kg/m^3)	(lb/ft^3)	0·062 lb/ft^3	16·02 kg/m^3

Application. To convert metric to imperial, multiply metric values by the conversion factor.

EXAMPLE

1 metre to feet = 1 metre × 3·28 = *3·28* ft

To convert imperial to metric, multiply imperial value by the conversion factor

EXAMPLE

1 gallon to litres = 1 gal × 4·54 = *4·54* litres

Note 1 N/m^2 = only 0·000145 lbf/in^2. Pressure measurements will therefore be more commonly expressed in kN/m^2 (kilonewtons/m^2)

The measuring and setting out of sheet metal work and pipework will not present many difficulties in use of the S.I. system.

Using a suitably divided metric rule, one will simply mark off metric dimensions as the work in hand demands. However, practice in this will help to make you 'metric-minded' and more ready to accept metric terms in plumbing calculations.

If you look at a metric rule, you will find it sub-divided into centimetres and millimetres.

You will find that 1 m = 100 cm
and 1 cm = 10 mm
clearly 1 m = 1000 mm
furthermore 1 cm $= \frac{1}{100}$ or 0·01 m
therefore, 3 cm = 0·03 m (3/100ths of a metre)
and 13 cm = 0·13 m (13/100ths of a metre)
also 1 mm = 0·001 m (1/1000th of a metre)
21 mm = 0·021 m (21/1000ths of a metre)
305 mm = 0·305 m (305/1000ths of a metre)
2500 mm = 2·5 m that is, 2 metres plus 500 mm which are equal to 500/1000ths of a metre which equals 0·5 metres.

Lengths will be written in metres and millimetres or in millimetres entirely. For example, 4227 mm could be written as a dimension. Alternatively, it could be written 4·227 m (4 m and 227 mm).

The centimetre is not a *preferred* unit and should not be used in mensuration.

From the above, it will be seen that 1 m² contains 100×100 cm = 10 000 cm² or 1000×1000 mm = 1 000 000 mm².

And that 1 m³ = $100 \times 100 \times 100$ cm = 1 000 000 cm³.

However, since the metre is the primary unit of length from which areas and volumes are derived, we can expect areas to be in metres² or fractions of a metre², and volumes too in metres³ or fractions of metres³.

Units of measurement

The primary, fundamental or basic units of the S.I. metric system are as follows—

Quantity	*Unit*	*Unit Abbreviation*
length	metre	m
mass	kilogram	kg
time	second	s
temperature	degree Kelvin	°K = °Celsius
current	ampere	A
luminous intensity	candela	cd

We need only concern ourselves with the first four at this stage.

Derived units are those which can be expressed in terms of the primary units so as to provide more units to work with.

For example, there is a primary unit for length, but not for area or volume. However, it is possible to derive such units as these from the primary unit as follows:

Area = length × breadth and if the length of these is given in metres, then Area = $m \times m = m^2$, that is, the derived unit of area is the *squared metre* (or *square metre*) which is written as m^2.

$$\text{Similarly, with volume which} = L \times B \times D$$
$$= m \times m \times m = m^3$$

and this would be referred to as cubic metre(s) or m^3, the derived unit for volume.

A list of some commonly used derived units is given below:

DERIVED S.I. UNITS

Physical quantity	*S.I. Units*	*Unit symbol*
area	square metre	m^2
volume	cubic metre	m^3
mass	kilogram	kg
density $\left(\frac{\text{mass}}{\text{volume}}\right)$	kilogram/cubic metre	kg/m^3
capacity	litre	l

DERIVED S.I. UNITS—*cont.*

Physical quantity	*S.I. Unit*	*Unit symbol*
		Note. 1 litre water = 1 kg i.e. 1000 grams at 4°C at normal atmospheric pressure, 760 mm mercury
intensity of pressure	newton per square metre	N/m^2 (1000 N/m^2 = 1 kN/m^2)
quantity of heat	joule	J
power	watt	W W = J/s

MULTIPLES AND SUB-MULTIPLES OF UNITS

Multiplication factor		*prefix*	*symbol*
1 000 000	10^6	mega	M
1000	10^3	kilo	k
100	10^2	hecto	h
10	10^1	deca	da
0·1	10^{-1}	deci	d
0·01	10^{-2}	centi	c
0·001	10^{-3}	milli	m

For example, 1 millimetre (mm) = 0·001 metre
= 1/1000 metre

and 1 kilogram (kg) = 1000 gram
1 litre = 1000 cm^3

then 1 cm^3 = 1/1000 litre

and the prefix m, meaning 1/1000 should be used to describe 1/1000 of a litre

thus: 1 ml = 0·001 l = 1 cm^3

The application of the newton as a unit of pressure is dealt with in the chapter entitled *Water*. (See also page 25.)

Temperature will now be measured against the *Celsius* scale. The temperature of any substance will be referred to as degrees Celsius, or °C.

A *temperature difference* which may exist between any two substances is written in the form, deg C. *Then*:

°C indicates temperature or degree of hotness, e.g. 60°C
deg C indicates temperature difference or interval of temperature, e.g. 80°C − 40°C = 40 deg C

See also Fig. 12 (*Plumbing 1*) and note that the freezing point of water is 0°C on the Celsius scale but 273°K on the Kelvin scale. Also that the boiling point of water is 100°C or 373°K.

The Kelvin scale is used in scientific work and 0°K which is 273 degrees below 0°C is referred to as *absolute zero*, the lowest temperature theoretically obtainable.

Note too that any 1°K covers the same interval of temperature as 1°C.

In plumbing work, as with everyday use, the Celsius scale will be used with its more familiar freezing and boiling point 0°C and 100°C respectively as defined on page 83 of *Plumbing 1*.

Heat and work are forms of *energy* (see Chapter 7, *Plumbing 1*) and in the S.I. system the unit of energy is the joule (J).

Quantity of heat

4·186 J = 1 gram calorie
1 gram calorie will raise the temperature of 1 gram (g) of water 1 deg C from which

1. 4·186 J will raise 1g water 1 deg C
   and
2. 4186 J will raise 1 kg (1 litre) water 1 deg C
   or
3. 4·186 kJ will raise 1 kg (1 litre) water 1 deg C

EXAMPLE 1

Calculate the quantity of heat energy in joules required to raise the temperature of 100 litres of water from 20°C to 100°C (boiling point)

using (1) joules = 4·186 × grams × temperature rise
= 4·186 × (100 litres × 1000 gram/litre) × (100°C − 20°C)
= 4·186 × 100,000 g × 80°C
heat energy = *33 488 000 joules*
= 33 488 kJ

using (3) kJ = 4·186 litres of water × temperature rise
= 4·186 × 100 litres × (100°C − 20°C)
= 4·186 × 100 litres × 80°C
heat energy = *33 488 kJ*

And this, you will note, produces the same result but with less working when the quantity of water is expressed in litres, as will generally be the case in calculation involving heat energy requirements for domestic hot water supply.*

EXAMPLE 2

Calculate the number of joules required to raise 140 litres of water from 10°C to 60°C.

kJ = 4·186 × litres of water × temperature rise
= 4·186 × 140 × (60 − 10)
= 4·186 × 140 × 50
= 29 302 kJ
= *29 302 000 joules*

EXAMPLE 3

Calculate the number of kilojoules required to raise 86 litres of water through 10 deg C.

* Since 4·186 may be taken as 4·2 for all practical purposes, this would further simplify the arithmetic involved.

$$\text{kJ} = 4{\cdot}186 \times 86 \times 10$$
$$= \textit{3601 kilojoules}$$

The joule has already been defined as the S.I. unit of energy. The foregoing examples show its application in the calculation of *heat energy* input requirements to water.

Mechanical energy (work done) is also measured in joules. In this sense 1 joule represents the work done when a force of 1 newton moves through a distance of 1 metre.
or joules = newtons × metres = Nm

The *newton* is the basic unit of force in the S.I. system. A newton is that force which gives a mass of one kilogram an acceleration of one metre/s^2 (1 kgf = 9·81 newtons).

force = mass × acceleration
1 newton = 1 kg × 1 m/s^2
Gravitational acceleration in S.I. units is 9·81 m/s^2.

EXAMPLE 4

How much work is done when a pump raises 100 litres of water to a height of 10 metres?

force = mass × acceleration
= 100 kg × 9·81
force = 981 newtons
work done = force × distance moved = N × m = joules
= 981 × 10
= 9810 Nm = joules
9810 joules

Power has to be expended to produce energy and the S.I. unit of power is the *watt* (W)

one watt = one joule/second

(4) $$\text{or watts} = \frac{\text{joules}}{\text{seconds}}$$

that is:

$$\text{power required in watts} = \frac{\text{joules (of heat or work energy)}}{\text{seconds (time power supplied)}}$$

EXAMPLE 5

From *Example 4* we find that 9810 joules of work energy were required to raise 100 litres of water through a height of 10 metres. What *power* would be required to do this work in one minute?

$$\text{from (4) watts (Power)} = \frac{\text{joules (energy required)}}{\text{seconds (time power applied)}}$$

$$\text{then power required in watts} = \frac{9810 \text{ joules}}{60 \text{ seconds in one minute}}$$

$$= \textit{163.5 watts} \ (0.1635 \text{ kW})$$

Power and Heat Energy. Boiler, and all water heating equipment, output ratings will be in *watts.* More conveniently, the output ratings will be given in kilowatts (1 kW = 1000 W).

EXAMPLE 6

From *Example 3* we find that 3601000 J or 3601 kJ of heat energy are required to raise 86 litres of water through 10°C.

What *power* would be required to heat this water in

(a) 1 second
(b) ½ hour
(c) 1 hour
(d) 2 hours?

Answers

$$\text{(a) from (4) watts (power)} = \frac{\text{joules (energy required)}}{\text{seconds (time power applied)}}$$

$$\text{then watts} = \frac{3\,601\,000 \text{ joules}}{1 \text{ second}}$$

$$= \textit{3 601 000 watts}$$

$$= 3601 \text{ kW}$$

Clearly, this is like using a steamroller to crack a nut. In practice, we shall be concerned with power applied to water for longer periods than 1 second.

The following examples illustrate that as the time factor (in seconds) increases, so the power requirement proportionately decreases:

(b)
$$\text{watts} = \frac{\text{joules}}{\text{seconds}}$$
$$= \frac{3\,601\,000 \text{ joules}}{1800 \text{ seconds in } \frac{1}{2} \text{ hour}}$$
$$= 2000 \text{ watts}$$
$$= 2 \textit{ kW power rating for water heater}$$

similarly,

(c)
$$\text{watts} = \frac{3\,601\,000 \text{ joules}}{3600 \text{ seconds in 1 hour}}$$
$$= 1000 \text{ watts}$$
$$= 1 \textit{ kW}$$

and

(d)
$$\text{watts} = \frac{3\,601\,000 \text{ joules}}{7200 \text{ seconds in 2 hours}}$$
$$= 500 \text{ watts}$$
$$= 0{\cdot}5 \textit{ kW}$$

EXAMPLE 7

An oil-fired boiler is required to raise 140 litres of water from 10°C to 60°C in three hours. Calculate its required power rating.

Answer Duty, heat load $= 4{\cdot}186 \times 140 \text{ litres} \times (60°C - 10°C)$
or heat energy requirement $= 4{\cdot}186 \times 140 \times 50$
in kilojoules
(See *Example 1 (3)*) $= 29\,302$ kJ

$$\text{power (kW)} = \frac{\text{kJ}}{\text{Time (seconds)}}$$
$$= \frac{29\,302}{10\,800 \text{ seconds in 3 hours}}$$
$$\simeq 2{\cdot}7 \textit{ kW boiler rated output}$$

and a 3 kW boiler might be selected.

By imperial units

140 litres $= 30\cdot8$ gal (1 litre $= 0\cdot22$ gal)
50 deg C $=$ 90 deg F (deg C $\times \frac{9}{5} =$ deg F)

then Btu

heat $= 30\cdot8$ gal $\times$ 10 lb $\times$ 90°F
load $=$ 27 720 Btu/h

$$\text{or} \quad \frac{27\ 720\ \text{Btu/h}}{3\ \text{hours}}$$

$=$ 9240 Btu/h at 3 hour 'heat-up' or recovery period

by conversion: Btu/h $\times 0\cdot293 =$ watts
power $= 9240 \times 0\cdot923 = 2707$ W
or *2·7 kW* (as in the metric worked example)

EXAMPLE 8

An instantaneous gas water heater raises the temperature of 9 litres of water per minute through 50 deg C. Calculate its power rating.

heat energy required $= 4\cdot186 \times 9$ litres $\times$ 50°C
in kilojoules $= 1883\cdot7$ kJ

$$\text{kW} = \frac{\text{kJ}}{\text{s}} = \frac{1883\cdot7\ \text{kJ}}{60\ \text{seconds—1 minute}}$$

$=$ *31·3 kW power rating*

By imperial units

9 litres $= 1\cdot98$ gal
50 deg C $=$ 90 deg F
then Btu
heat load $= 1\cdot98$ gal $\times$ 10 lb $\times$ 90°F
$=$ 1782 Btu/min
$= 1782 \times 60 =$ 106,920 Btu/h
and Btu/h $\times 0\cdot293 =$ watts
106 920 $\times 0\cdot923 =$ 31 327 watts
$=$ *31·3 kW power rating* (as in the metric worked example)

Note. In all these examples power rating is given in kW. There is no reference to time as in the imperial Btu/h. In the S.I. metric method you will see that the time factor is taken into consideration when calculating the power rating.

Also note that if the heat energy required in *Example 8* had been put in at a uniform rate over 1 hour (3600s) then power rating would have been:

$$\frac{1883 \text{ kJ}}{3600\text{s}} \simeq 0{\cdot}5 \text{ kW}$$

However, since the heat energy must be produced in one minute (1/60th of an hour) then 60 times more power must be expended in that shorter time to provide the same hot water output, that is, $0{\cdot}5 \text{ kW} \times 60 = 30 \text{ kW}$.

So far we have been concerned with calculating how much power is required to heat a given amount of water through a given number of C° using the formula

$$\begin{array}{c}\text{Power} \\ \text{(kW) or (W)}\end{array} = \frac{\text{Heat energy (kJ) or (J)}}{\text{Time (seconds) power applied}}$$

Note. If power is in kW, then energy must be in kJ.
If power is in W, then energy must be in J.

This can be re-written to read

(1) power = heat energy ÷ time

Then, remembering the simple rule of transposition of formula:

'change the side, change the sign' we have:

(2) heat energy = power × time

Note. The time factor has been changed to the power side of the = sign and the ÷ time becomes × time

or

$$(3) \quad \text{time} = \frac{\text{heat energy}}{\text{power}}$$

Note. Here the power × has been changed to the other side of the = sign and the × now becomes ÷ (the horizontal line between heat energy and power means 'divided by')

EXAMPLE 9

Calculate the power, in watts, to raise 9 litres of water through 50°C in 1 hour.

Answer heat energy required $= 4{\cdot}186 \text{ kJ} \times 9 \text{ litres} \times 50°\text{C}$

$= 1883{\cdot}7 \text{ kJ}$

$= 1\,883\,700 \text{ joules}$

Formula (1)

$$\text{power in watts} = \frac{\text{joules}}{\text{seconds}} = \frac{1\,883\,700}{3600 \text{ seconds}}$$

$= \textit{523 watts}$

or $0{\cdot}523$ kW

EXAMPLE 10

Calculate the heat energy, in joules, produced by 0·523 kW maintained for 1 hour formula (2)

$$\text{energy in joules} = \text{power in watts} \times \text{time in seconds}$$

$= 523 \text{ W} \times 3600\text{s}$

$= \textit{1 883 700 joules}$

or $1883{\cdot}7$ kJ

EXAMPLE 11

Calculate the time taken for 0·523 kW to produce 1 883 700 joules of heat energy

Formula (3)

$$\text{time in seconds} = \frac{\text{joules (heat energy)}}{\text{watts (power)}}$$

$$= \frac{1\,883\,700}{523}$$

$= \textit{3600 seconds}$

or 1 hour

EXAMPLE 12

How long will it take a 3 kW electric immersion heater to raise 140 litres of water through 55°C?

Answer heat energy required in kJ $= 4{\cdot}182 \times 140 \times 55$

$= 32\,122$ kilojoules

time in seconds $= \dfrac{\text{joules}}{\text{watts}}$ or $\dfrac{\text{kJ}}{\text{kW}} = \dfrac{32\,122}{3}$

$= 10\,707$ seconds

$\dfrac{10\,707}{3600}$

or $\simeq$ *3 hours*

Water pressures

Pressure may be defined as force/unit area.

In the S.I. system the unit of force is the newton (N) and the unit of area is the square metre (m^2)

Then: *intensity of pressure* $= \dfrac{\text{force (N)}}{\text{area (m}^2\text{)}}$ giving N/m^2

Owing to the smallness of the newton unit, the kilonewton (kN) will generally prove more convenient in use since smaller figures will be involved in calculations.

The 'Head of Water' idea as a pressure calculation device will still apply in the S.I. system where:

1 metre head of water $= 9{\cdot}8\ kN/m^2$

or: intensity of pressure in (kN/m^2) = head of water in (m) $\times 9{\cdot}8$

The relationship of pressure to m^2 unit area as compared with the in^2 of the imperial system should present no problems, as the following examples will show.

EXAMPLE 1

Calculate the intensity of pressure on a tap fixed 3 m below the water level in a cold feed cistern.

Answer intensity of pressure in (kN/m²) = head of water in (metres) × 9·8

$= 3\ \text{m} \times 9{\cdot}8$

$= \mathit{29{\cdot}4\ kN/m^2}$

total pressure (in kN) = intensity of pressure (in kN/m²) × area acted upon (in m²)

This applies to total force acting on areas larger or smaller than 1 m² and, since the force unit is the newton or, as sometimes more convenient, the kilonewton, the results of total pressure calculations will be in newtons or kilonewtons, *not* in N/m² or kN/m². (Note: $\frac{\text{kN}}{\not{\text{m}^2}} \times \not{\text{m}^2} = \text{kN}$).

EXAMPLE 2

Calculate the total pressure on the base of a cold store cistern 3 m long × 2 m wide and filled with water to a depth of 2 m.

Answer intensity of pressure in kN/m² = head (m) × 9·8

$= 2\ \text{m} \times 9{\cdot}8$

$= \mathit{19{\cdot}6\ kN/m^2}$

Total Pressure in kilonewtons = Intensity of Pressure in kN/m² × Area acted upon in m²

$= 19{\cdot}6\ \text{kN/m}^2 \times 2\ \text{m} \times 3\ \text{m}$

$= \mathit{117{\cdot}6\ kilonewtons}$

So much for areas larger than 1 m². What happens for areas less than 1 m²? This presents no difficulty since all that is done is to express the smaller area as a decimal fraction and then proceed as above.

EXAMPLE 3

A cold store cistern is fitted 3 m above a 50 mm (5 cm) diameter gate valve. Calculate (a) the intensity of pressure on the closed gate of the valve, and (b) the total pressure on the gate of the valve.

Answer (a) intensity of pressure in kN/m² = head (m) × 9·8

= 3 m × 9·8

= *29·4 kN/m²*

(b) total pressure in kN = intensity of pressure in kN/m² × area acted upon in m²

$= 29{\cdot}4 \times \pi \frac{D^2}{4}$ (*note* D in metres)

Note 50 mm

$= \frac{5\not{0}}{100\not{0}}$ m

= 0·05 m

= 29·4 × 0·7854 × 0·05 m × 0·05 m

= *0·057 kilonewtons*

$\frac{\pi D^2}{4}$ = area of circle

Here is a case where, owing to the smallness of 0·057 kN, one might convert the answer to newtons, thus:

$\frac{\pi}{4}$ = 0·7854

0·057 kN × 1000 = *57 newtons*

so 0·7854 × D² = area of circle

Sometimes the intensity of pressure is known and it becomes necessary to calculate what head, in metres, is creating this pressure. The following example shows clearly how this is done.

EXAMPLE 4

The pressure in a water main is 545 kN/m². What head, in metres, creates this pressure?

From *Example 1*

intensity of pressure (in kN/m²) = head of water (in metres) × 9·8 $\left(\frac{\text{kN/m}^2}{\text{m head}}\right)$

$$\text{or head (in metres)} = \frac{\text{intensity of pressure in kN/m}^2}{9{\cdot}8 \text{ kN/m}^2\text{/m head}}$$

$$\text{then head} = \frac{545 \text{ kN/m}^2}{9{\cdot}8 \text{ kN/m}^2\text{/m head*}}$$

$$= \mathit{55{\cdot}6\ metres}$$

Volume. The S.I. unit is m^3. However, the litre will continue in use as fluid capacity unit.

$$1 \text{ m}^3 = 100 \times 100 \times 100 \text{ cm}^3 = 1\ 000\ 000 \text{ cm}^3 \ (10^6 \text{ cm}^3)$$

$$= 1000 \times 1000 \times 1000 \text{ mm}^3 = 1\ 000\ 000\ 000 \text{ mm}^3 \ (10^9 \text{ mm}^3)$$

1 litre of water = 1000 cm^3

and 1 litre of water = 1000 g = 1 kg

$$\text{so 1 m}^3 \text{ of water} = \frac{1\ 000\ 000 \text{ cm}^3}{1000 \text{ cm}^3\text{/litre}} = 1000 \text{ litres}$$

$$= 1000 \text{ kg}$$

EXAMPLE I

A hot store cylinder is 1000 mm tall and 450 mm in diameter Calculate (a) its volume in m^3 and (b) its capacity in litres.

Note $1 \text{ mm} = \frac{1}{1000} \text{m} = 0{\cdot}001 \text{ m}$

then 1000 mm = 1 m

$$450 \text{ mm} = 0{\cdot}450 \text{ m or } \frac{45\not{0}}{100\not{0}} \text{ m} = 0{\cdot}45 \text{ m}$$

Answer (a)

$$\text{volume (m}^3) = \text{height (m)} \times \frac{\pi D^2\text{(m)}}{4}$$

$$= 1 \text{ m} \times 0{\cdot}7854 \times 0{\cdot}45 \text{ m} \times 0{\cdot}45 \text{ m}$$

$$= \mathit{0{\cdot}16\ m^3}$$

Answer (b)

$$\text{capacity in litres} = \text{volume in m}^3 \times 1000 \text{ litres/m}^3$$

$$= 0{\cdot}16 \text{ m}^3 \times 1000$$

$$= \mathit{160\ litres}$$

* Note that the kN/m^2 cancel out to leave the answer in metres head.

Volume water flow rates will be expressed in litres/s or m³/s. The former will generally apply to outflows from draw-off taps, and the latter to large water flows as in drains.

EXAMPLE 1

The desired outflow from a bath tap is 4 gal/min (imperial) Express this in metric terms

Answer 4 gal/min × 4·5 litre/gal = 18 litres/min

$$\text{and } \frac{18 \text{ litres/min}}{60 \text{ seconds/min}} = \mathit{0{\cdot}3\ litres/s}$$

EXAMPLE 2

Assuming a 4 in diameter drain discharges 140 gal/min (imperial) express this in metric terms.

140 gal/min × 4·5 litre/gal = 610 litres/min

$$\text{and } \frac{610 \text{ litres/min}}{60 \text{ seconds/min}} \simeq 10 \text{ litres/s}$$

$$\text{and } 10 \text{ litres/s} = \frac{10}{1000} \text{ m}^3\text{/s}$$

$$= \mathit{0{\cdot}01\ m^3/s}$$

Mass flow rate, as seen, in imperial form, as pounds of water circulating per hour in a hot water or heating circuit, for example will be expressed in kg/s (kilograms/second).

EXAMPLE 1

200 lb/h water circulation rate to be expressed in kg/s

Answer 200 lb/h × 0·45 kg/lb = 90 kg/h

$$\text{and } \frac{90 \text{ kg/h}}{3600 \text{ second/h}} = \mathit{0{\cdot}025\ kg/s\ mass\ flow\ rate}$$

or *0·025 litre/s volumetric flow rate*

Atmospheric pressure commonly expressed as 14·7 lbf/in² (imperial) becomes, by direct conversion, using 6895 × lbf/in² = newton/m²:

$14{\cdot}7 \times 6895 = 101\ 356$ newton/m²
or *101·3 kilonewton/m²*

Atmospheric pressure may be expressed in terms of mercury column:

(imperial) 30 inches mercury $\simeq$ 760 mm mercury (metric)
now, 760 mm mercury $\times\ 13{\cdot}6 = 10{,}366$ mm W.G.*
which represents a column of water 10·336 m high
and intensity of pressure = head of water × 9·8
in kN/m² — in metres

$= 10{\cdot}336 \times 9{\cdot}8$

Atmospheric pressure = *101·3 kN/m²*

Note This is the so called Standard Atmospheric Pressure. In practice, the atmospheric pressure varies according to its moisture content, air being less dense when moist than it is when dry, and it varies with elevation above sea level, the pressure being greater at sea level but getting less as height above sea level increases. (See Chapter 5, *Plumbing 1*.)

The bar† = 100 000 N/m² or 100 kN/m² which is very nearly the same as atmospheric pressure as it is defined above. The bar and its sub-multiples, particularly the millibar (mb) 100 N/m² will be used in heating and ventilation work. It will probably be used as a unit of pressure in pipes.‡

EXAMPLE I

The maximum working pressure for a pipe is given as 3·0 bars. Calculate the equivalent head of water in metres.

Answer 1·0 bar = 100,000 N/m²
3·0 bars = 300,000 N/m² = 300 kN/m² intensity of pressure

$$\underset{\left(\text{in kN/m}^2\right)}{\text{intensity of pressure}} = \underset{\left(\text{in metres}\right)}{\text{head of water}} \times 9{\cdot}8$$

$$\text{so, } \underset{(\text{m})}{\text{head}} = \frac{\text{intensity of pressure (kN/m}^2)}{9{\cdot}8\ (\text{kN/m}^2/\text{m head})}$$

$$= \frac{300}{9{\cdot}8}$$

$$\simeq \textit{30 metres head}$$

* See 'Manometer' page 391, *Plumbing 2*. † Strictly, not an S.I. unit (see page 63). ‡ 1 Bar = 14·5 lb f/in² = The Meteorological Atmosphere.

Clearly, if $3{\cdot}0$ bars $\simeq$ 30 metres head

then $1{\cdot}0$ bar $\simeq$ 10 metres head

therefore, pressure in bars $\times$ 10 $\simeq$ metres head

and, $\dfrac{\text{metres head}}{10} \simeq$ pressure in bars

EXAMPLE 2

Calculate the pressure, in bars, created by 3 metres head of water.

Answer Pressure in bars $= \dfrac{\text{head (m)}}{10}$

$$= \frac{3}{10} = 0{\cdot}3 \textit{ bar}$$

Note. $0{\cdot}3$ bar $= \dfrac{3}{10} \times 100$ kN/m² $= 30$ kN/m²

Compare with *Example 1* in section entitled 'Water Pressures' (page 25).

1 millibar (m bar) $=$ 1/1000th of $1{\cdot}0$ bar $=$ 100 N/m²

Rate of pressure drop in pipes due to frictional resistance to flow of water will be measured in m bar. The imperial system measures this in inches W.G./100 ft run of pipe.

EXAMPLE 1

A $\frac{1}{2}$ in nominal bore mild steel tube carries 792 lb/h water with a pressure drop of 42 in W.G./100 ft. Calculate this pressure drop in m bars/metre.

Note (1) $\frac{1}{2}$ in m.s. tube, nominal bore $=$ 15 mm nominal bore (metric)

(2) conversion lb/h to litres/s has been dealt with in *Example 1* in the section entitled '*Mass flow rate*'.

(3) feet $\times$ $0{\cdot}305$ $=$ metres

(4) 1 in W.G. $\simeq$ $2{\cdot}5$ m bar

Answer 42 in W.G./100 ft $= 42 \times 2{\cdot}5$ m bar/100 ft
$= 105$ m bar/100 ft
$= 105$ m bar/100 ft $\times\, 0{\cdot}305$ m/ft
$= 105$ m bar/$30{\cdot}5$ m

and $\dfrac{105 \text{ m bar}}{30{\cdot}5 \text{ m}}$ *3·4 m bar/metre run*

Incidentally, since 1 in W.G. $\simeq 2{\cdot}5$ m bar this sub-multiple of the bar may well find application in gas installation and air tests on drains and soil-waste pipe systems. 6 in W.G. (imperial) becomes 15 m bar, $1\frac{1}{2}$ in W.G. becomes $3{\cdot}75$ m bar which gives easily manipulated values in this particular unit.*

Density will be expressed in kg/m³. In imperial form it was given as lb/ft³. Conversion from imperial to metric values can be made by using the conversion factor:

1 lb/ft³ $= 16{\cdot}02$ kg/m³

EXAMPLE 1

1 ft³ water at 140°F (60°C) weighs 61·38 lb

61·38 lb/ft³ $\times$ 16·02 $= 983{\cdot}3$ kg/m³ at 60°C

* See also 'Manometer' page 391, *Plumbing 2*.

2

Tools of the trade

Hand tools

Good tools are indispensable to the plumber, but buying them can be an expensive business. It is advisable to get a few of them at a time, as your experience grows and as you need them. A kit of good-quality tools, built up in this manner, is a sound investment which will pay handsome dividends, as any experienced plumber will tell you.

Some employers will buy tools for their apprentices at a trade discount, which reduces the cost. These are then paid for by agreed weekly deductions from wages. The cost of the kit will, of course, depend upon the quality and number of tools bought. Manufacturers now produce such a wide variety of tools, and so many plumbers make or adapt others to suit their own special needs, that there is no limit to the possible contents of a kit. The list below, however, is agreed between the Master Plumbers' Federation and the Plumbing Trades Union to be a full kit of tools which should enable the plumber to carry out any reasonable job.

List of tools to be provided by plumbers

Bent pin or bolt
Brace and bits up to 10 mm
Blowlamp, excepting the replacement of spare parts, other than the body of the lamp
Bobbins, all sizes to 50 mm
Bradawl
Bossing stick
Boxwood dressers (large and small)
Boxwood setting in stick
Boxwood-bending dresser
Boxwood mallets (large and small)
Mandrels, 25 mm, 32 mm and 38 mm

Cardwire
Caulking tools for ordinary work
Chase wedges
Compasses
Copper bit
Cutting pliers
Draw-knife
Footprints, 140 mm, 230 mm
Fixing points (or clamps)
Flat chisel for wood
Gimlet for lead pipe
Gimlet for wood screws
Gouge for wood
Glass cutter and putty knife, if glazing is normally done by plumbers in the district
Hacking knife
Hacksaw frame
Hammers, small and large (1 kg maximum)
Hand dummy
Handsaw
Handbrush
Hand ladle
Lavatory union key
Pliers, two holes
Plumb-bob and chalk line
Pot hook
Rasp
Rule
Screwdrivers (large and small)
Scribing plate
Shave-hooks
Small brick drill
Snips
Soil pot and brush
Spirit level
Springs for bending 13 mm and 19 mm light gauge copper pipes
Square
Steel drip-plate
Steel chisels for brickwork up to 500 mm long
Stillson or other pipe wrench up to 300 mm
Tanpins up to 50 mm
Tool bag or box
Trowel, small
Wiping cloths

Plumbers able to provide tools from the list when they are needed for a job are eligible for a tool maintenance payment for every day during which plumbing work is done. This money is used for the upkeep of their kit. To qualify for this allowance one does not necessarily have to cart the whole kit around; but it is necessary to own all the tools, even though many of them might be at home or in the workshop.

You will have noticed that the list does not include such items as stocks and dies, pipe cutters, pipe vices, welding kits and several other of the larger and more expensive tools. These are provided by the employer. Neither does the list contain various tools used for specialized ways of working sheet copper and aluminium roof weatherings. Many of these are easily made by the plumber himself, and they will be described as the need for them arises.

The maintenance of hand tools is important both from the practical angle and from the point of view of safety. This applies to all tools, whether they be your own or provided by your employer for your use. Blunt, damaged and generally uncared-for tools will not produce good work, and they are liable to prove dangerous to the user.

All wood tools used for working lead should be kept quite separate from steel tools which could mark or bruise them. These blemishes would result in damage or scratching of the lead. Wood tools like, cricket bats, should be carefully preserved from damp, which would cause them to swell out of shape. On the other hand, they should not be allowed to become dry, since this would make them shrink and crack. An occasional dressing with linseed oil will keep bossing and other leadwork tools in good order.

The cutting surfaces of files and rasps must be kept clear by frequent cleanings with a wire brush.

Cutting tools must be kept sharp, and the cutting edges must be kept covered when the tool is not in use so as to protect them from damage and to avoid the risk of injuring anybody. For example, a short length of bicycle inner tube slipped over the handles of a pair of snips will keep the cutting blades closed and safe. Short lengths of hosepipe might be used to protect the cutting ends of chisels. You will think of several more ideas of this nature which will be effective.

'Mushroomed' or burred cold chisel edges should be dealt with quickly, since these burrs are apt to chip off. If they fly into the eye they can have tragic results. The simplest treatment is to grind off the burrs until the end of the chisel regains its original shape. Use goggles to protect the eyes from flying dust and metal particles when using the carborundum grinder.

Stock and dies, pipe vices, copper tube benders, pipe cutters, etc., all need an occasional rub over with an oily rag to protect them from rust. But be careful to keep the handles free from oil—it makes them slippery, and in building work a slip can be dangerous.

Wedges in the heads of hammer and mallet handles often work loose, and the tool becomes unsafe. Faulty wedges should be refixed or replaced as necessary. Split or damaged handles should be ruthlessly scrapped, and new ones fitted.

Always remember that well-kept tools not only produce better work; they also mean safer working. No one can afford to be careless in this important matter.

Electric-powered tools

Portable electric-powered tools help to reduce time and effort, but it is advisable always to seek advice from a competent electrician (or some other responsible person) before connecting or using them. They need constant care both when being used and when in store. Frayed or damaged leads must be replaced at once.

Many electrical tools are designed for use on a 250-volt supply. At this voltage an electric shock, caused by some defect in the tool or its earthing arrangement, could prove fatal. Electricity supply authorities strongly recommend that electric-powered portable tools should be of the 110-volt kind. These are supplied with current from the 250-volt main by way of a special centre-tapped transformer which reduces the voltage to 100, and also provides a safety device whereby, in the event of any earthing defect or fault in the tool, only a safe voltage of 55 volts would pass to the tool operator.

All portable electric equipment should be fitted with heavy duty three-core cable, properly connected through a three-pin plug and socket; and it should be used only when a proper earth connection has been made.

Electric extension-lead lamps should be waterproof, fitted with a three-core cable, and effectively earthed. Do not connect extension lamps to ordinary lampholders—they do not have an earth connection.

Oxy-acetylene welding equipment

This equipment is commonly used by the modern plumber for pipe welding and lead burning. If correctly stored and used the

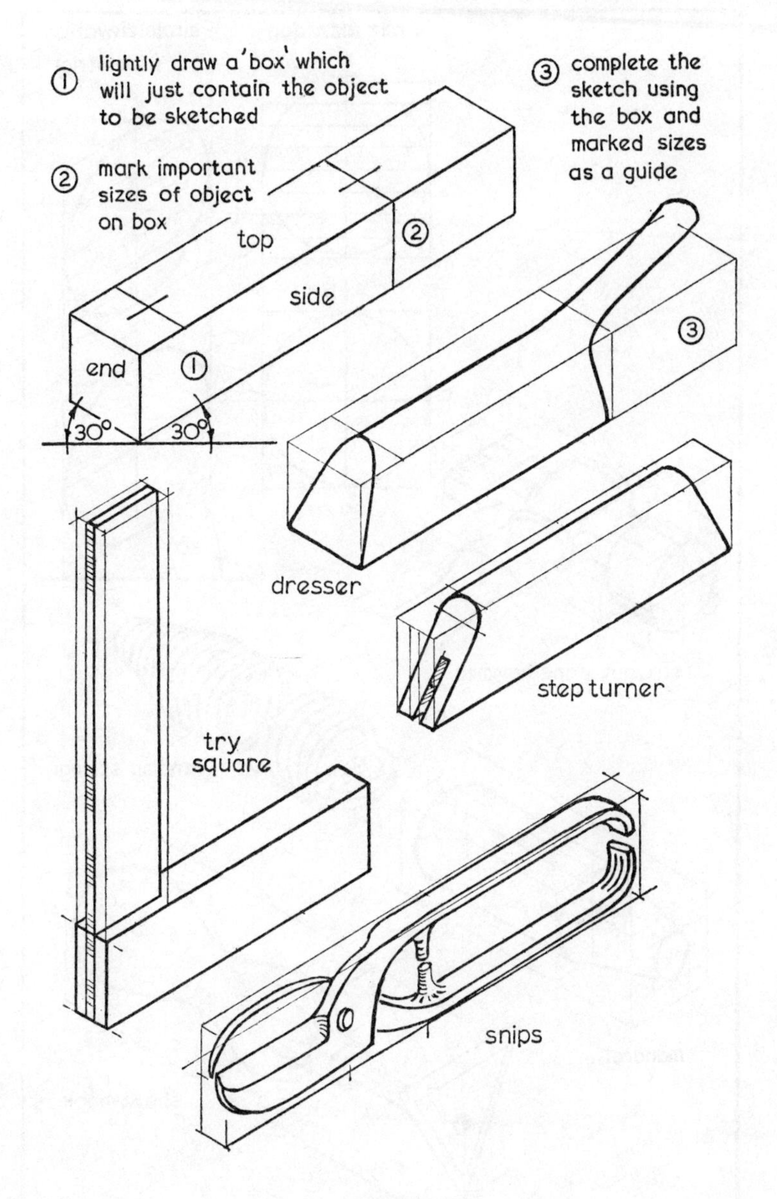

Sketching Plumbing Tools

FIG. 1

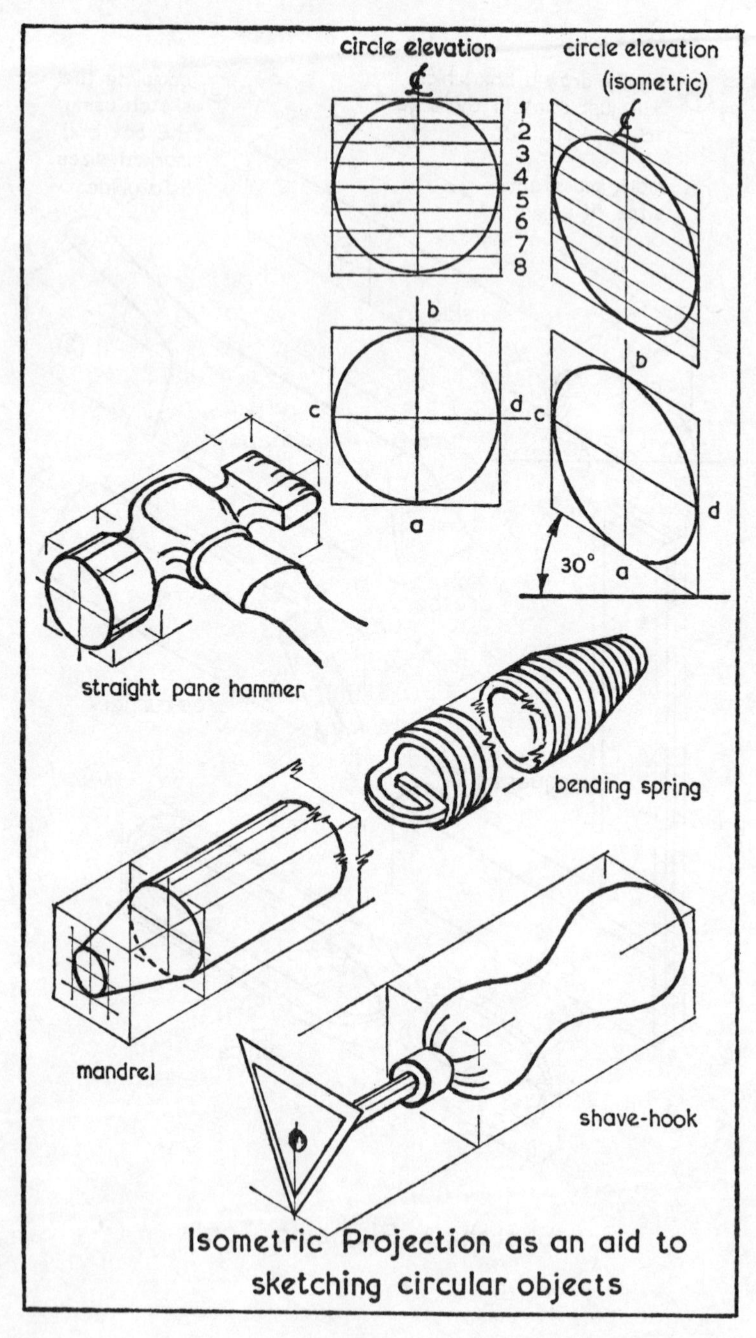
circle elevation
circle elevation
(isometric)
℄
1
2
3
4
5
6
7
8
℄
b
c
d
a
b
c
d
30°
a
straight pane hammer
bending spring
mandrel
shave-hook
Isometric Projection as an aid to
sketching circular objects

FIG. 2

gas and equipment are perfectly safe, but acetylene gas is inflammable and when mixed with air in a certain proportion is highly explosive. No one should meddle with oxy-acetylene equipment unless properly instructed in its use.

Sketching

A plumber often finds that the ability to sketch details of his work is very useful, since by this method he can express an idea more quickly, and often more effectively, than in a written or verbal description. It is not difficult to learn how to do it if one practises often enough. The plumber's tools shown in Figs. 1 and 2 have been specially chosen and drawn to suggest simple techniques of sketching. If you copy them, and apply the methods to drawing other tools, you will soon find it easy to sketch like this.

Job planning

Job planning is hardly a tool of the trade, but you will find that if some thought is given to the placing of materials on the site, and the order in which the work should proceed, there will be a considerable saving in labour and time. This sort of forethought is characteristic of the good plumber, who must very often think and act on his own.

3
Safety at work

The wise plumber works with a care for his own safety, and for the safety of those about him. Because not everybody is as careful, hundreds are killed and many thousands seriously injured at building work every year. The *Building Regulations* (for safety, health and welfare) exist to warn and protect building workers against these dangers. It is advisable to know some of the more common causes of accidents, since this may help you to avoid them.

Ladderwork is the cause of large numbers of accidents, mostly through ladders breaking or slipping. Ladders should be soundly constructed and well looked after; and the worker should always have an eye to the dangers involved in using them.

Ladders with missing or faulty rungs must never be used, and home-made ladders should be used only if they have been properly made. Even then they should be treated with great care. They must always stand on a firm, solid and level base. Bricks and other packings must not be used to level the ground; if the ground is uneven the high spots must be removed.

Ladders used to give access to a roof or a platform must be long enough to pass the roof or platform edge by at least four feet in order to give a good handhold. The top should if possible be securely tied to a good anchorage. If this cannot be done, a responsible person must 'foot' the ladder. Where a ladder is laid up a sloping roof with its foot thrusting against the main upright ladder, the weight and attention of a responsible person is essential to counteract this thrust and prevent the danger of slipping. A much safer method is to arrange that the top of the ladder on the roof should have some fixing over the ridge. This is sometimes done by tying on to it a short ladder, which will hand down the other side of the roof and act as a counterweight.

Heavy materials should not be carried up ladders; instead, a hoist of some kind should be used. Do not strain or overreach when working from a ladder. If necessary, come down and move it to a more convenient position.

Fragile roof coverings can seem deceptively safe. They are liable to give way suddenly under the weight of a person standing or walking on them with serious and often fatal results. Crawling-boards which can be attached to the ridge of the roof must always be used for large jobs on or over fragile roof coverings. Ladders suitably secured and arranged to give a good toehold can be used for smaller repair-jobs.

Falling materials and tools cause many accidents in building, and it is essential that one should take care when placing materials or tools anywhere, whether overhead or on ground level. Plumbing tools such as pipe wrenches and hammers, pipe fittings and tins of jointing compounds must never be left just balancing on a pipe or ledge.

Building rubbish often causes accident because people can easily trip over it. Rubbish should be cleared as it arises, so as to keep working spaces clear and safe.

Nails left sticking through timbers should be pulled out or knocked flat. It is not the funniest thing in the world to have a nail pass right through one's foot.

Poisonous materials. Lead taken into the human body can cause lead poisoning. The safeguard is to wash carefully after using it in any form. Always use plenty of hot water, soap and a nail brush.

Protective clothing helps to reduce accident risk. Avoid loose clothing, especially neckties, which might get caught in tools or machinery. And 'suède shoes with pink shoe-laces' may be fine for the odd night out with the lads, but good safety-sense demands stout boots or shoes for building work.

Eye protection. One instinctively 'ducks' when danger threatens. This is the natural reaction to seeing the danger and judging how near the body is to it. One cannot see one's own eyes, and one is apt to be more careless of their safety. It is wise to wear clear, unsplinterable goggles when working on a job such as cutting away brick or concrete where a chip could easily fly into, and damage the eyes.

Oxy-acetylene welding must not be carried out or watched without specially tinted goggles to protect the eyes against glare. Electric arc welding also produces an intensely bright light, which can be very damaging to the eyes if viewed without dark protective screens. Even the reflected light of arc welding can be irritating to the eyesight.

Suggested books for further reading:

Guide to the Building (Safety, Health & Welfare) Regulations, 1948, published by H.M.S.O.

Guide to the Safety Regulations, published by the London Regional Joint Committee for the Building Industry.

Building Safety Aids, published by the Building Safety & Welfare Committee, Southern Counties Federation of Building Trades Employers.

4
Water

Millions of years ago, when the earth was a white-hot mass hurtling through space, its hydrogen and oxygen gas content burned together. The result of this combustion was a new substance—water vapour or steam. Because of the intense heat, this remained as a gas, and mixed with the atmosphere enveloping the hot earth mass. In time the earth cooled, and so did the surrounding atmosphere. Its water-vapour content condensed to water, which fell as rain to settle in the surface depressions of the earth and form the seas. From these same seas water is still evaporated into the atmosphere by the heat of the sun. The water vapour cools, and forms clouds of condensed water droplets which eventually return to the earth as rain.

On a much smaller scale water can be manufactured in a laboratory by burning hydrogen in oxygen in the proportions of two volumes to one, as indicated by the chemical symbol for pure water, H_2O.

Properties of water

Water has weight, but this varies according to its temperature. The fact is important in the design and installation of hot-water systems. Moreover, it affects the placing of taps, for the weight of the water, combined with its height above the tap, affects the rate of flow. For normal every-day purposes, and for simple calculations of forces and pressures acting in water which is stationary in pipes and storage vessels, the weight of water is taken to be 1000 kg/m³. The following points are also worth noting:

1 *There are 1000 litres in 1 m³.*
2 *One litre weighs 1000 g.*

Pure fresh water is tasteless and without smell. At atmospheric pressure and between 0°C and 100°C it is a transparent and almost colourless liquid. It is essential to plant and animal life.

3 *The maximum density temperature of water is 4°C.* Most materials expand as their temperature rises, and water expands when heated above this temperature. It is unusual, however, in that it also expands when cooled below this temperature. The behaviour of water according to its heat content is shown in the diagrams on page 45. It will be seen that above the 'waistline' of its maximum density temperature it expands by $\frac{1}{25}$th or 4%, and that nearly threequarters of this expansion takes place between 45°C and 100°C. Below the 4°C 'waistline' water expands as its temperature falls. Note carefully that an equal volume of water weighs less at 0°C than at 4°C, since this provides proof that expansion has taken place.

4 *Pure water boils at 100°C* at standard atmospheric pressure (760 mm of mercury), and changes from the liquid to the gaseous state, expanding some 1600 times as it does so, as is shown in Fig. 3. It can be said that 1 m^3 of water will produce 1600 m^3 of steam.

Equally, 1600 m^3 of steam at 100°C, when condensed to water at the same temperature, produces 1 m^3 of water. This fact will be used later to explain in part why hot-water storage cylinders sometimes collapse inwards under the influence of atmospheric pressure.

5 *Pure water freezes at* 0°C, and changes from the liquid to the solid state (ice) with an immediate expansion of $\frac{1}{10}$th. Thus, if the water in a 100 mm tube were to freeze, and all expansion were lengthways, the column of ice formed would be 110 mm long, with 10 mm, or $\frac{1}{10}$th of the original volume of water, projecting beyond the end of the tube (see diagram, Fig. 3).

The force of this expansion is considerable, exerting great pressures on internal surfaces of pipes, which frequently burst under the strain. This stoppage of water supplies is not only irritating; it could also prove quite dangerous, especially if the hot-water supply system were dependent on it.

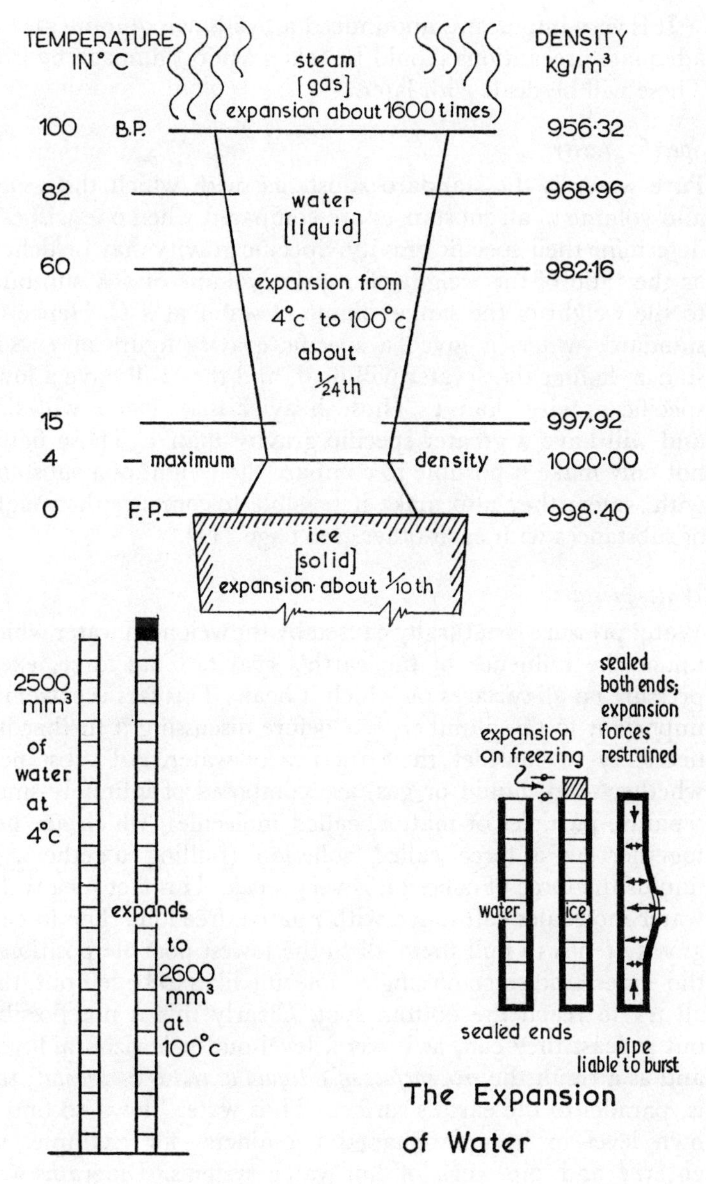
TEMPERATURE IN °C
DENSITY kg/m³
steam
[gas]
expansion about 1600 times
100
B.P.
956·32
82
968·96
water
[liquid]
60
982·16
expansion from
4°c to 100°c
about
1/24th
15
997·92
4
— maximum
density —
1000·00
0
F. P.
998·40
ice
[solid]
expansion about 1/10 th
2500 mm³ of water at 4°c
expands to 2600 mm³ at 100°c
expansion on freezing
sealed both ends; expansion forces restrained
water
ice
sealed ends
pipe liable to burst
The Expansion of Water

FIG. 3

It is very important, and indeed a bye-law requirement, that adequate precautions should be taken against damage by frost. These will be dealt with later.

Specific gravity

Pure water is the standard substance with which the weight and volume of all substances are compared when one wishes to determine their specific gravity. Specific gravity may be defined as the ratio of the weight of a given volume of any substance to the weight of the same volume of water at 4°C. Hence the standard, water, is given a specific gravity figure of 1. Substances lighter than water will float, and they will have a lower specific gravity than 1. Those heavier than water will sink, and will have a greater specific gravity than 1. Those figures not only make it possible to compare the weight of a substance with water; they also make it possible to compare the weights of substances with each other (see page 112).

Water pressure

Water pressure is naturally caused by the weight of water which, under the influence of the earth's gravitational force, exerts pressure on all surfaces on which it bears. This fact is extremely important to the plumber, but before discussing it further it is necessary to consider the structure of water. All substances, whether solid, liquid or gas, are composed of infinitely small, separate particles of matter, called molecules, which are held together by a force called cohesion (pulling together). In liquids the force of cohesion is very weak. This fact means that water molecules can move with relative freedom. The force of gravity tends to pull them all to the lowest possible position in the water, and since no single molecule likes to be left out, they all try to reach the bottom spot. Clearly this is not possible, but at least they can, as it were, level out in horizontal layers, and as a result the *free surface of a liquid at rest is horizontal*; that is, parallel to the earth's surface. Thus water 'flows' to find its own level in irregular-shaped containers—for example, the cisterns and pipework of hot-water systems. Diagrams (A), (B) and (C) in Fig. 4 will help to clarify this point.

Diagram (A) shows what would happen to a pyramid of

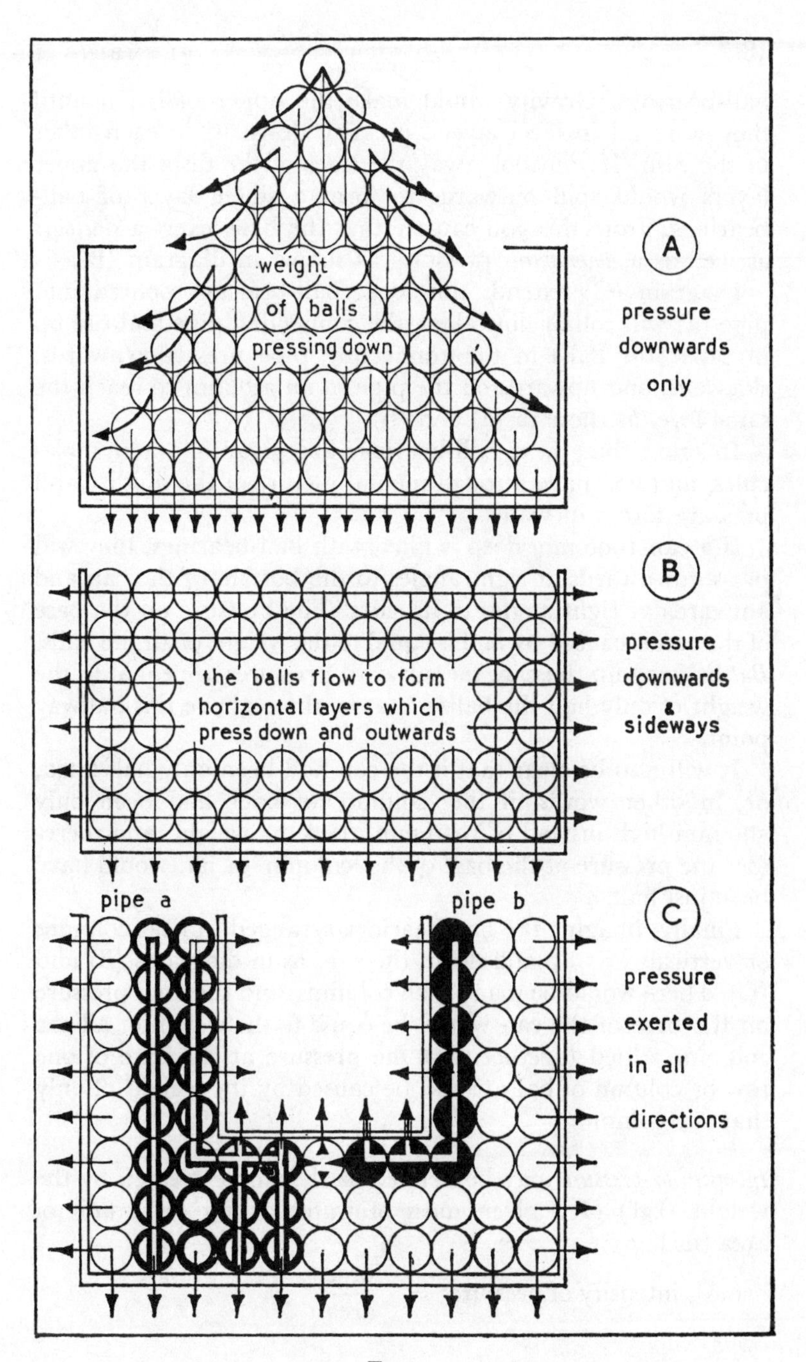
weight
of balls
pressing down
A
pressure
downwards
only
B
pressure
downwards
&
sideways
the balls flow to form
horizontal layers which
press down and outwards
pipe a
pipe b
C
pressure
exerted
in all
directions

FIG. 4

ball-bearings. Gravity would make the upper balls fall until they were all in level layers, pressing down upon each other in the dish. If you took away the sides of the dish, the upper layers would spill outwards to form a single layer of ball-bearings. From this you can see that the balls exert a *sideways* as well as a *downwards* pressure, as shown in diagram (B).

Diagram (C) extends this idea. Ball-bearings poured into pipe (a) will roll or flow along the connecting pipe and rise up in pipe (b). Balls in the connecting pipe press downwards, sideways, and upwards on the pipe in an attempt to reach the same level as those in (a) and (b).

Imagine that these ball-bearings are gigantic water molecules, and you have a rough mental picture of the fluidity and pressure forces in water.

If a can 1000 mm deep is filled with ball-bearings, they will press downwards at right angles to the bottom of the can and outwards at right angles to its sides. The pressure on the base of the can is caused by and is equal to the weight of all the balls. *But* halfway up the can, the outward pressures are equal to the weight of only half the balls—that is, those above this halfway point.

It will also be clear that if the can had been only half filled, or, in other words, if the 'column' of balls had been only 500 mm high instead of 1000 mm, then the weight, and therefore the pressure at the base of the 'column' or jar, would have been just half.

Finally, imagine the ball-bearings arranged in neat columns or vertical rows up and down the can, as in diagrams (B) and (C). There would be many such columns, and the total pressure on the base of the can would be equal to the weight of all the columns added together. But the pressure at the base of one row or column of balls would be caused by the weight of only that one column.

Intensity of pressure may be defined as that force created by the weight (kgf) of a given mass of water acting on 1 unit of area (m^2)

$$\text{or,}\quad \text{intensity of pressure} = \frac{\text{force}}{\text{area}}$$

and, since force is measured in newtons

$$\text{intensity of pressure} = \frac{\text{newtons}}{\text{area}}$$

However, the kilonewton (kN = 1000 newtons) is a more convenient multiple of the basic unit of force than the newton (N) which is rather small in numerical value.

1 cm^3 water at 4°C weighs 1 gram (g)
1 m^3 contains $100 \times 100 \times 100$ cubed centimetres (cm^3)
or 1 m^3 contains 1 000 000 cm^3 or 10^6 cm^3
from which it will be seen that:
1 m^3 water = 10^6 cm^3 = 10^6 g weight

Now, there are 1000 g in 1 kg.

$$\text{and so 1 m}^3 \text{ water weighs } \frac{1000\,000 \text{ g}}{1000 \text{ g/kg}} = 1000 \text{ kg}$$

Now, *1 kgf*, the force produced by the mass of one kilogram, in this case 1 kg of water, *equals 9·8 newtons*

Note: We are imagining a cube of water measuring $1 \text{ m} \times 1 \text{ m} \times 1 \text{ m}$ (1 m^3), that is, 1 m high and 1 m^2 base area. This volume weighs 1000 kg hence *intensity of pressure* (newtons) on the *unit area* (m^2) is given as 9800 N/m^2 or 9·8 kN/m^2

thus, *intensity of pressure* on the m^2 base of 1 m^3 of water

$$= \frac{\text{force}}{\text{area}}$$

$$= \frac{1000 \text{ kgf} \times 9{\cdot}8 \text{ N}}{1 \text{ m}^2}$$

$$= \frac{9800 \text{ N}}{1 \text{ m}^2}$$

or *9·8 kilonewtons per square metre*
i.e. 9·8 kN/m^2

and so if we visualise a column of water 1 m tall and call this height *head* we can derive the following simple formula which will be very useful:

$$\text{intensity of pressure} = \text{head (m)} \times 9{\cdot}8 \text{ kN/m}^2$$

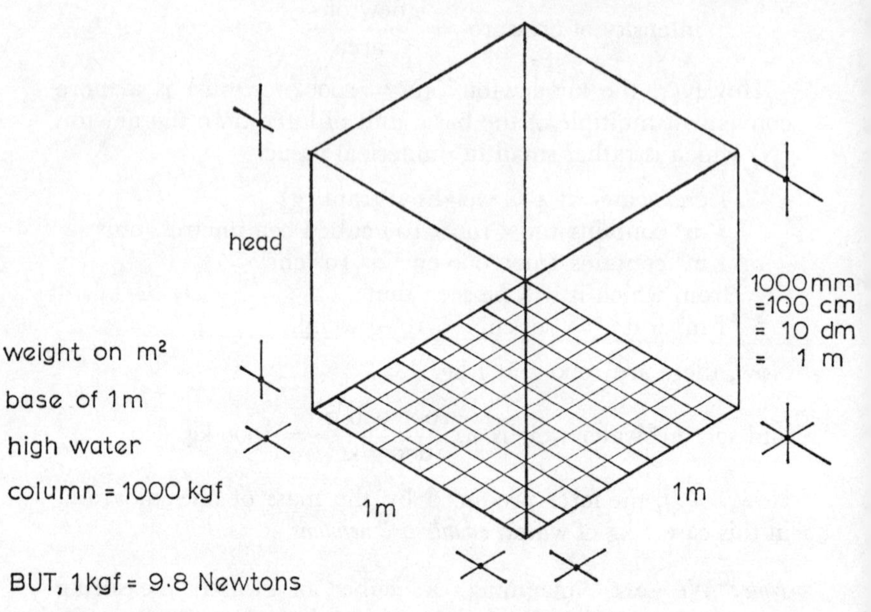

FIG. 5

EXAMPLE I

Find intensity of pressure on the base of the hot store tank in Fig. 6.

$$\begin{aligned} \text{intensity of pressure} &= \text{head (m)} \times 9{\cdot}8 \text{ kN/m}^2 \\ &= 3 \times 9{\cdot}8 \\ &= \mathit{29{\cdot}4\ kN/m^2} \end{aligned}$$

Now this is quite simple, but what happens if the area of the base is *more or less* than 1 m²?

This leads us to another simply derived formula whereby we can determine the *total pressure* acting on areas other than the m² unit.

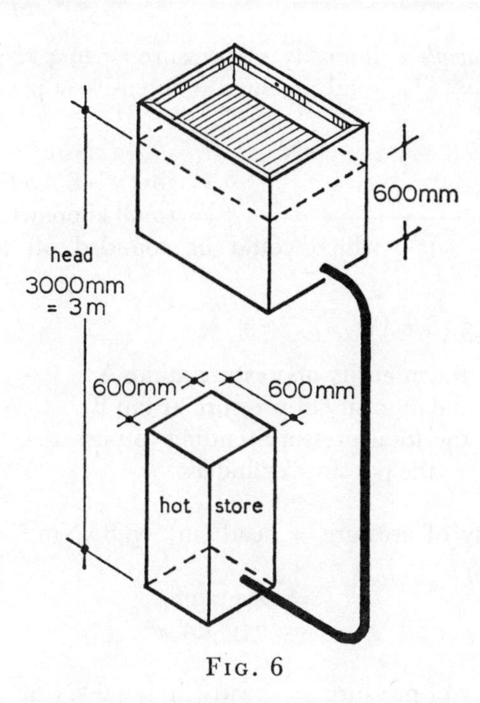

FIG. 6

Clearly from *Example 1*, the intensity of pressure due to a 3 m head is 29·4 kN/m² and if this pressure acted upon a base of two square metres area (2m²) then the total pressure acting upon the whole area would be 2 × 29·4 *kilonewtons* or 58·8 kN.

Note carefully that the m² part is missed out in the total pressure answer because we are not relating the pressure to one metre square but to an area bigger than 1 m². The same rule applies equally to areas less than 1 m². From which:

$$\underset{(kN)}{\text{total pressure}} = \underset{(kN/m^2)}{\text{intensity of pressure}} \times \underset{(m^2)}{\text{area acted upon}}$$

EXAMPLE 2

Find total pressure on the base of the hot store vessel shown in Fig. 6.

Note. $600 \text{ mm} = \dfrac{600}{1000} = 0{\cdot}6 \text{ m}$

From *Example 1* intensity of pressure = $29{\cdot}4$ kN/m²

total pressure = intensity of pressure × area acted upon

(kN) (kN/m²) (m²)

$= 29{\cdot}4 \times 0{\cdot}6 \times 0{\cdot}6$

$= 10{\cdot}58$ kilonewtons

which could be rounded off to *10·6 kN*

EXAMPLE 3 (See Fig. 7)

Find (1) the intensity of pressure at tap A

(2) the intensity of pressure at tap B

(3) the total pressure tending to push the manlid C of the hot store cylinder.

1. intensity of pressure at tap A = head (m) × 9·8 kN/m²

$= 2{\cdot}4 \times 9{\cdot}8$

$= 23{\cdot}5$ *kN/m²*

2. intensity of pressure at tap B = head (m) × 9·8 kN/m²

$= 4{\cdot}8 \times 9{\cdot}8$

$= 47$ *kN/m²*

Note. The head and the resultant pressure in this example are both exactly twice those in 3 (1).

From this we can deduce:

Intensity of pressure is directly proportionate to head that is, double the head and the resulting intensity of pressure is doubled, halve the head and the intensity of pressure is also halved.

This simple yet very important fact can be put to good practical use since, if we want good pressure to give good outflows at taps we would aim to get the head of water as high as possible above the draw-off tap or shower spray as the case may be.

3. *Note* area of circular manlid = πr^2

$\pi = 3{\cdot}142$

300 mm dia = 150 mm radius = 0·15 m radius

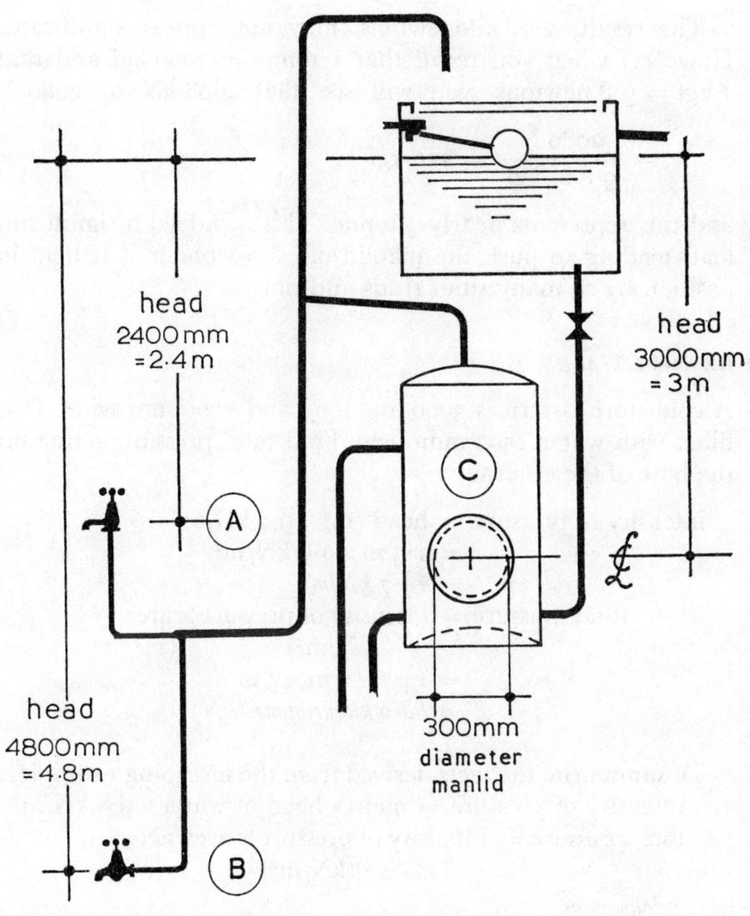

FIG. 7

$$\text{intensity of pressure} = \text{head (m)} \times 9{\cdot}8 \text{ kN/m}^2$$
$$= 3 \times 9{\cdot}8$$
$$= \mathit{29{\cdot}4\ kN/m^2}$$

$$\text{total pressure} = \underset{(\text{kN/m}^2)}{\text{intensity of pressure}} \times \underset{(\text{m}^2)}{\text{area}}$$
$$= 29{\cdot}4 \times 3{\cdot}142 \times 0{\cdot}15 \times 0{\cdot}15$$
$$= \mathit{2{\cdot}08\ kN}$$

The result, 2·08 kilonewtons, may not appear significant. However, when you recall that 1 tonne = 1000 kgf and that 1 kgf = 9·8 newtons, you will see that 2·08 kN or 2080 N

$$= \frac{2080\ \text{N}}{9{\cdot}8\ \text{N/kgf}} \simeq 210\ \text{kgf}$$

and this represents nearly $\frac{1}{4}$ tonne. This is indeed a significant load tending to push the manlid off. No wonder it is held in position by so many stout studs and nuts.

EXAMPLE 4

A cold store cistern is 3000 mm long and 2000 mm wide. It is filled with water 1500 mm deep. Find total pressure acting on the base of the cistern.

$$\begin{aligned} \text{intensity of pressure} &= \text{head (m)} \times 9{\cdot}8\ \text{kN/m}^2 \\ &= 1{\cdot}5\ \text{m} \times 9{\cdot}8\ \text{kN/m}^2 \\ &= \mathit{14{\cdot}7\ kN/m^2} \\ \text{total pressure} &= \underset{(\text{kN/m}^2)}{\text{intensity of pressure}} \times \underset{(\text{m}^2)}{\text{area}} \\ &= 14{\cdot}7 \times 3\ \text{m} \times 2\ \text{m} \\ &= \mathit{88{\cdot}2\ kilonewtons\ (kN)} \end{aligned}$$

To summarise the facts derived from the foregoing examples:

1. intensity of pressure = metres head of water × 9·8 kN/m²
2. total pressure = intensity of pressure × area acted upon
   (kN/m²) (m²)
3. $\frac{\textit{kilonewtons}}{1000}$ = newtons
4. metres head of water = $\frac{\textit{intensity of pressure}\ (kN/m^2)}{9{\cdot}8}$

Furthermore:

1. head of water is that *vertical* distance from the free surface of water in a storage vessel to any point of consideration below.
2. when water is still, it is said to be in a static condition and the head might be referred to as *static head.*
3. water pressure is directly proportionate to its depth, or static head.

4. water pressure is exerted on all surfaces upon which it bears, and it pushes at right angles to those surfaces.

EXAMPLE 5

An altitude gauge fitted on a central heating boiler indicates a head of water of 13 metres. What pressure is applied to the boiler as a result of this height of water column?

$$\underset{(\text{kN/m}^2)}{\text{intensity of pressure}} = \underset{(\text{m})}{\text{head}} \times 9{\cdot}8$$

$$= 13 \times 9{\cdot}8 \text{ kNm}^2$$

$$= \mathit{127{\cdot}4\ kN/m^2}$$

EXAMPLE 6

Assume the gauge in *Example 5* to be a pressure gauge reading 127·4 kN/m². What head of water is creating this pressure?

$$\underset{(\text{m})}{\text{head}} = \frac{\text{intensity of pressure (kN/m}^2)}{9{\cdot}8}$$

$$= \frac{127{\cdot}4}{9{\cdot}8}$$

$$= \mathit{13\ metres}$$

EXAMPLE 7

The pressure in a water main is 500 kN/m². Ignoring frictional resistances, etc., to what height would this water rise in a vertical supply pipe?

$$\underset{(\text{m})}{\text{height or head}} = \frac{\text{intensity of pressure}}{9{\cdot}8} \ (\text{kN/m}^2)$$

$$= \frac{500}{9{\cdot}8}$$

$$= \mathit{51{\cdot}02\ metres}$$

Measurement of water pressures

The intensity of pressure in plumbing systems is measured on a Bourdon gauge, a simple device containing a near-circular loop of flattened phosphor-bronze tube. This tube—the Bourdon

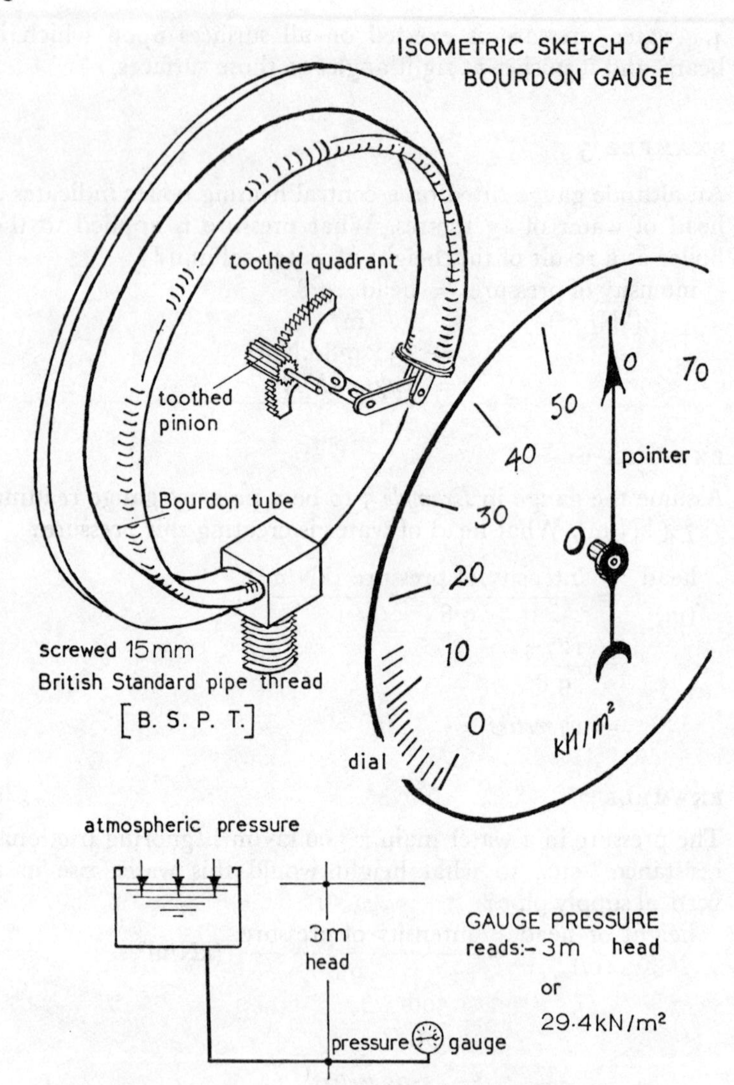

ABSOLUTE PRESSURE = gauge pressure + atmospheric pressure

= $29{\cdot}4\,\text{kN/m}^2 + 100\,\text{kN/m}^2$

Pressure Measurement

FIG. 8

tube—tends to open out when pressure is applied to its inner surfaces. In doing so it pulls on a toothed quadrant which rotates a toothed pinion. This pinion spindle carries a pointer, which moves round a suitably divided scale on the dial of the gauge to indicate the pressure within the Bourdon tube, and hence the pressure within the system of pipework to which the gauge is attached. Fig. 8 shows these working parts and their arrangement in the gauge. The dial scale of the Bourdon pressure gauge is calibrated, or divided off to register kN/m^2.

A similar gauge with a scale divided to register m head instead of kN/m^2 would be called an altitude gauge. It would be used to indicate the head of water in a central-heating system, so that if there were any loss of water from the system from evaporation or some other cause, and the cold-feed ball-valve failed to supply more water, the deficiency would be shown by loss of head on the gauge.

Gauge pressure indicates the pressure in a system caused by the head of water. It should be noticed that the atmospheric air also has weight, and exerts a pressure of some 100 kN/m^2* on the free surface of water. However, the same pressure also acts on the outside of the system, and so cancels itself out. Normally, therefore, one only refers to gauge pressure.

Absolute pressure indicates the sum of the gauge pressure+the atmospheric pressure. Fig. 8 illustrates this.

Absolute pressure = gauge pressure+atmospheric pressure

In the given example:

$$= 29{\cdot}4\ kN/m^2 + 100\ kN/m^2$$
$$= \mathit{129{\cdot}4\ kN/m^2}$$

* See Chapter 1, page 30.
$14{\cdot}7\ lbf/in^2 = 101{\cdot}3\ kN/m^2$
$14{\cdot}5\ lbf/in^2 = 100\ kN/m^2 = 1$ bar

5
The atmosphere

The atmosphere is a zone of air which envelops the earth, and extends for many miles above its surface. At high altitudes the atmosphere becomes rarefied or 'thin', and its pressure becomes less, so that high-flying aircraft have to be fitted with artificially pressurised cabins in order that passengers and crew should have air conditions near those to which they are accustomed on earth.

Composition of air

Air is a mixture of gases, and for the plumber's purpose the average composition of air may be taken to be as follows:

	%
Nitrogen	78·09
Oxygen	20·94
Argon	0·94
Carbon dioxide	0·003
	100·00

Water vapour, dust and a very small amount of other gases will also be present.

Oxygen is a clear, colourless gas without taste or smell and, besides being essential to plant and animal life, it is chemically an active gas. It will readily combine with other elements to form chemical compounds; that is, new substances with properties quite different from those of the original substances. The combination of oxygen with iron, for example, produces rust, a substance quite different from metallic iron.

The substance formed is the *oxide* of iron, since no other element is involved. If, however, there were three or more elements involved—perhaps iron, sulphur and oxygen—the substance formed would be called iron sulphate, the presence of oxygen being indicated by the ending 'ate' (see also page 76).

The fact that oxygen will combine with the exposed surfaces of metals, and moreover that oxides are non-metals, is very important to the plumber. It is particularly significant in the welding process, and will be fully dealt with later.

Oxygen is essential to combustion, and even highly inflammable materials will not burn without it. For example, a blowlamp will not burn if used in a small, airless space such as the inside of a hot-water tank.

Nitrogen is a clear, colourless gas without taste or smell. It is chemically inert, and can be obtained as a by-product of the manufacture of oxygen gas. It is compressed into grey cylinders, and is used in the welding of plastic sheet and pipe materials. This again will be more fully dealt with later.

Argon has similar properties to nitrogen, and is used in the argon-arc method of welding aluminium by electric arc.

Carbon dioxide is clear and colourless, but has a slight, not unpleasant smell. It is about one and a half times heavier than air, and so it is found in greater quantities at ground level than in the upper regions of the atmosphere.

Carbon dioxide is produced and poured into the air by the breathing processes of animals and plants. It is also produced whenever a substance is burned, and so the atmosphere in towns may be expected to have a larger carbon dioxide content than that of country districts, because of the larger number of fuel-burning appliances that are to be found there. On the other hand, carbon dioxide is also produced by the decay of plants, so that in large areas of heathland or moorland the concentration of carbon dioxide from this source may be relatively high.

Carbon dioxide will not support combustion; indeed, it is

well known as a constituent of some fire extinguishers. It will, however, readily dissolve in water, producing a dilute carbonic acid which has a corrosive effect upon metals.

Properties of air

Air occupies space—a fact illustrated each time it is pumped into a car tyre.

Air has weight—a fact not so readily observed largely because one is accustomed to it. Air can be weighed in the craft science laboratory, but it will be sufficient to know that it is about 800 times lighter than water.

Weight exerts pressure, and atmospheric weight or pressure has considerable influence on the design and working of plumbing appliances and the systems in which they are used. It is put to advantage in the working of all siphonic appliances —for example, flushing cisterns—but it can cause trouble. For instance, it may produce air-locks in plumbing systems, or siphonage which can unseal the traps of sanitary fitment waste pipes.

Atmospheric pressure can be measured on a barometer (*baro* means 'weight', *meter* means 'measure'). A simple barometer such as is shown in Fig. 9 is easily made in the craft science laboratory and will illustrate just how much pressure the atmosphere exerts.

It will be seen that the column of mercury, a substance weighing 13·6 times the weight of water, stands some 760 mm high in the glass tube. The weight of the mercury is supported or balanced by the weight of a column of the atmosphere of the same diameter as the barometer tube. The same atmospheric weight or pressure would support a column of water 13·6 times taller. That is, $13{\cdot}6 \times 760$ mm mercury $= 10350$ mm or 10·35 m of water column.

Since each m head of water exerts a pressure of 9·8 kN/m^2 on its base, the pressure exerted on the base of a water column 10·35 m will be $10{\cdot}35 \text{ m} \times 9{\cdot}8$, or 101·3 kN/m^2. It follows that atmospheric pressure will balance a pressure of $\simeq$ 100 kN/m^2. Its pressure, therefore, must also be $\simeq$ 100 kN/m^2.

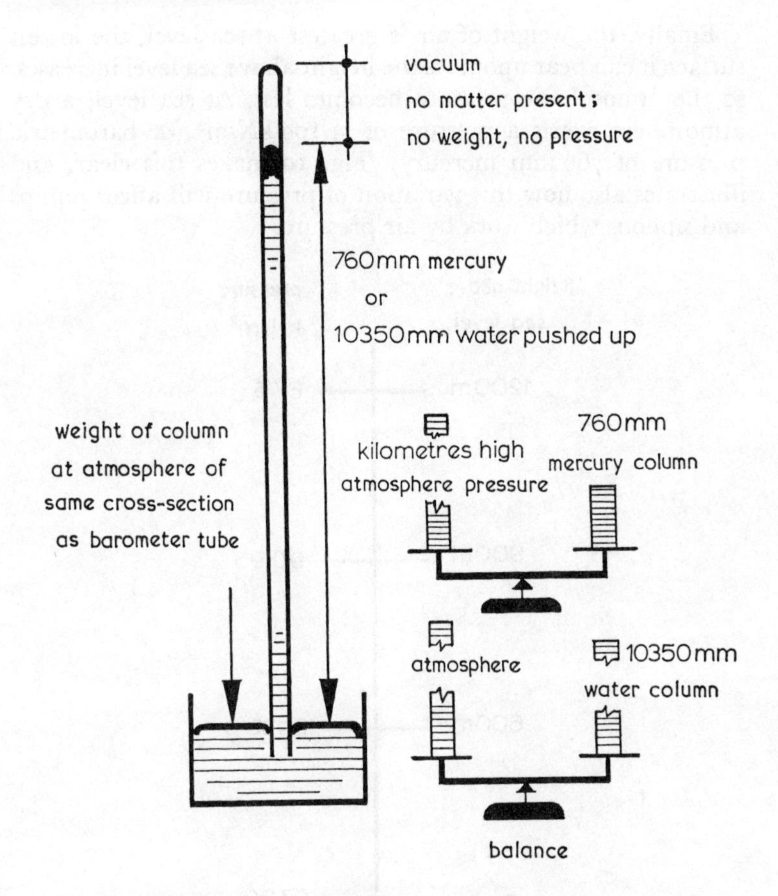

Atmosphere Weight or Pressure

FIG. 9

Pressure variations may be caused by variations of temperature, for cold air weighs more than warm air. This is an important factor in the design and placing of appliances for heating buildings by convection currents of warm air. It may also be caused by variations in humidity (moisture content of air), for moist air weighs less than dry air. This fact may be shown on a barometer, where the mercury column falls a little when the atmosphere is humid, and therefore suggests that rain is likely.

Finally, the weight of air is greatest at sea level, the lowest surface it can bear upon. As the height above sea level increases, so the atmospheric pressure becomes less. At sea level, a dry atmosphere exerts a pressure of $\simeq$ 100 kN/m² (or barometric pressure of 760 mm mercury). Fig. 10 makes this clear, and illustrates also how this variation of pressure will affect pumps and siphons which work by air pressure.

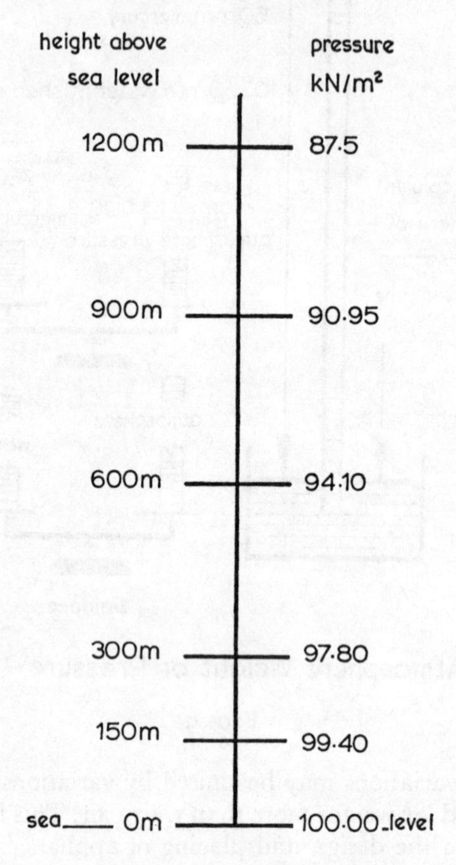

Atmospheric Pressure;
variations with altitude above sea level

FIG. 10

Atmospheric pressure and the bar

Since the bar (b) and its sub-multiple, the millibar (mbar or mb), are widely used to express fluid pressure, it is useful to examine their relationship.

At *normal temperature* (0°C) and *pressure* (760 mm of mercury column) one *standard atmosphere* or standard atmospheric pressure at sea level and latitude 45° has commonly been taken to = 14·7 lbf/in² = 34 ft head water column = 30 in mercury column. These figures have been slightly corrected from the following for convenience in use.

$$\frac{760 \text{ mm mercury}}{25{\cdot}4 \text{ mm/in}} = 29{\cdot}9 \text{ in mercury column}$$

and $29{\cdot}9 \times 13{\cdot}6$ (s.g. mercury) = 406·6 in water column
= 33·9 ft water column

and 33·9 ft head × 0·434 lbf/ft head = 14·71 lbf/in²

now, by conversion factor lbf/in² × 6·894 = kN/m²

Atm: pressure 14·7 lbf/in² × 6·894 = *101·3 kN/m²*

One bar = *100 kN/m²*

Therefore, standard atmospheric pressure = *101·3 kN/m²*

$$\text{or} = \frac{101{\cdot}3 \text{ kN/m}^2}{100 \text{ kN/m}^2} = \mathit{1{\cdot}013\ bar}$$

You will note, 1·013 bar = 1 standard atmosphere and this is so nearly 1 = 1 that it might well be taken that

1 bar = 1 atmosphere or, conversely,
1 atmosphere = 1 bar.

Furthermore, although the bar is not, strictly speaking, an S.I. metric unit, it is most likely to be widely used as a unit of pressure in steam, water, and gas.

Looking at this another way and remembering:

2·2 kg = 1 lb
and $(2{\cdot}45)^2$ cm² = 1 in²
and taking atmospheric pressure = 14·7 lbf/in²

then $\dfrac{14{\cdot}7 \text{ lbf/in}^2}{2{\cdot}2 \text{ kgf/lbf} \times (2{\cdot}45)^2 \text{ cm}^2/\text{in}^2}$

note, the lbf and in² cancel out leaving:

$$\frac{14{\cdot}7}{2{\cdot}2 \text{ kgf} \times 6{\cdot}45 \text{ cm}^2} = 1{\cdot}03 \text{ kgf/cm}^2$$

now $1 \text{ m}^2 = 100 \text{ cm} \times 100 \text{ cm} = 10\,000 \text{ cm}^2$

so $1{\cdot}03 \text{ kgf/cm}^2 = 10\,300 \text{ kfg/m}^2$
but $1 \text{ kgf} = 9{\cdot}8$ newton
so 1 atmosphere $= 10{,}300 \text{ kg/m}^2 \times 9{\cdot}8 \text{ N/kgf}$
$= 100{,}940 \text{ newtons/m}^2$
or $= 100{\cdot}9$ say *101 kN/m²*

again, this result is so nearly 100 that you will see that the bar (100 kN/m²) might conveniently be used as an alternative to atmospheric pressure (101·3 kN/m²).

See also the chapter entitled *The International Metric System S.I.*

Atmospheric corrosion

The moisture content of atmospheric air; the solubility, of carbon dioxide and sulphur gases which make this moisture acidic; and the readiness of oxygen to combine with and convert metal surfaces to non-metallic oxides, are corrosive influences which attack some metals exposed to them, with costly and damaging effect.

In the case of non-ferrous metals; that is, metals other than iron or steel, exposure to moist air will result first in the formation of a surface oxide. Following closely upon this, acidic atmospheric moisture in the form of dew, mist, or rain will act on the surface of the non-metallic oxide, producing a 'skin' of basic oxy-carbonate, and losing its own acidity in the process. This skin is adherent; that is, it sticks to the metal, and forms a protective film which resists further corrosion. The underlying metal will therefore last for a long time; that is, it will be durable.

This is not the case with the ferrous metals—iron and steel—for the oxides of these (rusts) expand as they form and crack off the metal in doing so. Fresh metallic surfaces are therefore revealed for further corrosive attack by moisture and oxygen.

It will be seen that non-ferrous metals protect themselves by developing a skin from the products of initial corrosion, and suffer no further attacks in consequence. The atmospheric corrosion of iron and steel is progressive, and unless precautions are taken these metals will be completely destroyed by oxidation, or rusting.

Protective coatings applied to metals

To prevent this, it is necessary to protect metals, and particularly iron and steel, with some form of resistant coating. Some of these in common use are listed in Table I below.

TABLE I

Type of coating	*Protective component*	*Application*
Paints (applied by brush)	Oil base or bitumen base	Cast-iron and mild steel pipework, and fittings exposed to atmospheric air or to soil corrosion
Dr Angus Smith's Solution (applied by hot dip)	Bitumen base. N.B. The term bitumen is correctly applied only to a naturally occurring mineral substance which exudes from the ground, as in Trinidad. The term is loosely used to describe pitch—a by-product of petrol refining from crude oil; or coal-tar pitch—a by-product of coal-gas manufacture	Cast-iron drain-pipes and fittings; cast-iron soil pipes and fittings

TABLE I—*cont.*

Type of coating	*Protective component*	*Application*
Liquid coal tar or bitumastic base modified by addition of anthracine oil (applied by brush or hot dip)	These preparations are the present-day counterparts of the older Angus Smith's Solution (see BS 1211)	
Galvanising (applied by hot dip)	Adherent zinc film	Mild steel tubes and fittings; mild steel hot-store vessels; mild steel cisterns
Plastics (extruded on application)	Adherent, inert polythene 'sheath'	Copper tubes for use in severely corrosive soils
Electro-plating (applied by electrolysis)	Adherent electro-deposition of corrosion-resistant metal; chromium, for example	Taps, copper tubes and fittings for towel rails, etc. Generally applied more for appearance than as true protection against corrosion
Prepared wrappings (applied by hand or machine)	Petroleum jelly (vaseline) or alternatively a bitumen base on hessian, cotton or a glass silk bandage	Pipes and structures of all kinds exposed to severe atmospheric or soil corrosion

6

The make-up and behaviour of plumbing materials

How does one decide which is the right way to do a job, and which are the right materials to use for it? Experience helps, of course, but the real key to these problems is technical 'know-how'. Know-how merely means a sufficient knowledge and understanding of the basic scientific principles which govern all plumbing design and materials and which must be applied in even the simplest plumbing operation. It helps one to learn more quickly from experience, and, even more important, it tells one *why* a job should be done in a certain way, so that a reasoned choice of method and material is possible.

Physical properties of matter

The skilful plumber needs to know a good deal about the properties of matter, and the basic rules which govern them. There is some likeness between the meaning of the words 'properties' and 'characteristics', and sometimes one is used for the other, but we shall stick to 'properties'.

'Matter' is the scientist's word for materials of any kind, and different materials can be identified by the qualities or properties which they possess. Four important properties possessed by all matter are listed here:

1 *All matter occupies space.* Air and water occupy space and so they are materials just as much as are bricks and other solid objects. Before a new hot-water system is filled with water the boiler, pipes and hot-storage vessel are full of air, which occupies these apparently empty spaces.

2 *No two pieces of matter can occupy the same space at the same time.*

Air and water are both forms of matter, and you know that the new hot-water system is full of air. Since air and water cannot be in the same space at the same time, the system cannot be filled with water until the air has been got out. This explains why a vent pipe is put at the top of the hot-store vessel to allow the air to move out as the water enters.

All very simple—when you know how. Everything which follows is just as simple, and just as useful.

3 *All matter has weight.* The weight of lead is a well-known property of that material. If one wanted a piece of material that was small but heavy, one would choose lead because of its property of heaviness. If one wanted something light one would look for a material which had the property of lightness—aluminium perhaps.

Air and water have weight, but water is 800 times heavier than air. This explains why hot-water pipes are graded to rise to the vent which is fixed at the highest point of the hot-water system. Since water is heavier, it will fall and fill the lower parts of the system first, pushing the air to the highest point in the system where it can escape through the vent pipe.

4 *All matter is inert;* that is, no matter or material can move of its own accord. The air in the emptied down hot-water system did not move out until the force of the greater weight came to push it out. It is important to remember that matter—air and water in particular—cannot move unless some force or forces make it do so.

Force is not a property of matter. It is something which moves or tries to move an object. You have seen an example of this where the greater weight of water forced the air out of the hot-water system. Force will also change, or try to change, the direction in which a moving body is going. When a cricket ball leaves the bowler's arm it moves in the direction of the wicket. When the ball strikes the bat its direction is changed by the weight force of the bar, or by the force power exerted by the batsman.

The action of forces upon air and water have a very important bearing upon the design of hot- and cold-water, sanitary and central-heating systems, and the way they are installed.

5 *Matter can exist in three forms, or 'states'.* These, and the ways of telling them apart, are given in Table 2.

When matter changes from one state to another, this is called a *change of state* or a *physical change*. Water provides a clear illustration of this kind of change. At normal temperatures

TABLE 2

Physical state	*Volume or size*	*Shape*	*Remarks*
Solid	Constant (always the same)	Constant	Rigid (resists alteration of shape by pressure)
Liquid	Constant	Variable (according to shape of container)	Fluid (opposite to rigid); will flow to take up shape of containing vessel
Gas	Variable	Variable (tends to fill space in which it exists)	Fluid

it is a liquid. Cooled to 0°C, it will freeze and change to the solid state (ice). Heated to 100°C at standard atmospheric pressure (100 kN/m^2), water turns to steam; that is, to the gaseous state. Later the explanation of the physical change and its effect on plumbing materials will be given, but for the moment it is enough to remember the following points which help to identify it:

1 The substance itself is not changed (ice, water and steam are all H_2O [water], in different states).

2 The change of state is temporary (the ice can be melted back to the liquid water, and the steam can be condensed back to the liquid water.)

Notice that to freeze water heat has to be taken away from it, and to turn it into steam heat has to be added to it. Clearly, heat can affect the state of a material, and this will again be dealt with later.

A note on pressure

Pressure can also greatly influence the state of matter; but before discussing its effects it is necessary to understand the structure of matter, and particularly the reason why it can exist in three different states. It has been said already, on page 46, that matter is composed of separate molecules which are held or 'pulled' together by a force called cohesion. However, they are not steady, but are always vibrating to and fro. The chemist calls this 'molecular vibratory motion'. How then do substances like iron, solder and so on, which appear to be so firm and solid, keep together or even keep their shape if all their molecules are moving?

Each molecule attracts or pulls its neighbours to itself by what is called the inter-molecular force of cohesion. If this force is very great, the molecules will hold together tightly and rigidly, and the substance will be firm, rigid and, in fact, solid. Should the cohesive force weaken because of the chemical or physical condition of the substance, then the molecules can more easily be pushed out of place. The material will become less rigid and will tend to flow; in other words, it will become liquid.

Further loss of cohesion between the individual molecules will allow them to move considerable distances apart. The material will not merely be fluid; it will be so fluid that it will flow in all directions to fill its containing space. In short, it will become a gas.

If a gas is compressed by some mechanical means, its widely spread molecules will be pressed closer and closer together. As they come closer, the cohesive force between them becomes stronger. Eventually, if sufficient pressure is applied, the inter-molecular cohesive force will become strong enough to hold the molecules reasonably close together, and the gas will change into a liquid. If the compression force (pressure) is released, the liquid will return to its gaseous state.

This fact is used to separate the gases which make up the air. If atmospheric air is sufficiently compressed under carefully controlled conditions of temperature and pressure, it will change from a gaseous to a liquid state. When the compression forces are released, the air is able to revert to a gas. Since the elements which make up air become gases at different temperatures, it is possible to collect them separately, and store them in cylinders for use in welding, and so on. In this way oxygen, nitrogen and argon gases are obtained.

Pressure can also affect the actual process of a physical change. For example, water, as you know, boils and changes from a liquid to a gas (steam) at 100°C at Standard Atmospheric Pressure (760 mm mercury column or 100 kN/m^2). If a pressure greater than this is acting on the water, it will not boil until a much higher temperature is reached. In fact, it will not boil at all until it has gained sufficient heat energy to do so. The extent to which pressure will affect the ability of water to change from liquid to gaseous state is shown in Table 3.

TABLE 3

Absolute pressure; i.e. gauge pressure plus atmospheric pressure kN/m^2	*Gauge pressure in* kN/m^2	*m head of water column*	*Boiling point at base of water column in °C*
100	0	0	100
129·4	29·4	3	107
158·8	58·8	6	113
217·6	117·6	12	123

(Absolute and gauge pressures are fully described under the heading 'The Measurement of Water Pressures', on page 57).

It will be clear that water in a boiler some metres below the feed cistern of a domestic hot-water or central-heating system can be heated much above its boiling temperature at atmospheric pressure. Should some fault arise in the system, so that the heated water could not circulate freely and so lose some of its heat, then the water would overheat, expand, and possibly exert such outward pressure on the boiler walls that they would burst. If this happens, the high-temperature water immediately flashes to steam which fills the room in which the boiler is fitted. The actual bursting of the boiler is less of a menace than this scalding steam.

It is interesting to notice that as increased pressure on water raises its boiling point, so reduced pressures upon it reduce its boiling point. At 600 m above sea level the atmospheric pressure is about 94 kN/m^2. In this reduced atmospheric pressure water boils at 98°C.

Chemical properties of matter

If one takes a ball-valve to pieces and puts it back together again one is better able to understand just how and why it works. It would help considerably if one could take the various plumbing materials to bits to examine how they are made up, for the chemical composition of a material has an influence on its durability, strength and fitness for use. The plumber cannot easily do this but the chemist can, and so you must rely upon what he knows and is able to tell you about the chemical structure of materials.

He will tell you that the smallest separate particle of material that it is possible to visualize, which would be far too small to be seen, is called a *molecule*. A molecule is composed of *atoms*. Sometimes it is made up of atoms all of the same kind, so that every single atom in the material is of the same type. Such materials are called *elements*. The following table gives a list of some of the elements which are commonly used in plumbing. The chemical symbols by which these elements are known are also given.

The chemical symbol is the initial letter of the element's name. If two or more elements have the same initial letter, this is used together with the next, or some other letter in the name. For example, carbon and calcium both begin with 'C'. In this case carbon is known by the initial 'C' and calcium by 'Ca'. Sometimes the name of the element is in Latin. For example, lead, copper and tin have the Latin names *plumbum*, *cuprum* and *stannum*, and their chemical symbols are Pb, Cu and Sn, in that order.

Some elements have only one atom in each molecule, in which case a molecule would be referred to simply by the symbol of the element. Others have two or more atoms in each molecule, and then the symbol of the element is followed by a

TABLE 4

Element	Chemical symbol	Remarks
Aluminium	Al	metal
Argon	A	gas
Calcium	Ca	metal
Carbon	C	non-metal (solid or gas)
Copper	Cu	metal
Hydrogen	H	gas
Iron	Fe	metal
Lead	Pb	metal
Nitrogen	N	gas
Oxygen	O	gas
Sulphur	S	non-metal (solid or gas)
Tin	Sn	metal
Zinc	Zn	metal

small figure showing how many atoms are present. For example, H_2 and O_2 indicate that one molecule of hydrogen has two atoms of hydrogen, and one molecule of oxygen has two atoms of oxygen. If a figure is placed in front of a chemical symbol it indicates how many molecules are present; thus $2H_2$ is read as two molecules of hydrogen.

At first this chemical 'shorthand' will seem difficult, perhaps hardly worth bothering about, but it will soon become clear that it is useful and, indeed, essential.

Chemical composition of matter

As has already been said, elements are substances in their simplest chemical form. If atoms, from two or more different elements join chemically together in definite proportions, a new kind of molecule, and a new kind of substance, is formed. There are two important points to notice. First, the substance formed has quite different properties from any of the elements from which it is made. Second, it is formed by what is known as a *chemical change*. A chemical change can be distinguished from a physical change in that:

1 A new substance is produced.
2 The properties of the new material are different from those of the elements from which it was made.
3 The result of the chemical change is permanent in that the new material cannot be changed back to its constituent elements by mechanical means—for example, crushing or straining. (Later you will see how a second chemical change can reduce the now complex compound to its simple constituent elements.)
4 Heat is usually produced when a chemical change takes place. (It is interesting to note that heat will hasten a chemical change or will help to bring one about. Again, this knowledge is extremely useful in plumbing work.)
5 The compounded elements are in definite and unalterable proportions.

Here is a simple chemical change which will illustrate how usefully plumber's chemistry shorthand can be applied:

Water is a chemical compound of hydrogen and oxygen. It can be made by burning the gaseous element hydrogen in the gaseous element oxygen in the proportions of two volumes to one.

Notice

i The definite proportions of the two elements going into the compound
ii The production of heat in this chemical change by the burning of H_2 in O_2
iii The new substance, liquid water, formed from the chemical compound of two gaseous elements.

This could be written as follows:

two molecules of hydrogen (two volumes) gas	plus	one molecule of oxygen (one volume) gas	(burns and produces heat)	to become	two molecules of steam which condenses to form two molecules of liquid water

Or, in pluming chemistry 'shorthand',

$2H_2+O_2$ heat→$2H_2O$ (steam) or 2 molecules of water (H_2O)

Mixtures

In considering the composition of plumbing materials, remember that elements can also exist together in simple mixtures, which are quite different from chemical compounds.

Sand stirred into water forms a mixture. The sand particles do not chemically unite with the water, but 'float' in suspension in it. The sand can be clearly seen in the water, and if the mixture is allowed to stand, the tiny sand particles will settle at the bottom of the container, leaving clear water above. Whatever the proportions of sand and water mixed together, you would still have a mixture. Notice how this differs from a chemical compound, in which the constituents are present in definite and unalterable proportions.

Solder is a mixture of lead and tin in proportions differing according to the purpose for which it is to be used. Concrete used in building work is a mixture of cement, sand and coarser material called aggregates. When the concrete is set, the sand and aggregates are clearly visible, and if the concrete is crushed they come apart and can be separated. Notice how this differs from the chemical compound, where the constituent parts lose their individual qualities in the new substance that is formed, and cannot afterwards be separated by such mechanical means as crushing.

Mixtures, then, have the following characteristics:

i No new substances are formed although the mixture may have somewhat different properties from its ingredients. For example, a concrete wall is very different from its ingredients of sand and aggregate. These ingredients have, however, in no way changed; they have simply been stuck together by cement.
ii The proportion of ingredients in a mixture can be varied.
iii The ingredients of a mixture are not chemically combined, and are each visible in the mixture.
iv The ingredients of a mixture can be separated mechanically; that is, without chemical means.
v Heat is not essential to the making of a mixture. Although

lead and tin have to be heated to make them melt and mix, no such heat is needed to mix concrete, or linseed oil and whiting putty.

Reduction of metallic ores

Through the ages, metallic elements in the earth's crust have chemically combined with the other elements, usually oxygen, carbon or sulphur. In this way, new substances have been formed which are generally rock-like in character, and are called metallic ores, though in this state they do not resemble metals at all. It is from these ores, however, that metals are obtained. The process by which they are extracted is called 'reduction', and is a special and important kind of chemical change.

First it is necessary to understand the make-up of the ore. It may be mixed with earthy matter and other material which does not bear metal, and these waste substances are called 'gangue' (pronounced 'gang'). Before anything else is done, the ores are crushed and the worthless gangue separated from the parts that are rich in metal.

Metallic ores may be oxides, carbonates, sulphides or sulphates, according to the substances with which the metal has combined:

Oxide ores are metal plus oxygen
Carbonate ores are metal plus carbon and oxygen
Sulphide ores are metal plus sulphur
Sulphate ores are metal plus sulphur and oxygen.

Table 5 illustrates the chemical composition of some of the ores of the more common plumbing metals. One can readily work out the composition by using chemical shorthand together with the above information. For example, aluminium, it will be seen, is obtained from an ore called bauxite. This is a hydrated aluminium oxide. 'Hydrated' means that water is chemically combined in the compound ore. More simply, one might say that bauxite is a chemical compound of aluminium, oxygen and water. The chemical symbol for this ore is $Al_2O_3.3H_2O$. This formula states that one molecule of the compound contains 2 atoms of aluminium, 3 atoms of oxygen

TABLE 5

Metal	*Name of ore*	*Chemical composition*		*Chemical form*
		Chemical name	*Chemical symbol*	
Aluminium	Bauxite	Hydrated aluminium oxide	$Al_2O_3.3H_2O$	Hydrated oxide
Copper	Copper pyrites	Copper and iron sulphide	$CuFeS_2$	Sulphide
Iron	Haematite	Iron oxide	Fe_2O_3	Oxide
Lead	Galena	Lead sulphide	PbS	Sulphide
Tin	Cassiterite	Tin oxide	SnO_2	Oxide
Zinc	Blende	Zinc sulphide	ZnS	Sulphide
	Calamine	Zinc carbonate	$ZnCO_3$	Carbonate

and 3 molecules of water. Once you grasp this chemical build-up, you can see that if these elements are separated then the metallic aluminium can be removed and processed into sheets and equipment for plumbing work.

The chemical composition of ores of other metals are perhaps easier to follow. Do try to separate them into their constituent parts, for then it is easier to understand how the reduction process works (see also Table 5).

Oxygen combines readily with metals—hence the oxide, carbonate and sulphate ores—but under suitable conditions, usually at high temperatures, the oxygen content of a metallic ore prefers to re-combine with some other element, for example carbon. On page 74 it was mentioned that heat could hasten or produce a chemical change, and that there is the possibility of a second chemical change reversing an earlier one. Thus, when an oxide ore of a metal is put together with sufficient carbon in the presence of enough heat to bring about a chemical change, the oxygen content of the ore will leave it to form a new compound with the carbon. The molten metal will be left behind, and can be collected and used.

The extraction of tin from its oxide ore (SnO_2) may be taken as an example:

tin oxide (a rocklike ore)	plus	carbon (in the form of coke)	(heat is produced by the burning of the coke)	the oxygen content of the ore combines with the carbon to form carbon dioxide gas, and leaves the metal tin behind

Or

$$SnO_2+C+heat \rightarrow CO_2 \text{ (escapes as gas)}+Sn \text{ (tin)}$$

Notice how the coke (carbon) serves a double purpose; it burns to produce the temperature necessary to bring about a chemical separation of the elements tin and oxygen, and it acts as a 'blotter', soaking up the oxygen to form a new compound of carbon and oxygen (CO_2). The molten tin is collected at the bottom of the furnace.

One can sum up by noting that oxidation is the result of a chemical change in which the element oxygen combines with another element, a metal. Reduction is the result of a chemical change artificially brought about to separate the oxygen from the metal. By this reduction, or reverse chemical change, a chemical change can quickly be undone, and useful metal extracted from seemingly useless pieces of ore.

The extraction of *iron* from its ores follows much the same pattern. *Lead*, which is derived from a sulphide ore, presents additional problems. Sometimes the sulphur content of the ore is replaced by oxygen in a preliminary treatment with heat. The ore is then reduced in the same way as those of tin or iron, in a blast furnace. Thus

$$PbS+O_4 \rightarrow PbO_2+SO_2\text{; and } PbO_2+C \rightarrow Pb+CO_2$$

The blast furnace in which the reduction process, called smelting, takes place, is a tall circular furnace of steel plates lined with firebrick. It works non-stop, and charges of ore,

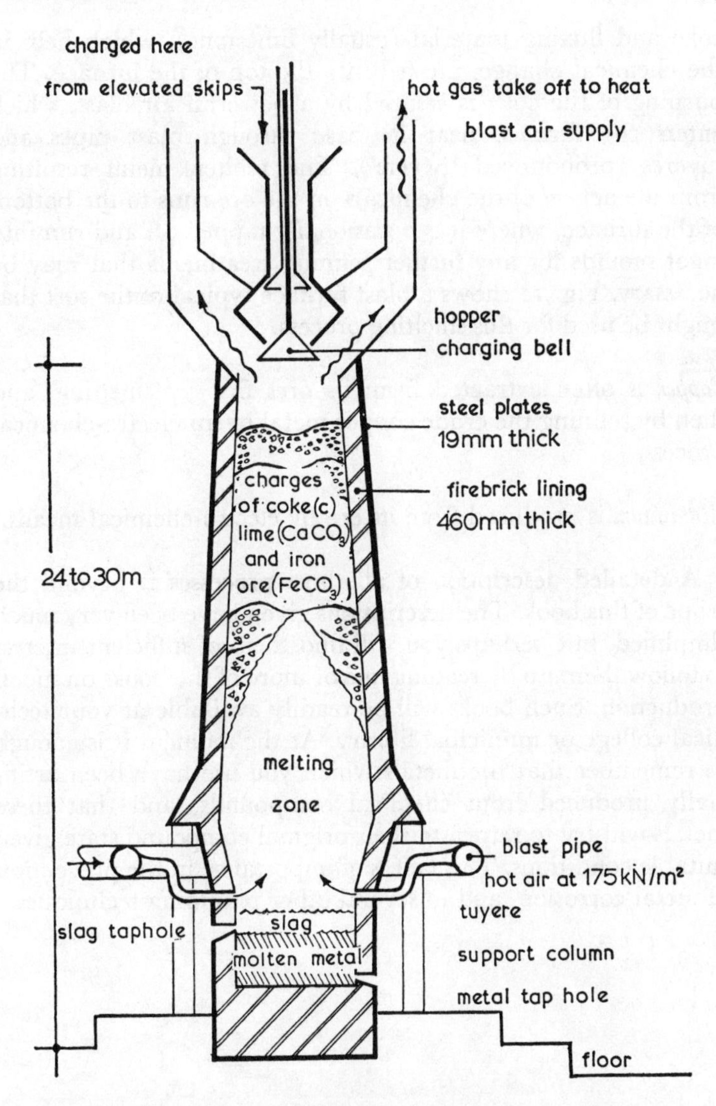
charged here
from elevated skips
hot gas take off to heat
blast air supply
hopper
charging bell
steel plates
19mm thick
charges
of coke(c)
lime($CaCO_3$)
and iron
ore($FeCO_3$)
firebrick lining
460mm thick
24 to 30m
melting
zone
blast pipe
hot air at 175kN/m²
tuyere
slag taphole
slag
molten metal
support column
metal tap hole
floor
Diagram of blast furnace as used
in the reduction of metal

FIG. 11

coke and fluxing material (usually limestone), which help in the chemical change, are fed into the top of the furnace. The burning of the coke is assured by a powerful air blast, which enters the furnace near its base through blast pipes and tuyeres (pronounced 'twyers'). The molten metal resulting from the action of the chemicals on the ore runs to the bottom of the furnace, where it is occasionally tapped off and run into ingot moulds for any further refining treatments that may be necessary. Fig. 11 shows a blast furnace typical of the sort that might be used for this smelting process.

Copper is often extracted from its ores first by smelting, and then by refining the crude copper metal by an electro-chemical process.

Aluminium is produced from its ore by electro-chemical means.

A detailed description of all these processes is beyond the scope of this book. The descriptions given have been very much simplified, but perhaps you will find them of sufficient interest to follow them up by reading one or more of the books on metal production. Such books will be readily available at your technical college or municipal library. At the moment it is enough to remember that the metals which you use have been artificially produced from chemical compounds, and that these metals will try to return to their original compound state given suitable conditions. This fact is of importance in the prevention of metal corrosion, and in several other plumbing techniques.

7
Heat and its measurement

Heat is a form of energy. 'Energy' means capacity for doing work. But what work goes on in boiling water?

You will see (page 90) how the molecules of substances are always vibrating to and fro. They need energy for this 'work', and heat provides it. The more energy the molecules possess, the more vigorously and further apart they will be able to vibrate. Given sufficient heat input or heat energy, the molecules can so step up their vibratory rate, and so weaken their cohesive bonds, that they can change from the solid to the liquid, or the liquid to the gaseous state. For example, ice is water in a solid state. It has so little heat content that its molecules hardly vibrate at all. Since they are close together, the cohesion between them is great and ice is therefore rigid, or solidified. If ice is heated, the molecules gain energy, vibrate farther apart, and weaken their cohesive bonds. The solid ice changes into liquid water. Further heating of the water increases the heat energy until the molecules vibrate so strongly that they actually 'jump' out of the water to form gaseous steam.

In the same way solder, which is normally a solid metal, becomes liquid when sufficiently heated. Mild steel pipes are not easily bent when cold, unless a bending machine is available. When the pipe is heated, however, it is much more bendable because its molecules vibrate further apart, lose some of their cohesive force, and do not 'pull together' so hard. The hotter the pipe, the easier it is to bend. But care must be taken that it is not heated too much, or the molecules will vibrate so far apart that the metal will melt and become liquid.

These are a few illustrations of the effect heat can have on the physical state of materials. No doubt you can think of many more. As might be expected, the taking away of heat from a material has the opposite effect to the addition of heat. As materials cool they lose heat energy, their molecules come closer together, and greater cohesion exists between them. As a result the material changes its physical state. For example, if steam is cooled it changes from gas to liquid. Further cooling of water to 0°C causes it to become solid ice. If molten solder is allowed to cool, it solidifies. When the hot, bent mild steel pipes are cooled their molecules slow up their vibratory rate, come closer together, and the bend in the pipe becomes rigid again.

All materials contain heat. Even ice, which one normally regards as cold, is very hot in comparison to liquid air, which exists at some 166°C below the freezing point of water.

Temperature or 'degree of hotness' of a material is measured on a thermometer.

Thermometers

If the end of a fine-bore glass tube is heated to a dull red heat, it becomes less solid and can be 'blown' to form a bulb. When cold this bulb is filled with mercury, a metallic element which is liquid at normal temperatures. The bulb is warmed, and the mercury expands up the fine bore of the tube, pushing out all air as it does so. When all the air has been expelled, the end of the tube is sealed off by its edges being melted together. The mercury cools down and contracts to a smaller size, dropping down toward the bulbous reservoir at the bottom of the tube as it does so. The result is an instrument which, when marked in degrees, can be used to measure temperatures.

Thermometer degree markings are etched in the glass between two fixed points. The first fixed point is the temperature at which ice melts, and the second the temperature at which pure water boils at standard atmospheric pressure ($101 \cdot 3 \simeq 100$ kN/m^2). There are 100 degree divisions between the two fixed points. The degrees are numbered according to whether the thermometer is to measure in the Celsius or the Kelvin scale. Fig. 12 shows this more clearly.

Sometimes it is necessary to convert a temperature given in

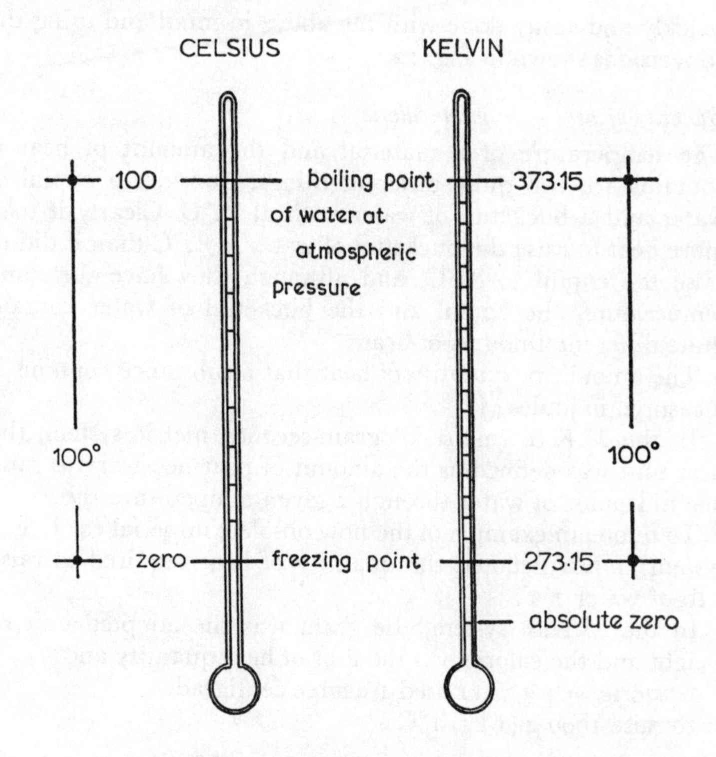

FIG. 12

one scale to the equivalent temperature on the other. It is helpful to remember that each single degree Celsius equals the same temperature interval as each single degree Kelvin. Furthermore, whereas the freezing point Celsius is 0° or zero, it is 273·15° on the Kelvin scale.

The conversion of temperature from one scale to the other is quickly and easily done with the above in mind and using the conversions shown in Fig. 12.

Quantity of heat—the gram calorie

The temperature of a material and the amount of heat it contains are two quite different things. Consider a cupful of water and a bucketful of water both at 82°C. Clearly it took more heat to raise the bucketful of water to 82°C than it did to raise the cupful to 82°C, and although they have the same temperature, the cupful and the bucketful of water contain quite different amounts of heat.

The amount or quantity of heat that a substance contains is measured in joules (J).

In the M.K.S. (metre-kilogram-second) metric system, the heat unit was defined as the amount of heat necessary to raise one unit mass of water through a given temperature rise.

To quote an example of the now obsolete imperial (or U.K.) system; 1 Btu denoted the quantity of heat required to raise 1 lb of water 1°F.

In the M.K.S. system, the gram was the adopted unit of weight and the calorie was the unit of heat quantity and:

1 calorie = 1 gram raised 1 degree centigrade
or to raise 1000 g (1 kg) 1°C

quantity of heat in calories = 1000 g × 1°C
= 1000 calories
or 1 kilo calorie would raise 1 litre of water 1°C

Note the use of the term *Centigrade* which was correct for the M.K.S. system. In the S.I. system, however, the correct term is *Celsius*.

Now, 4·186 joules = 1 calorie
or, we need 4·186 × more joules than calories for a given duty

And so, to raise 1 gram of water 1°C
we need 4·186 joules × 1 gram × 1°C
= 4·186 joules

Or, since 1 litre = 1000 grams and is the unit of liquid capacity in the S.I. metric system, quantity of heat required to raise 1 litre of water 1°C

$= 4{\cdot}186 \text{ J} \times 1000 \text{ g} \times 1°\text{C}$
$= 4{\cdot}186 \text{ J or } 4{\cdot}186 \text{ kJ}$

Then, 4·186 kJ will raise 1 litre of water 1°C

and this forms the basis of simple calculations of heat required for raising given amounts of water through given rises in temperature.

Thus,

quantity of heat energy required in kilojoules

$= 4{\cdot}186 \times \text{kg} \times (t_1 - t_2)$

where kg = kilograms of water being heated*
t_1 = final heated water temperature
t_2 = water temperature before heating

Conversely, the statement can be used to find the quantity of heat given off by a known amount of water when it cools down a known number of degrees Celsius.

Thus,

heat energy emitted in kilojoules

$= 4{\cdot}186 \text{ kJ} \times \text{kg} \times (t_1 - t_2)$

where kg = kilograms of water involved
t_1 = the hot water temperature
t_2 = the ultimate cooled water temperature

For examples in application see Chapter 1 (International Metric System) and Chapter 30, *Plumbing Calculations*, also pages 86 to 89.

Specific heat

The specific heat of a material is the amount of heat necessary to raise 1 kg of that material 1 deg C. The specific heat of water is 4·186 kJ/kg deg C.

Not all materials require this amount of heat to raise 1 kg by 1 deg. C, and different materials have different specific heat values as Table 6 shows. Furthermore, the Specific Heat Capacities of all substances vary slightly as their temperature varies. The following Table is sufficiently accurate for our present needs.

* For practical purposes 1 kg water = 1 litre.

TABLE 6 THE SPECIFIC HEAT VALUES OF SOME PLUMBING MATERIALS

Material	*Btu/lb deg F*	*kilojoules/kilogram deg C*
Water	1·000 (one)	4·186
Aluminium	0·212	0·887
Cast iron	0·13	0·544
Mild steel	0·12	0·502
Zinc	0·095	0·397
Copper	0·092	0·385
Tin	0·056	0·234
Lead	0·03	0·125
Mercury	0·03	0·125
Air	0·25	1·046

(Arranged in descending order of 'heat capacity' or specific heat values, except the last, Air).

It will be seen that only a small amount of heat is required to raise 1 kg of air 1 deg C. Mercury, too, has a low specific heat value and is consequently very sensitive to heat variation. This is one reason why it is used in thermometers.

Lead has a low S.H., and so has tin although, as will be seen, tin requires more heat to raise its temperature 1 deg C than does lead. In consequence tin takes longer to cool down than lead, and this explains why one is sometimes troubled with 'tin runs' out of the bottom of a wiped soldered joint. The more rapidly cooling lead solidifies in the solder joint whilst the tin is still liquid and can run over the surface of the joint.

A knowledge of specific heats and the Heat Energy requirement to raise a given mass of substance through a given rise in temperature is necessary in work on domestic hot-water and central-heating systems. One or two examples will show how simple it is to use them.

EXAMPLE 1

How many joules will be required to raise 150 litres of water from 10°C to 65·5°C?

Note: 1 litre = 1000 g = 1 kg   S.H. water = 4·186 kJ/kg

Heat energy required = mass × S.H. × rise in temperature)
(kJ) (kg) (kJ/kg) (deg C)
= 150 × 4·186 × 55·5
= 34 848·45 kJ
= *34 848 450 joules*

Checking this by working the same example in imperial values:

$$\frac{150 \text{ litres}}{4\cdot54 \text{ litres/gal}} = 33 \text{ gal or } 150 \text{ litres} \times 0\cdot22 = 33 \text{ gal}$$

55·5°C = 100°F

Heat energy required = 33 gal × 10 lb × 100°F
(Btu) = *33 000 Btu*

Now, 1 Btu = 1055 joule
and 33 000 Btu × 1055 J/Btu = *34 815 000 joules*
which gives virtually the same result.

EXAMPLE 2

10 kg of copper is heated through 80°C. How many joules of heat energy is taken up by the copper?

Heat energy = 10 kg × 0·385 (S.H. copper) × 80°C
= 308 kJ
= *308 000 joules*

Checking, using imperial values:

10 kg × 2·2 lb/kg = 22 lbs of copper
S.H. copper (imperial) = 0·092 Btu/lb
80°C temperature difference = 144°F temperature difference
Heat energy = 22 lb × 0·092 × 144°F
= 291·5 Btu
and 291·5 Btu × 1055 J/Btu = *307 533 joules*

EXAMPLE 3

Calculate the number of joules required to raise 140 litres of water from 10°C to 60 C. The water is held in a copper container which weighs 5 kg.

Answer Heat energy for water = 140 litres × 4·186 kJ/kg × 50 deg C (= 140 kg)

= 29 302 kJ = *29 302 000 joules*

Heat energy for copper container = 5 kg × 0·385 kJ/kg × 50 deg C (S.H. from Table)

= 98·25 kJ = *98 250 joules*

Then total heat energy required = 29 302 000 J + 98 250 J (water) (copper)

= *29 400 250 joules* or 29 400 kJ

Note 1 joule = 1 watt × 1 second and the amounts of heat energy derived in examples 1, 2 and 3 above are those which would be required if the substances involved were to be heated through the stated temperature rise in *one second.* However, in practice the 'heating-up' periods will be of much longer duration. Supposing in example 3 the heating up period was to be 1 hour (3600 seconds) then the *power* required for the stated duty would be:

$$\text{kW} = \frac{\text{heat energy required in kJ}}{3600 \text{ seconds}}$$

$$= \frac{29\,400 \text{ kJ}}{3600 \text{ s}}$$

$$= 8{\cdot}166 \; kW$$

See example 6 in the section entitled 'Heat and Work' in the chapter entitled 'The International Metric System' (page 20).

EXAMPLE 4

9 litres of water cool from 82°C to 60°C. How many joules of heat energy will be given off?

This is a problem involving specific Heat but in the opposite direction of heat flow to the previous examples. Here, heat is being given off from a cooling substance whereas previously we were considering heat taken up by substances being heated.

However, the solution is found in much the same way, as follows:

Heat energy given off in kJ $= 9 \text{ kg} \times 4{\cdot}186 \text{ kJ/kg} \times 22 \text{ deg C}$
$= 828{\cdot}828 \text{ kJ} =$ *828 828 joules*

If the heat is liberated over a period of 1 hour (3600 seconds) it would be liberated at the rate of

$$\frac{828\ 828 \text{ joules}}{3600 \text{ seconds}} = 230 \text{ J/s} \begin{array}{l} = 230 \text{ watts} \\ = 0{\cdot}23 \text{ kW} \end{array}$$

In the above examples, no allowance has been made for the thermal efficiency of the water heating apparatus or for continuous loss of heat from the hot water container to the cooler surrounding air. These considerations are, of course important points, which affect the design and installation of domestic hot water supply and central heating systems. They are dealt with in *Plumbing 2*.

8

Heat: its effects on plumbing materials

The following effects of heat on materials have now been noted:

i It can bring about a change of state (page 82).
ii It can hasten or bring about a chemical change (page 74).
iii As a result of i or ii above, the appearance, properties and behaviour of the material are very much altered.

There is another equally important effect of heat on materials; they change in size according to their heat content. It is well known that when a substance or material is heated it gets larger, or expands; and when it cools it gets smaller, or contracts. In other words, material 'moves' as its temperature varies, and this is called the heat movement or the thermal movement of materials.

Why do materials expand when heated and contract on cooling? Imagine the space taken up by a group of people standing fairly still. Now, if all these people started to rock-'n'-roll, they would jostle and push one another about, and in doing so they would take up more space. The separate molecules which go to make up a material are always on the move, vibrating to and fro at a rate depending upon the amount of heat energy they possess. When cold, their heat energy is small and they keep fairly still and close to one another, taking up as little space as possible. When the material is heated, its molecules gain energy and start to 'dance' with increasing vigour and so, of course, they take up more space. The result is an increase in the size of the material; that is, the material expands on being heated.

When the material cools down it loses heat energy, the molecules slow down the rate of their vibratory 'dances', and a reduction in size, or contradiction of the material, results.

Coefficient of thermal expansion

The amount of material expands when it is heated depends upon the material itself; whether it is in a solid, liquid or gaseous state; and, of course, upon its heat content. The amount that solids will expand for each °C of temperature rise is easily measured and is fairly constant. To whatever extent a solid expands or contracts for 1° change in temperature, it will expand or contract ten times as much for 10° change in temperature. Liquids and gases do not behave quite so conveniently, and water in particular behaves in a most unexpected manner, as has already been shown (page 44 and Fig. 3).

For the time being, consider the effect of thermal movement on solid materials such as pipes, boilers, and sheet metal roof coverings. Thermal movement affects all the dimensions of a material. Length, width and thickness all increase as the temperature of the material increases. Since most of the materials the plumber deals with—for example, pipes, and bays of sheet metal on roofs—are so much longer than they are wide or thick, it will be convenient to examine only the more readily seen effects of heat on the length of material.

Reference will be made to the 'linear coefficients of thermal expansion'. The word 'linear' means lengthwise and 'coefficient' means fraction; so the term really describes that fraction by which a given or unit length of material will expand when its temperature is increased by 1°, or will contract when its temperature decreases by 1°.

This unit length can be measured by any convenient scale, generally in metres, but whatever the unit length used, then the fraction that the material expands for 1° change in temperature will also be measured as a fraction of that unit.

Table 7 gives the coefficients for the commonly used plumbing materials.

The table shows that lead 'moves', or expands and contracts, nearly twice as much as does copper. Plumber's solder also expands less than lead. There is very little difference between them, but it is enough to set up damaging stresses between sheets of lead and the solder used to join them. This occurs especially where such solder joints are exposed to a wide variation in temperature; for example, that between day and night summer temperatures on a roof. The difference in the expansion coefficients of solder and copper are even greater. For this reason it is not considered good practice to use solder to join sheets of metal used for roof weathering.

Notice that mercury expands a good deal more than the other materials. For a small change in temperature there will be a visible change in the length of a mercury column, and this

TABLE 7 LINEAR COEFFICIENTS OF THERMAL EXPANSION

Material	*Variation per unit length for one degree Celsius temperature variation*
Solid metal elements	
Lead	0·000029
Zinc	0·000029
Aluminium	0·000026
Tin	0·000021
Copper	0·000016
Iron	0·000011
Solid metal mixtures or alloys	
Plumber's solder	0·000025
Cast iron	0·000011
Mild steel	0·000011
Liquid metal element	
Mercury	0·00005
Non-metal	
Polythene ('plastics') over its normal working range, 5°C to 60°C	0·00018

Note:

Superficial (area) coefficient of expansion = linear coefficient × 2

Cubic (volumetric) coefficient of expansion = linear coefficient × 3 is another reason why mercury is so useful for measuring temperatures in a thermometer.

Now that 'plastic' materials are more widely used in plumbing, it is worth while noting the comparatively high coefficient of expansion of polythene.

Using the coefficients given in Table 7, and the following rule, it is possible to find out just how much a given length of material will alter for a given change in its temperature:

Change in length	= length of material	× temperature	× coefficient of linear expansion
(heated, it expands and increases in length. Cooled, it contracts and reduces in length)	(in convenient unit lengths, usually m)	change in material	

A few examples will show how this rule can be usefully applied to plumber's work.

EXAMPLE 1

A copper hot-water pipe, 30 m long, is filled and tested with water at 10°C. By how much will the length of this pipe increase when it carries hot water at 60°C.

The question tells you the length of the pipe and the temperature change to which it is subject. From Table 7 you know that copper's linear coefficient of expansion is 0·000016 deg. C.

$$\text{Increase in length} = 30\,000\ \text{mm} \times (60°\text{C} - 10°\text{C}) \times 0{\cdot}000016$$
$$= 1\,500\,000 \times 0{\cdot}000016$$
$$= \textit{24 mm increase in length}$$

EXAMPLE 2

If the same hot-water pipe is cooled from 60°C to 10°C, how much will it shorten or contract?

$$\text{Decrease in length} = 30\text{m} \times (60°\text{C} - 10°\text{C}) \times 0{\cdot}000016$$
$$= 1500 \times 0{\cdot}000016$$
$$= 0{\cdot}024\ \text{m}\ \textit{or 24 mm decrease in length}$$

Example 2 shows that temperature change will make a material contract at the same rate as it will expand.

EXAMPLE 3

A polythene waste pipe at 15°C receives a discharge of hot water at 50°C. If the waste pipe is 3 m long, how much will this temperature increase cause it to lengthen?

$$\text{Increase in length} = 3\text{m} \times (50°\text{C} - 15°\text{C}) \times 0{\cdot}00018$$
$$= 135 \times 0{\cdot}00018$$
$$= \mathit{0{\cdot}024\ m} \text{ or } \mathit{24\ mm}$$

Work out how much a 3 m copper waste pipe would have expanded for the same temperature rise, and you will appreciate just how much more polythene expands than copper. You will also realise how important a knowledge of thermal movement in materials is in the design and fixing of pipes used to convey hot water.

Thermal movement of pipework

Small as these changes in material size may appear to be, the fact remains that they are irresistible and can exert a considerable push or pull on anything which tries to restrain their movement. Unless suitable precautions are taken to accommodate thermal movement in pipework, one or other of the following troubles will arise:

i The tube will buckle or 'snake' if its ends are restrained from movement by being solidly built into walls.
ii If the tube is of such diameter and wall thickness that it resists the tendency to buckle, then the expansion and contraction forces will impose severe and damaging strains upon walls into which the pipe is built.
iii Thermal movement of tubes through tight wall openings will give rise to unpleasant scraping noises, which will be transmitted throughout the pipework system.

The precautions taken generally include:

i The provision of purpose-made pipe sleeves. These are cut from odd lengths of mild steel tube, and are built into the wall.

The smaller pipe is then passed through the sleeve, and is thus given freedom of movement.

ii The provision of some form of expansion absorber at some point on a long pipeline which is liable to be affected by large variations of temperature. In some cases a simple loop is formed in the pipe. This can frequently be done by means of fittings and a simple diversion of the pipe route. Sometimes special expansion loops in the shape of a horseshoe are employed. Notice that all expansion loops should be fitted in the horizontal plane; that is, in line with the ceiling or floor, in order that sludge or air pockets should be avoided. As an alternative to loops, specially made expansion fittings can be inserted in the pipelines where necessary.

Cast-iron smoke pipes from boilers are often wrongly built solid into the flue wall. When the boiler is lit the smoke pipe gets very hot and expands, often to a degree sufficient to fracture the brickwork of the flue. A useful tip in such cases is to wrap the cast-iron pipe with two thicknesses of stout paper before building it into the brickwork. As the pipe heats up the paper scorches away, and the small gap which it leaves will generally be enough to accommodate the increase in pipe diameter on heating up.

Fig. 13 illustrates the effects of heat movement on pipes and fixings, and shows some types of precautionary measures against expansion which are commonly employed.

Thermal movement of sheet-metal roof coverings

Temperatures of up to 60°C have been recorded on the surface of metal roof coverings in this country. The difference between this temperature, caused by heat from the summer sun, and the freezing and below freezing temperatures of winter are considerable. Clearly, some precautions against the effect of thermal movement in metal roof coverings is essential. Even the inevitable variation of temperature between day and night can give rise to serious troubles unless simple, common-sense measures are taken to accommodate the expansion and contraction of roofing materials.

What thermal movement should be expected and how can

one best guard against its bad effects? Here is an example. Imagine a roof 30 m by 15 m covered by one sheet of lead. Then suppose that this imaginary roof acquires a temperature of 50°C during the day but cools to 15°C at night. What sort of movement might one expect?

Its length increase would be: 30 m × 35°C temp. diff. × 0·000029

$$= 0{\cdot}03 \text{ m}$$
$$= \text{say } \mathit{30\ mm}$$

Its width, being half the length, would expand 15 mm. Its superficial (area) expansion could be found by multiplying its area in square metres by the temperature rise, and again by the linear coefficient of expansion × 2 (see footnote, Table 7, page 92) as follows:

$$30 \text{ m} \times 15 \text{ m} \times 35°\text{C} \times 0{\cdot}000029 \times 2$$
$$= 31\,500 \times 0{\cdot}000029$$
$$= \mathit{0{\cdot}9\ m^2\ increase\ in\ area}$$

What would happen in the case of such expansion? Either the metal would push the extra length over the edge of the roof, if it were flat and there were no restrictions to such a movement; or, if it could not move along its length or width, the sheet of lead would rise up to form a hollow mound in its centre. If it were free to move in one direction but not the other, then expansion in the free direction would take place over the roof edge, but the expansion the other way would make the lead rise up all along the centre-line of the sheet, parallel to the fixed edges.

If the sheet were free to move in all directions it would push outwards as it expanded, but on cooling it would have to pull back again. The friction between it and the roof decking, and the weight of the lead, would resist this backward pull, so that the lead would be subject to severe tensile strains. Eventually it would stretch, become thin, and tear.

Where restriction of movement causes the sheet to rise up when expanded by heat, it will return to flat on cooling. If these movements are repeated, as they would be with each

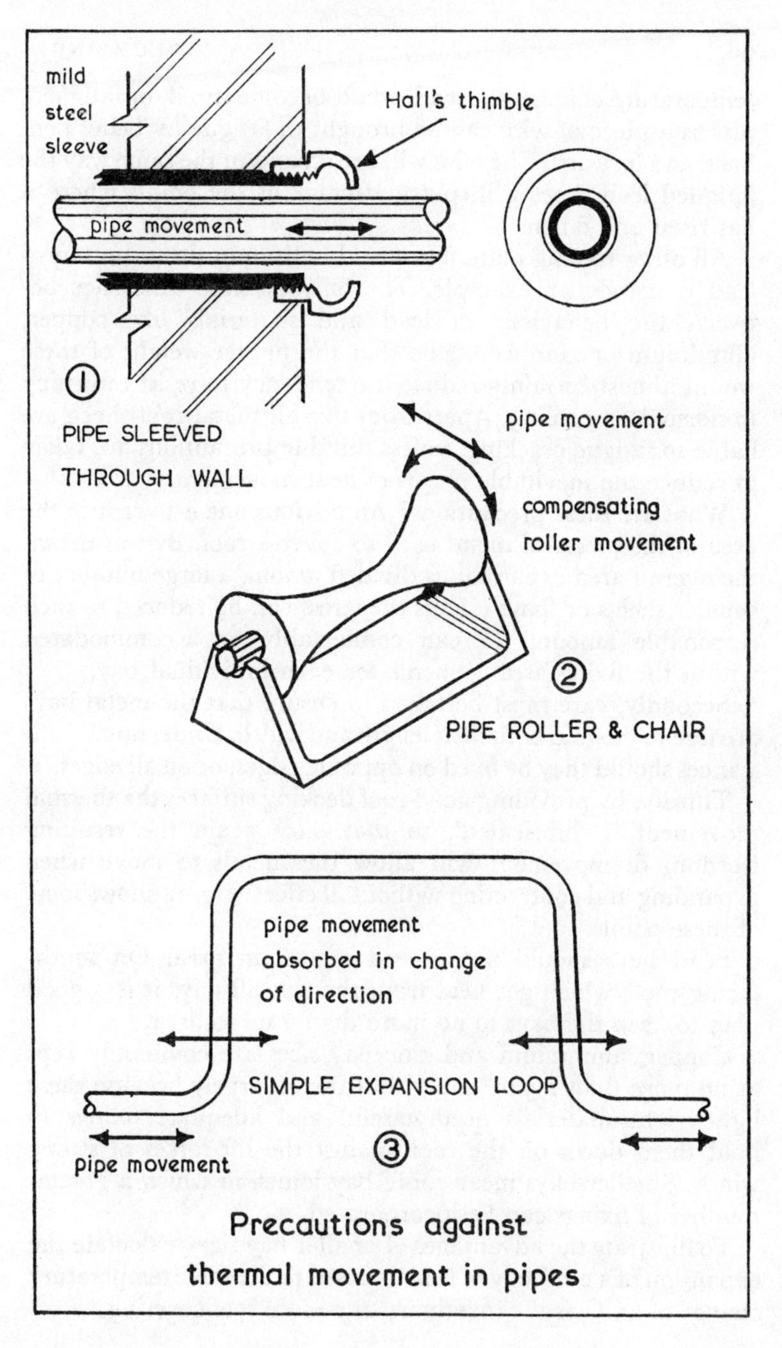
mild
steel
sleeve
Hall's thimble
pipe movement
①
PIPE SLEEVED
THROUGH WALL
pipe movement
compensating
roller movement
②
PIPE ROLLER & CHAIR
pipe movement
absorbed in change
of direction
SIMPLE EXPANSION LOOP
③
pipe movement
Precautions against
thermal movement in pipes

FIG. 13

temperature change, the lead would become 'tired' or fatigued, just as a piece of wire can be brought to fatigue by being bent back and forward. The wire will break, and in the same way the fatigued lead sheet will split and crack at the points where it has risen and fallen.

All other roofing materials would behave in the same way—lead is merely an example. The only possible difference between the behaviour of lead and materials like copper, aluminium or zinc would be that the lighter weight of these would almost certainly reduce the tendency to resist returning to normal on cooling. Apart from this all metal roof sheets are liable to fatigue cracking, unless suitable precautions are taken to reduce the inevitable effects of heat movement.

What are these precautions? An obvious one is to reduce the area of the pieces of metal used to cover a roof. By this means the overall area expansion is divided among a large number of smaller sheets or 'bays'. Thus the stress can be reduced to such reasonable amounts as can comfortably be accommodated within the fixing arrangements for each individual bay.

Secondly, care must be taken to ensure that the metal bays are free to move in both their length and width. Under no circumstances should they be fixed on opposite edges, or on all edges.

Thirdly, by providing good roof decking surfaces the thermal movement is 'lubricated', so that once again the resulting freedom of movement will allow the metals to move when expanding and contracting without ill effect. Fig. 14 shows some of these points.

Lead bays should not exceed 2·3 m² in area. On south-facing roofs which get heat from the sun all day, it is a good plan to keep the bays to no more than 2 m² in area.

Copper, aluminium and zinc bay sizes are commonly kept at no more than 1·5 m² super, but this is largely because these lightweight materials need careful and adequate fixings to hold them down on the roof against the lift forces of strong winds. Smaller bays mean more bay joints, in which a greater number of fixings can be incorporated.

To illustrate the advantages of smaller bay sizes, calculate the expansion of a 2 m² bay of lead exposed to the same temperature change of 35°C as the imaginary one-piece roof covering.

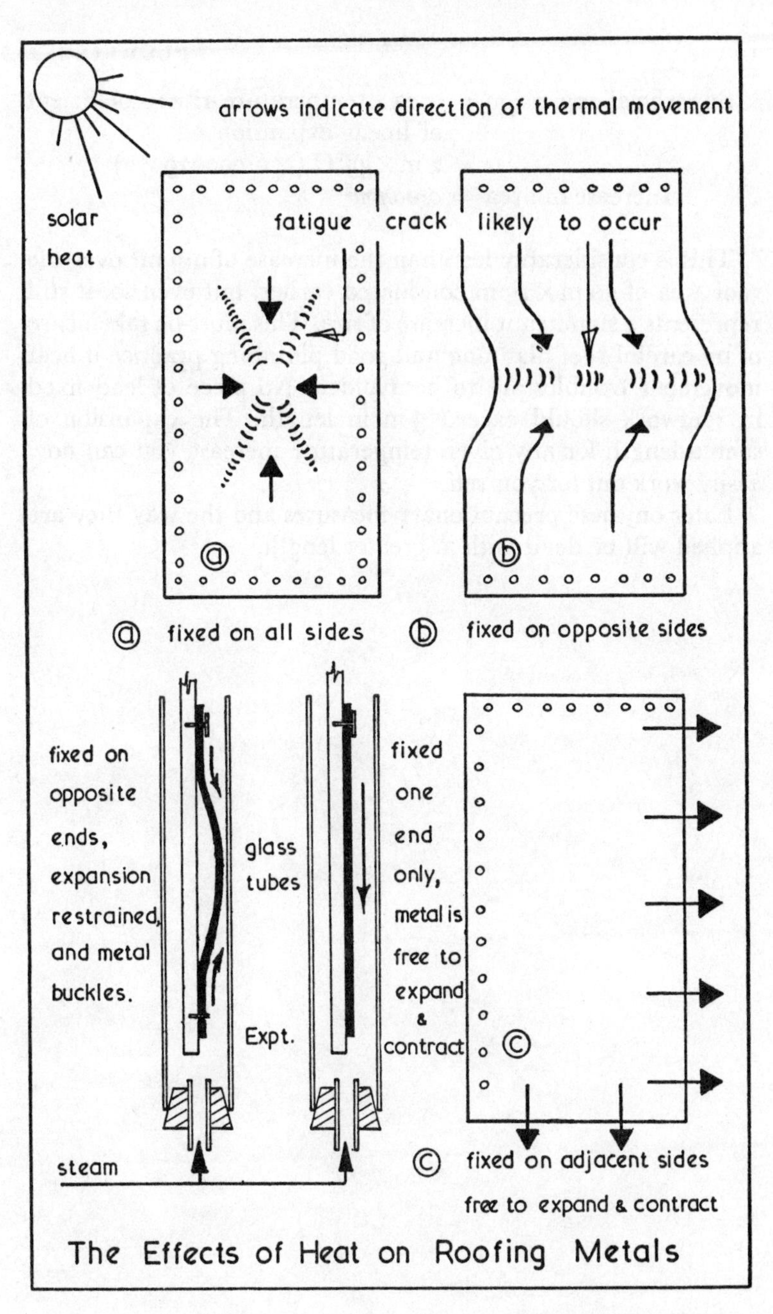
arrows indicate direction of thermal movement
solar
heat
fatigue crack likely to occur
a
b
a fixed on all sides
b fixed on opposite sides
fixed on
opposite
ends,
expansion
restrained,
and metal
buckles.
glass
tubes
Expt.
fixed
one
end
only,
metal is
free to
expand
&
contract
c
steam
c fixed on adjacent sides
free to expand & contract
The Effects of Heat on Roofing Metals

FIG. 14

$$\begin{aligned}\text{Superficial expansion} &= \text{area} \times \text{temperature rise} \times \text{coefficient of linear expansion} \times 2\\ &= 2\ \text{m} \times 35^\circ\text{C}\ (\times 0{\cdot}000029 \times 2)\\ \text{Increase in area} &= \mathit{0{\cdot}004\ m^2}\end{aligned}$$

This is considerably less than the increase of 0·9 m^2 over the roof area of 30 m × 15 m considered earlier, but even so, it still represents a significant increase of size. This must be taken care of by careful roof planning and good plumbing practice if heat movement troubles are to be avoided. No piece of lead fixed in roofwork should exceed 3 m in length. The expansion of such a length for any given temperature increase you can now easily work out for yourself.

Later on these precautionary measures and the way they are applied will be dealt with at greater length.

9
Heat: its effects on air and water

On page 91 it was said that whereas solids expand and contract at a regular rate for each deg C of temperature change, gases and liquids do not. Moreover, a solid material is rigid and self-supporting. Its change in size can therefore be directly measured by comparing dimensions before and after any temperature change in the material.

Liquid expansion and contraction cannot be so easily measured. Since they are fluid, liquids must be held in a container of some kind. When heated, measurement of their rate of expansion is upset by the fact that the container expands at the same time, and therefore increases its capacity. By using containers whose increase in capacity is known, allowances can be made and the actual amount by which the liquid expands can be calculated. The result will, of course, be greater than the amount by which the liquid appears to have expanded and so, when dealing with the thermal expansion of liquids, one refers to their 'actual' and their 'apparent' expansions.

By this means, a coefficient of expansion for water could be found for a temperature difference of 1 deg C, but since the expansion of water varies for each rise in temperature of 1 deg C, there is no constant coefficient of expansion for water.

It is useful to understand the following points:

i Water expands on heating, and its volume or bulk increases.
ii This being so, 1 kg of water at a high temperature will occupy more space than 1 kg of water at a lower temperature.
iii To make the volumes of the hotter and cooler waters the same, a small amount of water must be taken away from the hotter and bigger volume.

iv When this has been done, the hot and cooler volumes of water are the same. But clearly, the hotter volume will now weigh less than the equal but cooler volume.

This can be summed up in one short statement: 'Hot water is less dense, or weighs less bulk for bulk, than cooler water'. This explains why heated water circulates in hot-water supply and central-heating systems. The cooler, heavier water in the system falls by gravity to the lowest point, the boiler. In doing so it pushes the lighter, heated water out of the boiler, up and around the circulatory pipework.

The movement of warmed air in a room is explained in the same way. Cool air weighs more than warm air, and so it falls by gravity, pushing the lighter, warmed air up and around the room as it does so. So long as there is a difference of temperature between the falling and rising liquid or gas streams, there will be a difference of weight in those streams, and circulatory movement will result. When the temperatures become equal, there will be no such lack of balance in the weights.

It follows, therefore, that the greater the temperature difference, and the more vigorous will be the circulation.

These important facts find good practical use in the installation and design of hot-water supply systems and central-heating systems.

Transmission of heat

On page 53 it was shown that water will flow from a higher level to a lower level, and that it will continue to do so as long as there is a difference of head. It was also indicated that the greater the difference in head, the greater will be the pressure, and therefore the rate of flow of water. In just the same way, if two materials at different temperatures are placed in contact with one another, heat will flow from the hotter to the cooler material, and will continue to do so until the difference in temperature disappears. Furthermore, the greater the difference in temperature, the faster will be the rate of flow of the heat. Thus heat will flow from a hot radiator to warm up the air around it; but on the other hand it will also flow through the walls of a warm building to be wasted in the cooler air around it. These important points are illustrated in Fig. 15.

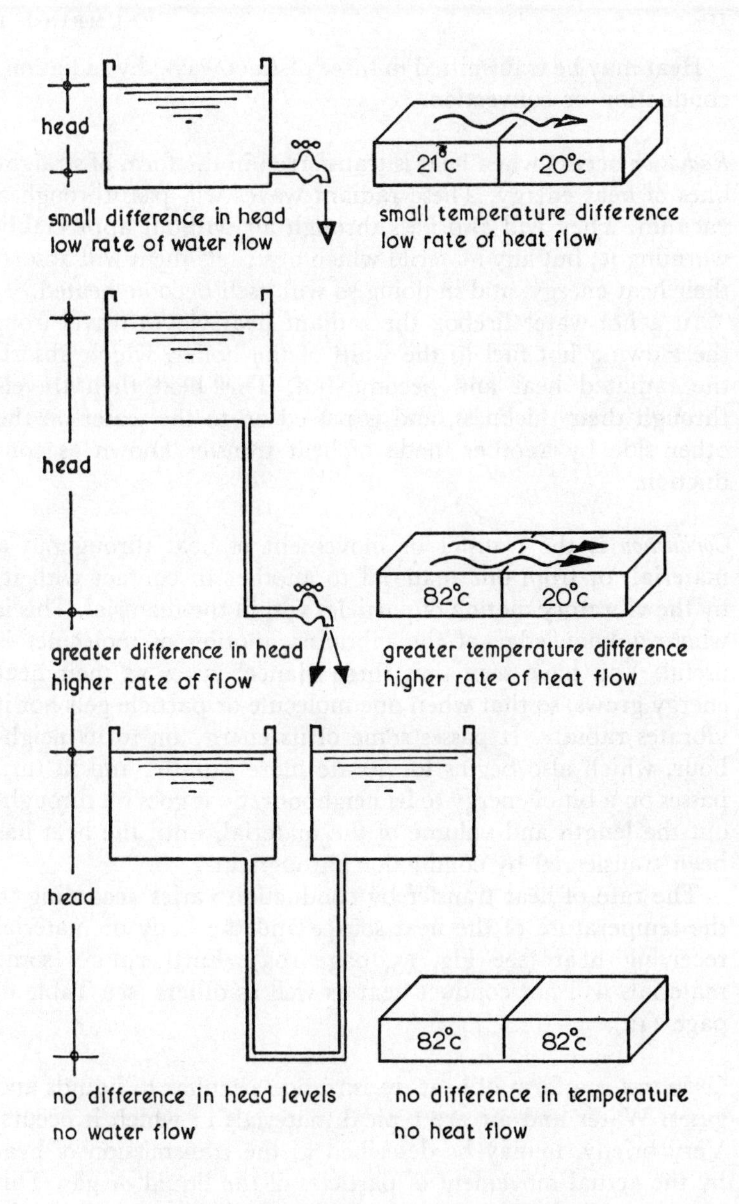

Heat Transfer by Conduction

FIG. 15

Heat may be transmitted in three distinct ways; by radiation, conduction or convection.

Radiation occurs when heat is transferred in the form of straight lines of heat energy. These radiant waves will pass through a vacuum. They will also pass through air without appreciably warming it; but any material which obstructs them will absorb their heat energy, and in doing so will itself become heated.

In a hot-water firebox the radiant heat waves travel from the glowing hot fuel to the walls of the boiler, which absorb the radiated heat and become hot. This heat then travels through their thickness, and is passed on to the water on the other side by another mode of heat transfer known as conduction.

Conduction is the transfer or movement of heat throughout a material, or from one material to another in contact with it, by the vibratory motion of particles within the material. This is where a knowledge of the vibratory motion of molecules is useful. You have seen that these 'dance' faster as their heat energy grows, so that when one molecule or particle gets hot it vibrates rapidly. It passes some of its energy on to its neighbour, which also begins to vibrate more rapidly, and in turn passes on a bit of energy to its neighbour. So it goes on throughout the length and volume of the material, until the heat has been transferred by conduction throughout.

The rate of heat transfer by conduction varies according to the temperature of the heat source and the body or material receiving heat (see Fig. 15, page 103). Furthermore, some materials will not conduct heat as well as others (see Table 8, page 112).

Convection is a form of heat transmission peculiar to liquids and gases. Water and air are typical materials in which it occurs. Very briefly, it may be described as the transmission of heat by the actual movement of particles of the liquid or gas. This movement is caused by the change in the particles' weight brought about by a variation in their temperature.

This form of heat transfer explains the movement of heated

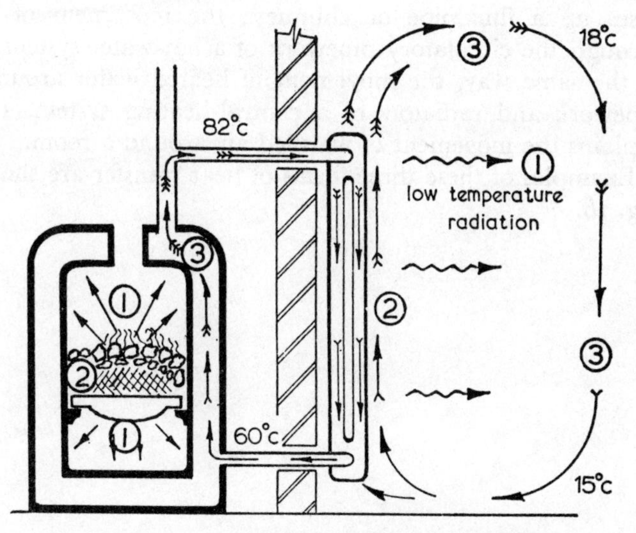

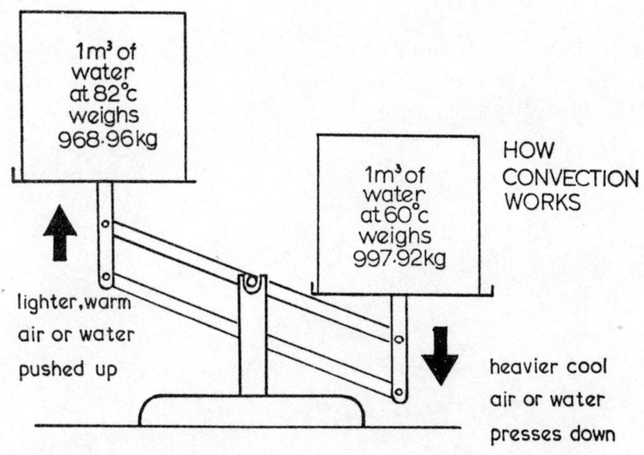

Heat transfer in solids, liquids & gases

FIG. 16

gases up a flue pipe or chimney; the movement of water through the circulatory pipework of a hot-water system; and, in the same way, the movement of heated water around the pipework and radiators of a central-heating system. It also explains the movement of warmed air around a room.

Examples of these three forms of heat transfer are shown in Fig. 16.

10
Properties of plumbing materials

Here is a list of the most important properties of matter, which the plumber must consider when selecting his material.

Specific gravity denotes the weight of a substance in comparison to the weight of the same volume of water at 4°C (see page 46).

Specific gravity figures, shown in Table 8, make possible quick and easy comparisons between the weights of equal volumes of different metals, since it is known that the higher the specific gravity figure, the heavier is the metal. For example, the specific gravity of lead is 11·3 and that of aluminium is 2·7. This shows that lead is 11·3 times heavier than water, and aluminium is 2·7 times heavier than water. Therefore lead is over four times heavier than aluminium.

Fusibility, or melting point, denotes the temperature at which a metal changes from its solid state to a molten liquid. It should be noted that cast iron melts at a much lower temperature than mild steel, which is a purer form of iron. The lower fusibility of cast iron enables that metal to be easily melted and poured into casting moulds, and thus permits cast-iron goods to be quickly and cheaply made. Mild steel has the highest melting point of all the plumber's metals. Although unyielding at normal temperatures, when heated to red heat it can easily be bent.

Malleability denotes the ability of a metal to be 'bossed' or worked to shape without breaking. Lead has this property to a remarkable degree, whereas ordinary cast iron is not appreciably malleable.

Malleable cast-iron fittings used for mild steel pipework are specially treated with heat to make them less brittle than they would otherwise be. The malleability of these fittings is such that they can be squeezed flat in a vice without breaking, and this would not be possible with ordinary cast iron. However, this degree of malleability is useful more for the fact that it can relieve strain than for the working properties which are normally associated with malleability.

Thermal conductivity denotes a material's ability to transmit heat from particle to particle throughout its mass or length. All metals possess this property, though to different degrees.

Electrical conductivity is a property possessed by all metals, though some are better conductors than others.

Of the plumber's metals, copper is the best conductor of electric current. Aluminium is next best; it is about half as effective as copper for the purpose. Because of its lightness it is much used, suitably reinforced with a central steel core wire, for high voltage cables on the electric grid system.

Coefficient of thermal expansion denotes the extent to which a metal will expand when its temperature is raised by 1 deg C. The effects of the expansion of metals must be carefully allowed for in work on roofs, and in work on hot-water and central-heating pipework systems (see Table 7, page 92).

Ductility denotes the property of stretchability, which enables a metal to be worked, and especially wires or tubes to be drawn, without breaking.

Annealed or softened copper and aluminium are very ductile. Lead is ductile to an extent; it can be stretched within limits. The stretching of lead or any other material reduces its thickness. In the case of leadwork the aim is always to leave the finished work of equal thickness throughout. For this reason the plumber must be careful not to rely on what ductility lead possesses to 'stretch' it; it is better 'bossed' into shape.

Tenacity denotes ability to resist pulling forces. As you may know, once a bulldog clamps his jaws on the seat of anyone's pants he resists all efforts to pull him off, and there is a well-known expression, 'as tenacious as a bulldog'.

The tensile strength of a material is a measure of its tenacity, and it is determined by clamping a short length of material—steel, copper or any other—between the jaws of a tensile testing machine. The jaws are made to pull in opposite directions and thus impose pulling stresses on the test piece, which stretches until it finally breaks.

The pull or weight force (N/m^2) at which the test piece breaks is a measure of its tensile strength. So that the tenacity of various metals can be compared, the tensile strength of each is fixed as the number of newtons pull needed to break a bar of the metal 1 m^2 in cross section (or it could be MN/m^2 or MN/mm^2).*

Elasticity denotes the ability of a material to resume its normal shape after being pushed or pulled out of shape.

Rubber is a good example of an elastic material. Certain steels, hard brass and hard copper can be made into springs which possess elasticity. Cast iron is not elastic to any appreciable extent, and neither are lead or 'dead soft' copper, or aluminium.

So far as roofing metals are concerned, what little elasticity they do have is generally referred to as 'spring back'. This is measured by bending sample strips of various metals to a right angle and then noticing how much they tend to spring back to their original flat form, if they do so at all. Sheet zinc has a measurable spring back, and so have copper and aluminium sheets in the 'hard' state. Lead, 'soft' copper and aluminium do not show a marked tendency to spring back, but stay in the shape to which they are bent.

Work-hardening denotes that a metal, though ductile in the normal or 'soft' state, will become gradually harder as it is worked upon by tools; for example, in bossing processes, or in

* M = Mega = 1 000 000
MN = 1 000 000 newtons.

drawing processes used in wire or tube manufacture. Lead does not work-harden to any noticeable extent. Copper and aluminium do, but can be restored to a soft, ductile state by annealing; that is, by heating the metal and then either cooling it in water or allowing it to cool in air.

Annealing is another way in which heat can affect the properties of metals, and it has practical applications in the working of sheet copper and aluminium, and the bending of tubes.

It is interesting to notice that ductility, tenacity or tensile strength, elasticity, and work-hardening are all closely connected. As cold-working on metals alters the shape of the crystals which make up the metal, so the property of ductility is lessened, but tenacity, elasticity and hardness increase.

Creep denotes the tendency of materials to 'flow' under the influence of a load. All metals tend to creep, and they do so when there is a change in shape of the metal crystals; when, for example, a heavy load tends to squash them, or a strong pull to stretch them. Creep is tied up with tenacity, hardness, and ductility, and an increase in a metal's temperature will increase its tendency to creep.

It is necessary to be very careful when fixing sheet lead on steep slopes or vertical surfaces, since otherwise the weight of the lead will impose a tensile stress upon itself. A hot summer sun will heat the lead and make matters worse. If lead movement down the slope is not restrained by proper design and fixing the lead becomes thin enough to tear. The lighter weight and greater stiffness of copper and aluminium sheeting combine to make them less subject to these problems.

Colour is produced as a result of a surface reflecting certain lights. As a property of metals it is important because it is a means of identification; one can often recognise a metal by its colour without even picking it up to judge its weight or other properties. Copper is clearly recognisable because it is reddish brown, and therefore not easily confused with, for example, newly cut lead, which is silvery in colour.

Durability denotes the quality of lasting, and is therefore very important. It has already been said, on pages 64 and 77, that metals are acted upon by the oxygen in the atmosphere. The oxygen combines with the surface of most metals to form a 'skin' of oxide which protects them against further attack. With iron and steel, however, the oxides crack and split, revealing fresh surfaces for attack beneath. In order for these metals to be durable, therefore, it is necessary to provide artificial protection in some form; for example, paint or galvanising (see Table 1, page 65).

Metal solvency. Water is a solvent; that is, it will dissolve substances. The dissolving of sugar in tea is a good example of this.

Metals tend to dissolve in water, some more than others, and in each case the dissolving rate or solvency of the metal will depend upon the chemical condition of the water. Waters collected from moorland gathering grounds absorb large amounts of carbon dioxide from the surrounding decayed plant life. This makes the water acidic, and more liable to dissolve metals. Such waters are usually 'soft' waters, but it is wrong to assume that only soft waters are metal solvents; cases of metal being dissolved by 'hard' waters also occur.

Lead is poisonous, and lead piping must never be used in districts where waters show a tendency to dissolve metals. If it were, *plumbo-solvency* might occur, with unpleasant results. Copper and zinc are similarly acted upon by metal solvent waters, giving rise to *cupro-solvency* and *zinc-solvency*. In the case of copper the danger to health is not great, but it does sometimes cause green staining of baths and sinks. Zinc-solvency is not particularly harmful, but metal solvent waters will quickly dissolve the protective zinc galvanising from mild steel tubes, fittings and cisterns, leaving the underlying mild steel open to attack by rust. The effect of this will be water discoloured by the rust and a very short life for the mild steel pipes and cisterns.

TABLE 8 SOME PROPERTIES OF PLUMBER'S METALS

Metal	*Chemical symbol*	*Specific gravity*	*Density or weight kg/m³*	*Melting point °C*	*Tensile strength MN/m²*	*Conductivity taking copper as 100*	
						Electrical	*Thermal (heat)*
Lead (milled)	Pb	11·3	11300	327	18	7·7	9
Copper	Cu	8·9	8900	1080	210 to 390	100	100
Aluminium	Al	2·7	2700	660	90 to 150	62	58
Zinc	Zn	7·1	7100	420	105 to 150	28	29
Tin	Sn	7·3	7300	215	15	15	17
Cast iron	Fe	7·2	7200	1526	75 to 270	17	17
Wrought iron	Fe	7·7	7700	2200	300 to 405	do.	do.
Steels	Fe	7·8	7800	1926	420 to 495	do.	do.
Mercury	Hg	13·6	13600	—	liquid at normal temperatures	—	—

Note: Properties vary according to condition; that is, whether the material is cold-worked, hot-worked, hard, or annealed. The above figures are therefore approximate, especially for tensile strength.

Note: Density of water = 1000 kg/m³ at 4° C. Hence lead is 11.3 times more dense than water, and so on.

Application

11

The weathering of buildings

Damp buildings are cold and unhealthy; moreover, dampness will hasten the decay of timber and other materials to such an extent that the useful life of the building will be seriously reduced. Thus, the exclusion of dampness is an essential part of good building work, and one in which the plumber is directly concerned.

There are many ways in which damp may enter a building, and they should all be looked for if one is called to investigate trouble of this sort:

1 Damp might enter in the form of moisture rising through porous walls or floor materials which are in contact with damp ground. Special waterproof membranes called damp-proof courses are used in good new work to prevent this.

2 The condensation of atmospheric moisture on cold, non-absorbent internal surfaces can give rise to serious dampness. There is little that the plumber can do about this except to advise the client to provide more ventilation to the room or building. The alternative, though this is, of course, more expensive, is to apply more heat to the room. By this method, either the temperature of the air will be raised to above the point at which its moisture condenses out, or the temperature of the walls will be raised so that local cooling of the air in contact with them is avoided.

3 Dampness in the form of rainwater may penetrate through walls, through projections from walls, or through faulty weatherings where walls and roofs meet. All weatherings should be provided with properly fashioned and fixed metal coverings at these points.

4 Dampness in the form of rain or snow may penetrate

through the roof covering. In the case of roofs covered with metal, this may be due to faulty design of roof weatherings, defective workmanship or defective materials.

Structure of roofs

Roofs may be either pitched or flat. A pitched roof is one which slopes at an angle of more than 10° to the horizontal plane. It may be 'double-pitched', so that from an end view it looks like a 'V' turned upside down; or it may be 'mono-pitched', with only one sloping surface.

A flat roof is one with a pitch of less than 10°, but generally a 'flat' roof is understood to be one which falls or slopes only enough to drain off rainwater. Such falls are usually about 1 in 60 (50 mm in 3000 mm).

The substructure or framework and decking for these roofs may be of timber, the size and strength of which will be designed to carry its proposed covering plus an allowance for the pressure and suction effects of strong winds. A flat roof substructure might be of concrete.

The design and preparation of any roof substructure which is to have a metal roof covering is very important. Faults in design or carelessness in preparation of the decking cannot be put right once the metal is laid, but they can seriously affect the efficiency and durability of the metal coverings.

Many roof designers are wise enough to get special advice, so that these faults can be avoided. Unfortunately, this is not always the case, and the plumber often meets roof preparations which are clearly unsuited to the metal that has to be put on. In such cases one must hope to convince some responsible person about what should be done to put things right before work is started. This needs a great deal of tact, but the plumber would be failing in his duty if he did not attempt to have these important matters dealt with.

Points to look out for:

1 Timber deckings to support sheet metal roof coverings should be of well-seasoned tongued and grooved boarding at least 25 mm thick.

2 It is false economy to use cheap boarding, and wherever possible the boards should be selected so that there will be as little warping as possible. The boards should be laid 'heart' side uppermost, so that if warping does occur no sharp board edges will turn up to cut the metal roofing.

3 On flat roofs the falls must be sufficient, and 1 in 60 is considered just right. To allow more is unnecessary and expensive, since more timber is needed to provide the 'firring' pieces which tilt the boarded surface. To allow less would result in poor roof drainage, and might cause leakage at the joints.

4 Boarding must be laid in the direction of, or diagonal to, the fall.

5 All nail heads must be punched below the surface of the boards.

6 All sharp edges of boarding with which metal will come in contact must be rounded off slightly with a plane or rasp.

7 'Box' gutters, which collect the rainwater from the flat roof before it is discharged to the rainwater pipes, should be wide enough to allow even a big-footed plumber to walk along with ease; 230 mm should be regarded as a minimum width, but 300 mm or 450 mm is better from the practical point of view; it will help in the laying of the metal gutter linings, and their subsequent cleaning and maintenance.

Metals conduct heat, and by themselves metal-covered roofs are poor heat insulators.

Felt underlays of flax fibre soaked in bitumen offer some measure of heat insulation, and should be laid butt jointed on roof deckings before any metal sheets are laid. The felt should be as described in BS 747, and is generally known as inodorous felt No. 1. It is brown, 'hairy' in texture, a little over 3 mm thick, and supplied in rolls. Its chief purpose is to act as an insulator against heat and sound, but it has other useful properties: the fact that it insulates heat reduces the effect of wide temperature variations. Furthermore, it reduces the damp and possibly corrosive results of condensation on the underside of the metal roof coverings. On boarded roof deckings the felt also acts as an insulator of sound, and helps to deaden the drumming noise of heavy rain-fall.

Metals are liable to corrosive attack when in contact with some timbers, notably oak; with concrete; and, particularly, with breeze concrete. The felt underlay 'insulates' the roofing metals from the substructure, and so removes this risk of corrosion. Finally, the felt does smooth out minor imperfections and roughnesses in the timber or concrete substructure surfaces, and so helps the metal to expand and contract freely.

It has already been said, on page 91, that a large roof area must not be covered with a single sheet of metal, but divided up into a number of smaller bays in order to allow for the movement of the metal caused by thermal expansion.

The fixing and jointing of bays forming a roof or gutter needs great care, since the fixings for the metal must be secure whilst at the same time allowance is made for its free expansion and contraction. This freedom of movement is provided for in the design of bay joints and the style of working the different metals at these points.

Fixings must provide for free movement of the metal sheets in at least two directions, and this means that no area of metal should be permanently fixed at opposite ends. To do so would result in the metal rising up in the middle when hot and falling flat again when cool. Continuous movement of this kind would cause the metal to get 'tired' or fatigued, and split. The lighter materials, copper, aluminium and zinc, require special care in fixing since they are more likely to 'lift' to the suction effects of strong winds. These points will be dealt with in detail in chapters 12 and 15.

Capillarity

Capillary attraction is the well-known phenomenon, or strange occurrence, which happens when water appears able to rise above its own level, either between two surfaces that are close together, or through fine-bore, tube-like passages such as are found in some building materials (see also page 130). Capillary attraction accounts for the fact that dampness will rise up walls from the ground. It can also work in reverse; when dampness appears at ceiling level in top-floor rooms it is because of the downward passage of water through chimney or parapet walls.

Damp-proof courses of some material through which water cannot pass are placed at suitable levels in the walls of buildings to prevent these upward and downward capillary movements. Lead, copper, and aluminium foil sandwiched between two layers of bitumen are often used for this purpose; and chimneys are sometimes built with damp-proof courses in line with the roof slope and a little below it, or at a horizontal level just below the point where the chimney-stack begins to pass out of the roof.

Capillarity can also cause rainwater to enter joints in metal roof coverings unless suitable precautions are taken. Capillary attraction lessens as the gap between the two surfaces gets bigger, and if the gap is wide enough it will stop. The provision of a gap between two surfaces fixed closely together will prevent capillarity. Fig. 26, page 140, shows a capillary gap or groove used at a vertical drip on a lead flat, and at a lapped joint between sheet lead ridge or hip weatherings.

Roof weathering details

Slate sizes are known by female titles of rank. A 'Duchess' slate is 610 mm long and 305 mm wide; a 'Countess' 510 mm by 255 mm, and a 'Lady' 406 mm by 203 mm.

Slates are fixed to 50 mm by 25 mm slate battens with two stout, large-headed slate nails. These are placed a little above the centre-line of the slate, and about 32 mm in from each side. This arrangement is called 'centre nailing', and is generally used because it holds the slates more securely against the wind than does 'head nailing', where the nails are fixed at the top corners of the slate.

Slates are fixed in rows or courses along the roof, beginning, as with all roof weathering work, at the lowest point of the roof. In this case it is the eaves. Each course overlaps parts of two courses below it. Fig. 17 (page 121) makes this clear, and explains how the courses are worked out and arranged.

Note carefully that the top edge of the fascia board is higher than the top line of the slate battens. This causes all the slates to tilt instead of lying flat. The tilt of the under slates makes the tails of the centre nailed upper courses press down on those below. This helps to keep the slates from lifting and blowing off in a strong wind. The widening gap between the slates prevents rainwater from 'climbing' between them and into the roof by capillary attraction.

Tilting fillets of triangular sectioned timber are used for the same job of tilting the slates at places on the roof where no fascia board is fixed; for example, behind chimney-stacks.

Note also that the 'lap' of a slate is the distance by which the tail of one slate passes beyond the head of the next slate but one, below. This is a bit of a tongue twister, but the diagram will make it clear.

You will also see that the first course of slates at the bottom or eaves of the roof are shorter than the rest by the length of the 'margin'. The length of these under-eaves slates is equal to the 'gauge' plus the 'lap'. For 510 mm long slates at 45° pitch this would be 217 mm plus 76 mm; i.e., 293 mm in all.

Abutment is a building term used to describe the junction or meeting line between a roof edge and the wall against which the roof edge 'butts'. You will see many forms of roof abutment; for example, where a low-level roof abuts the wall of a taller building; or where a chimney passes through a sloping roof. Weatherings to abutments are necessary to prevent the rainwater that falls on their vertical wall faces from running down into the building.

Sheet lead, copper, aluminium or zinc could be used for this work. The following description applies equally to all these metals, except in the case of certain setting out and working techniques which will be described separately as they arise. The procedure for this kind of work is as follows:

First, mark out and cut some soakers for the slaters to put in place as they finish off the slating along the roof abutments.

Soakers are pieces of sheet metal bent to form a right angle (Figs. 17 and 22, page 131), used to divert the water which

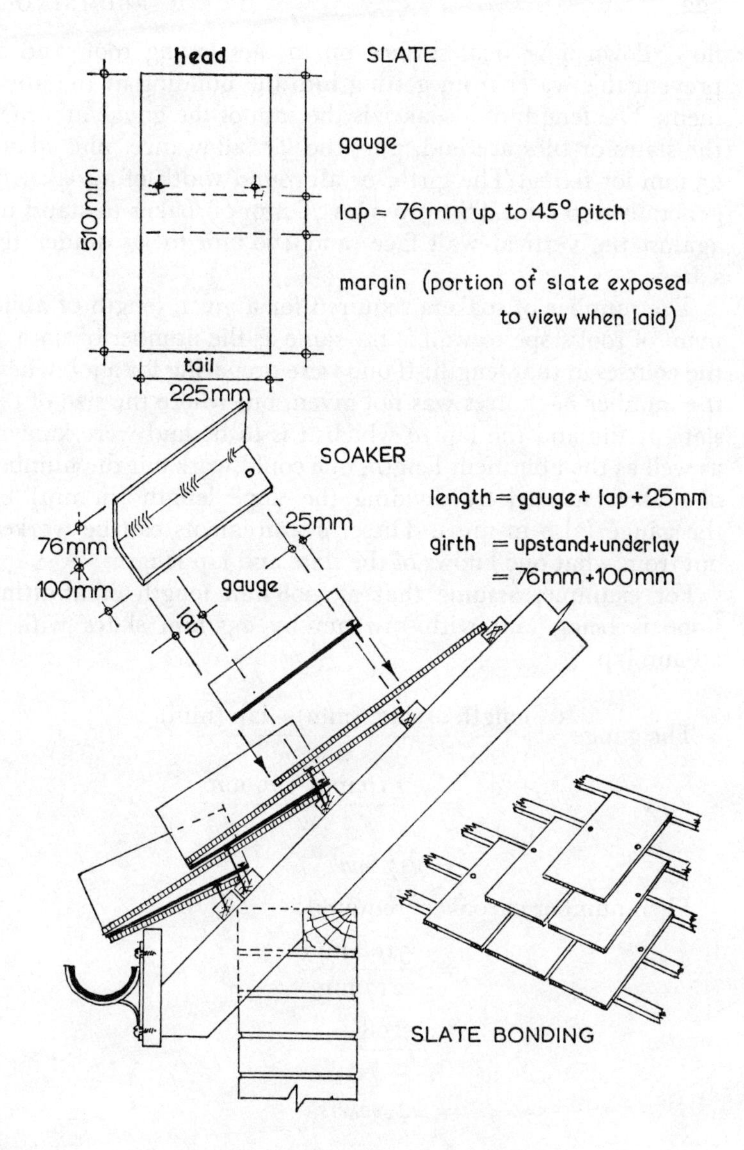

Slates, Soakers & Bonding

FIG. 17

flows down a vertical surface on to the sloping roof, and so prevent this water from getting into the building at the abutment. The length of a soaker is the sum of the gauge at which the slates or tiles are laid, plus the lap allowance, plus about 25 mm for fixing. The girth, or all-round width of a soaker, is generally 176 mm. This provides 76 mm of soaker to stand up against the vertical wall face, and 100 mm to lay under the slates.

The number of soakers required for a given length of abutment of roof slope to wall is the same as the number of slate or tile courses in that length. If one were preparing for a job where the number of courses was not given, but where the size of the slate or tile and the lap to which it is to be laid were known, as well as the abutment length, one could work out the number of soakers needed by dividing the slope length (in mm) by the gauge (also in mm). These measurements can be worked out from what one knows of the slate and lap lengths.

For example, assume that a 5208 mm length of abutting slope is being laid with 510 mm by 255 mm slates with a 76 mm lap.

$$\text{The gauge} = \frac{\text{length of slate (mm)} - \text{lap (mm)}}{2}$$

$$= \frac{510\text{ mm} - 76\text{ mm}}{2}$$

$$= \textit{217 mm}$$

Then number of soakers required:

$$= \frac{5208\text{ mm}}{217\text{ mm}}$$

$$= \frac{5208}{217}$$

$$= \textit{24 soakers}$$

Cover flashings are now needed to guide the water falling down the wall face over the upstand of the soakers, and on to the sloping surfaces which will take it to the gutter.

Where the abutment does not have a slope but is straight,

for example where the top end of a single-slope roof abuts a wall, the cover flashings will be cut in straight lengths of suitable width and length. Where necessary, joints might be seamed or welted in sheet copper, aluminium or zinc work, and for lead they would be lapped one over the other for some 100 mm or 125 mm (see Fig. 18).

Raking or stepped cover flashings are necessary where the abutment slopes, as for example at the sides of chimneys, or where a roof slope meets a vertical wall.

The setting out of stepped cover flashings is shown in Fig. 18, and a brief description of the procedure is as follows:

1 Carefully mark out the required length and width of metal sheet with a soft pencil in the case of copper, aluminium or zinc, and chalk or a chalked line for lead or Nuralite.

The width allowance for roofs with a pitch of 45° is 150 mm to 176 mm, but in order to get the 63 mm cover extension on each 'step', this width will have to be varied to suit other pitches. Allow extra metal at the bottom end if this is to turn round the front wall.

2 Cut the strip of metal, take it to the roof, and offer it into position so that the bottom line of the brick bed joints can be marked on and extended until they meet the 'water-line'. A line drawn from where a joint line cuts the water-line to the joint line above, and at the top edge of the strip, will give the line of 'cut back' for one step.

3 Carry on marking other steps in this way, and allow about 32 mm at the top of each to be turned into the brick joints. Cut away the unwanted triangular pieces, saving them carefully for scrap, or for use as fixing wedges. The 32 mm turns on each step are made to fit into the brick joints, and the step flashing is then ready to be finally secured in place with 'tags' or wedges of the same material. The bricklayer completes the weathering by filling, and pointing the brick joints with mortar.

This method is quite commonly used, and is the only one suitable where the walls are built of stones of many different sizes, and not laid to regular courses like bricks.

The prefabrication of flashings is an idea that saves time and

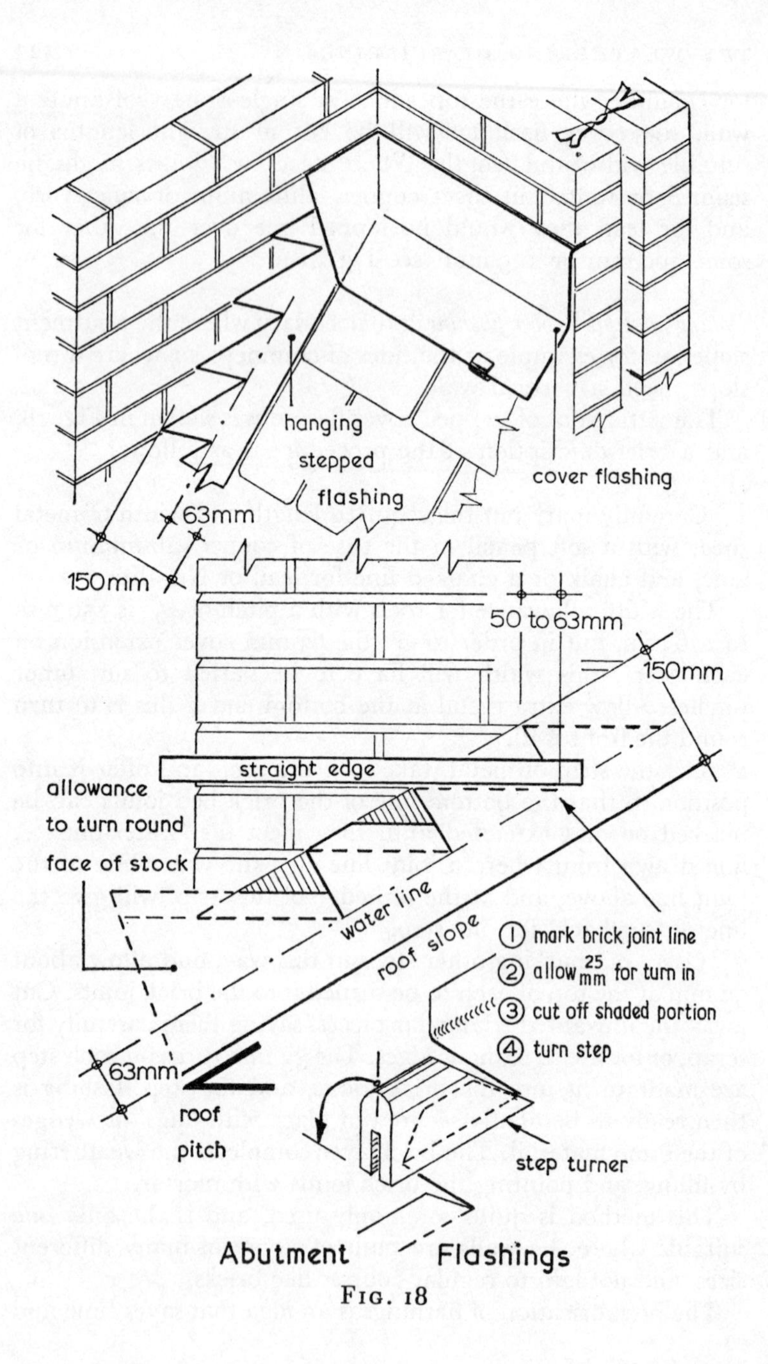
hanging
stepped
flashing
cover flashing
63mm
150mm
50 to 63mm
150mm
straight edge
allowance
to turn round
face of stock
water line
roof slope
1 mark brick joint line
2 allow 25 mm for turn in
3 cut off shaded portion
4 turn step
63mm
roof
pitch
step turner
Abutment Flashings

FIG. 18

money, especially where weathering details are repeated throughout a job. It can be helpful, too, if rain or other holdups prevent work on the roof.

It is perhaps wise to wait until the building part of the job is complete, and then to number and measure each detail separately. This information is taken back to the workshop, where if necessary it is reproduced as a full-size picture of the job.

A ridge piece is needed to weather the raking stepped flashings where they meet near the ridge of a roof abutting against the gable end wall of the main building. Fig. 19 shows the setting out and making of a lead ridge piece.

Weathering of chimney-stacks and parapet walls

If you look closely at chimney-stacks in your neighbourhood, you will see that they occur in three places in the roof; at the ridge, halfway down the slope, or, if the flue is built on to an external wall, partly cut into the roof at the eaves. Flues are best placed inside the building, well away from the outside walls. Clearly, a flue on an outside wall will lose valuable heat to the cold air outside, and will furthermore be more exposed to the weather than one passing through the roof halfway up its slope. The best protected is the chimney at the ridge. All flues must be carried to an adequate height above the roof—generally a metre or so above the ridge line—in order that they draw properly in any wind conditions (see Fig. 21).

All chimney-stacks suffer severely from exposure to rain and frost. Cement and sand flaunching is generally used to hold the pots in place and to shed off water. The combined action of rain and frost will crack these flaunchings. Water gets into the brickwork beneath them and frost action expands this saturated brickwork, opening up joints and cracks. Eventually, the top of the stack becomes quite unsafe, and expensive rebuilding is necessary.

Cap flashings, carefully designed and fixed by the plumber, will permanently weatherproof the top of a chimney-stack and keep it safe for all time. Fig. 20 shows examples of how this might be done.

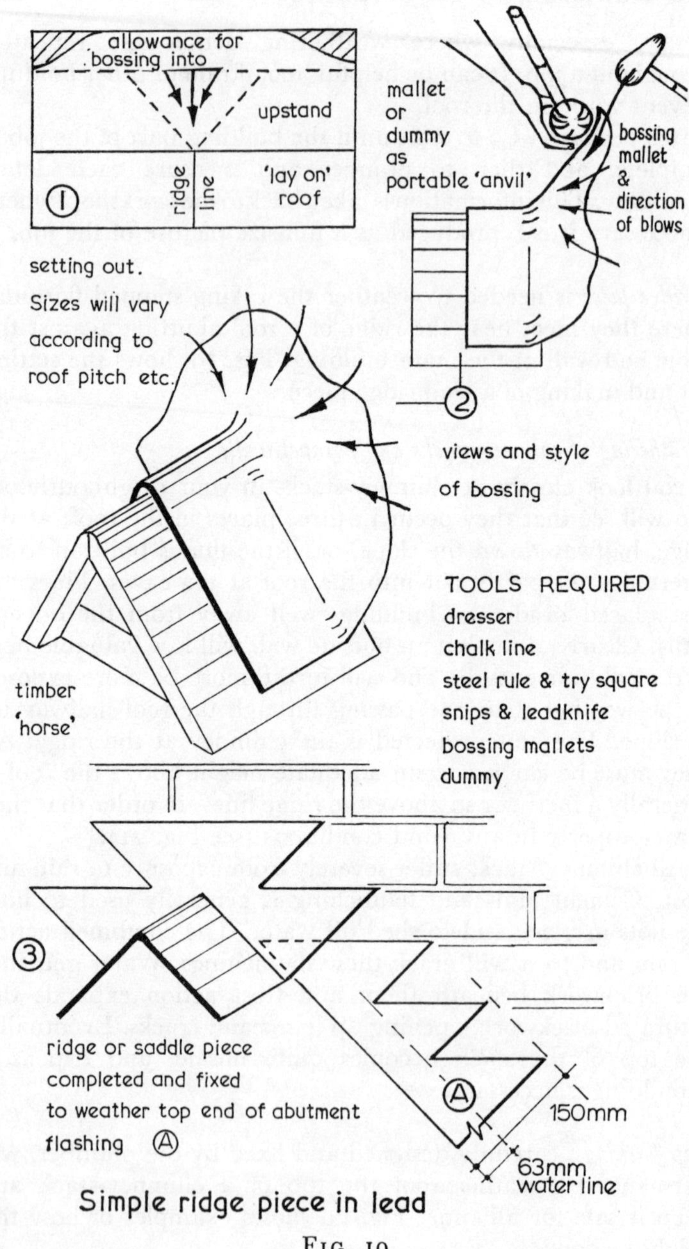

Simple ridge piece in lead

FIG. 19

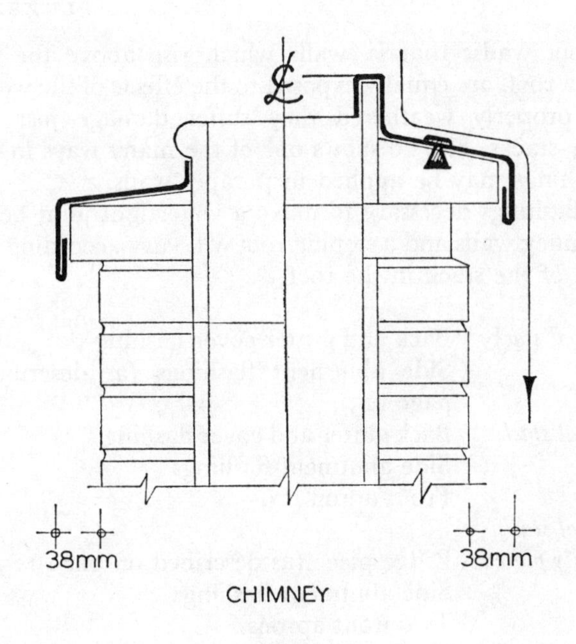

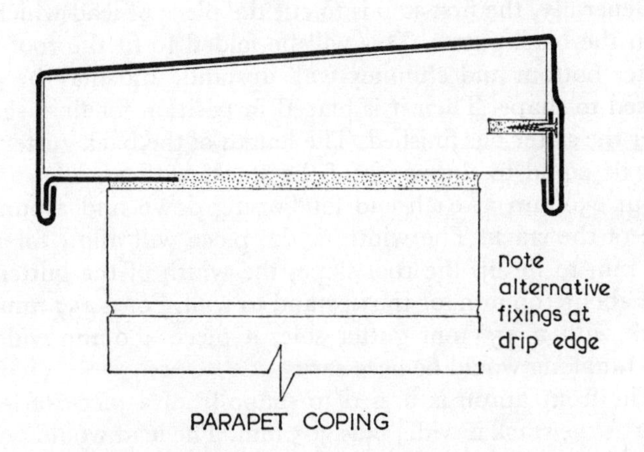

Cap Flashings or Weatherings

FIG. 20

Parapet walls; that is, walls which rise above the gutter level of a roof, are equally exposed to the effects of the weather. If not properly weathered they suffer damage just as do chimney-stacks. Fig. 20 shows one of the many ways in which cap flashings may be applied to parapet walls.

The flashings necessary to make a watertight joint between the chimney walls and a sloping roof will vary according to the position of the stack in the roof.

External stack: Back and gutter cover flashing
Side abutment flashings (as described on page 123).

Internal stack: Back gutter and cover flashing
Side abutment flashings
Front apron.

Internal stack centrally placed: Ridge piece (as described on page 125)
Side abutment flashings
Two front aprons.

You will meet variations of these, but Fig. 21, page 129, shows these three basic stack placings.

Generally, the first step is to cut the piece of lead which is to form the back gutter. This will be folded to fit the roof slope, gutter bottom and chimney wall upstand, and may be partly bossed to shape. Then it is placed in position for final shaping after the slater has finished. The length of the back gutter piece will be equal to the width of the stack, with an allowance of about 230 mm at each end for bossing down and around the side of the stack. The width of the piece will allow for about 230 mm to lie up the roof slope, the width of the gutter sole, and about 100 mm for the upstand to wall. For a 457 mm wide stack with a 150 mm gutter sole, a piece 480 mm wide and 915 mm long would be necessary.

The front apron is bossed to shape from a piece of lead as long as the stack is wide, plus 305 mm. The lead would be wide enough to provide 150 mm for laying over the slates plus an allowance for the upstand to front wall. This allowance will vary according to the way the brick courses are arranged at

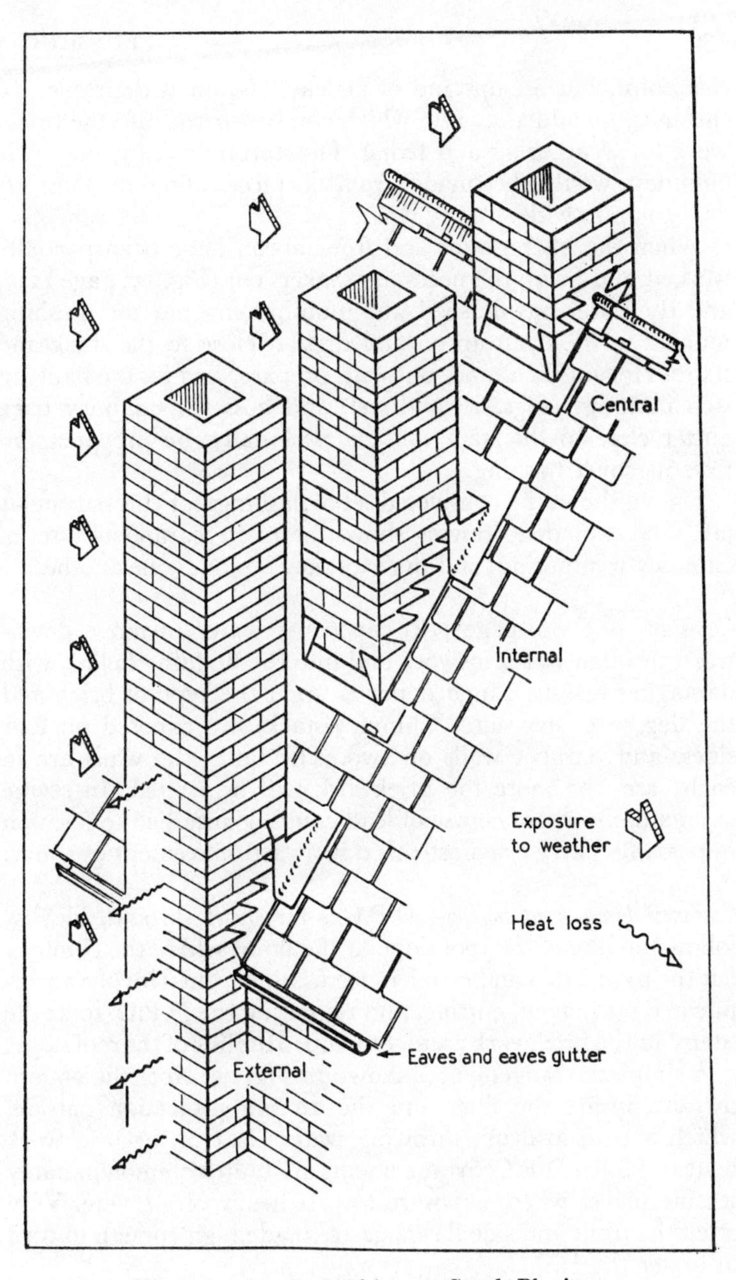

FIG. 21. Typical Chimney Stack Placings

this point, but an upstand of at least 76 mm is desirable. To this must be added 32 mm which can be turned into the brickwork for weathering and fixing. The apron for a 457 mm wide chimney would be made from a piece 760 mm long by 254–305 mm wide.

When the back gutter and front apron have been partially worked and fixed, the necessary soakers cut (Fig. 17, page 121), and the slater has finished, the plumber can put the finishing touches to the front apron and dress it close to the stack and slates. He will then mark out, cut, prepare, and fix the hanging step flashings (Figs. 18 and 22), and boss the chimney back gutter close to the stack and slates. Finally, he prepares and fixes its cover flashing.

Fig. 22 shows the completed set of flashings for this particular job. The exploded drawing shows the main features of a set of chimney weatherings and the way they lap over one another.

Capillary (see also page 118) can cause water to move downward through wet brickwork and into the building below, with damaging results. Much depends upon the type of brick and the degree of exposure. Chimney-stacks are exposed on four sides, and parapet walls on two. The taller the walls are in each case, the more the brickwork will be wetted. In severe cases a damp-proof course of lead could be installed to form an impassable barrier against this downward movement of water.

Chimney damp-proof courses—D.P.C.'s for short—would, ideally, follow the line of the roof slope at the point where the chimney left the roof. This can be and, indeed, is done, but it involves complicated setting out, cutting, and rejointing the D.P.C. to fit the 'steps' in the brickwork as it is built up to the line of the roof slope.

A simpler arrangement is shown in Fig. 23. Note the 25 mm turn-up inside the flue, and the 19 mm projection outside, which acts as a 'drip', throwing water clear of the brickwork beneath. Such D.P.C. arrangements are quite common in many seaside places where exposure to weather is very severe. Very often the front and side flashings are made high enough to turn in under the D.P.C.

Fig. 23 also shows the application of D.P.C.'s to a parapet wall.

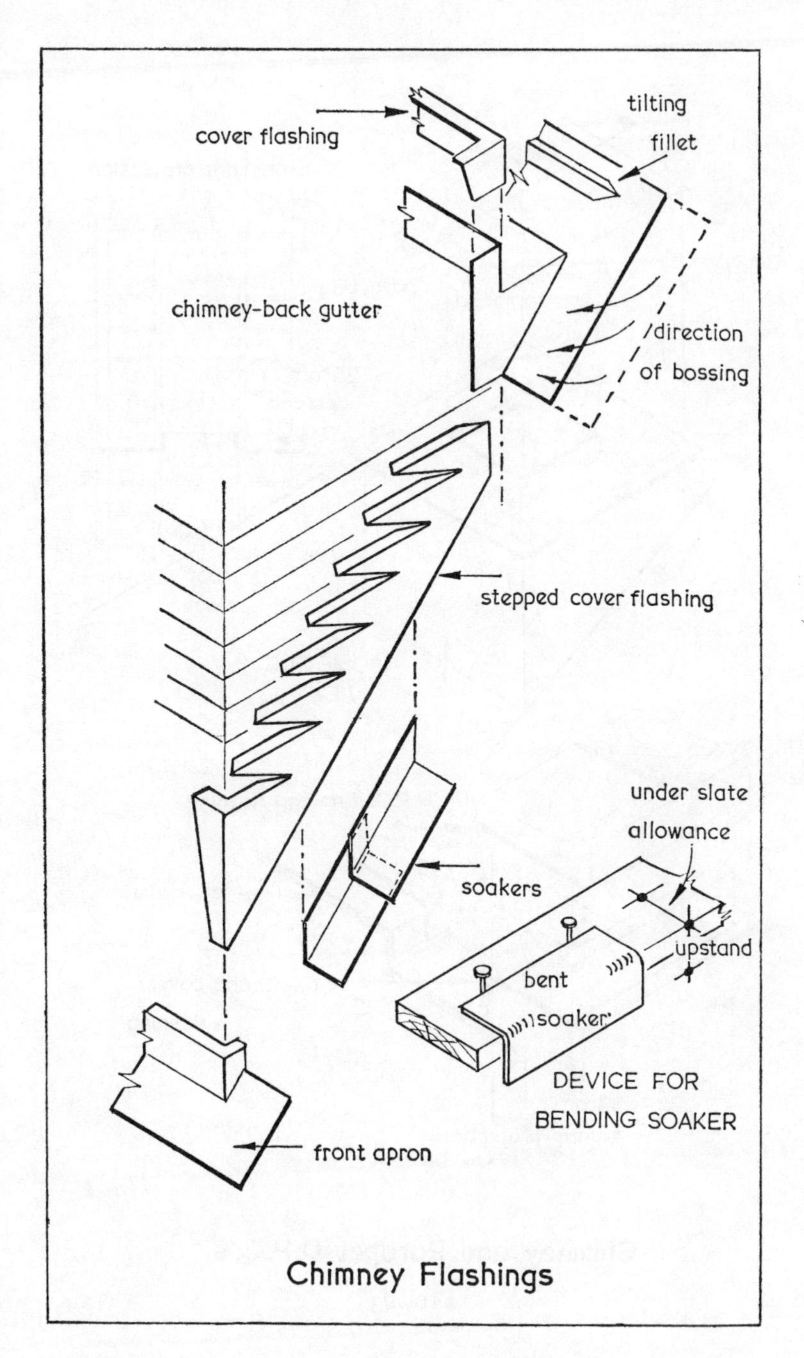
cover flashing
tilting
fillet
chimney-back gutter
direction
of bossing
stepped cover flashing
under slate
allowance
soakers
upstand
bent
soaker
DEVICE FOR
BENDING SOAKER
front apron
Chimney Flashings

FIG. 22

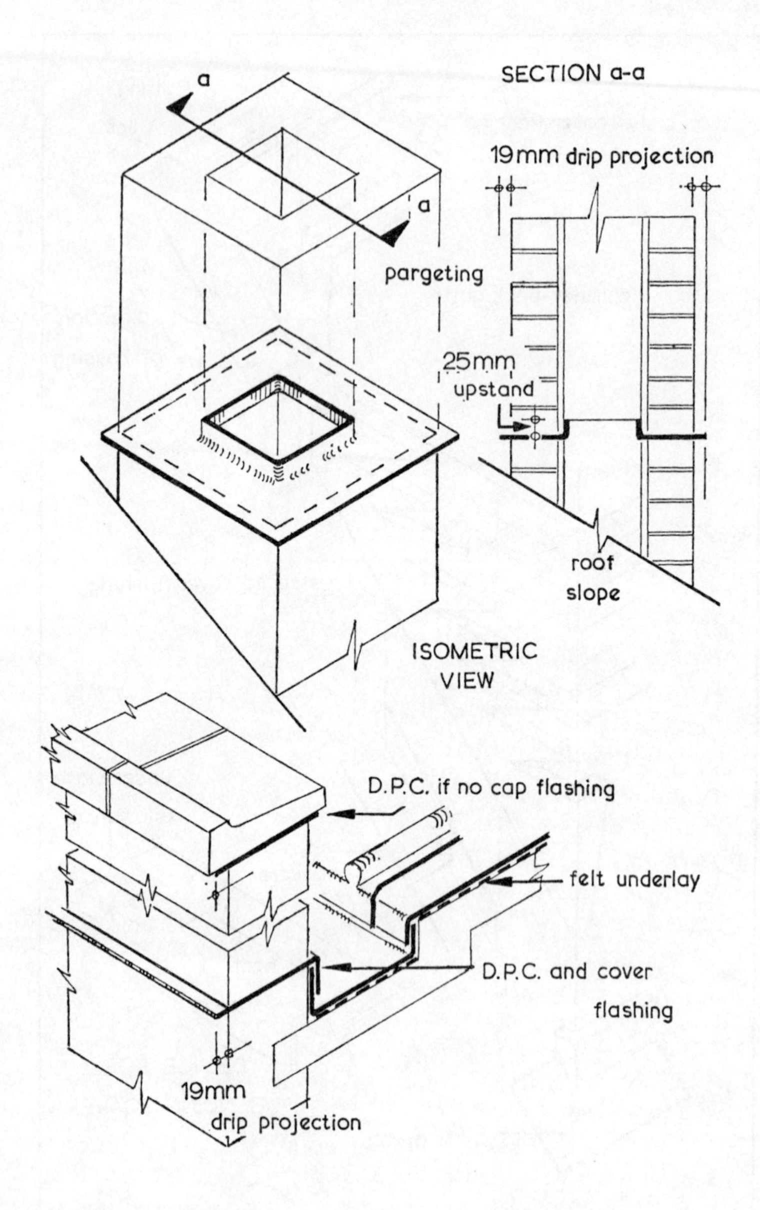

Chimney and Parapet D.P.C.'s

FIG. 23

12

Lead roofwork

The metals lead, copper, aluminium and zinc, and the non-metallic Nuralite are materials considered suitable for the weathering of roofs. The order in which they appear is not intended to suggest any order of suitability, but with the help of a knowledge of the properties of these metals (Chapters 8, 10, pages 90 and 107) one can make a reasoned choice of material and use it to the best practical advantage.

The basic requirements of a metal roofing or weathering material are durability and easy workability, so that details and joints may be worked in position on the roof. The appearance of a metal, particularly in relation to the other materials used in buildings nearby, and its resistance to the spread of fire, are other factors which must be considered. Finally, the matter of cost just cannot be ignored today.

Manufacture of milled sheet lead

Milled sheet lead for roofwork is made first by casting a slab of lead some 1500 mm square by 150 mm thick. When the slab has solidified and is cool enough to handle, it is hoisted by overhead cranes on to the roller bed of a milling machine. The milling machine, with its adjustable rolls, looks like a wringer used to squeeze water out of clothes. The slab of lead is moved backwards and forwards between the rolls, and at each passage the top roll is brought nearer the bottom one. Thus the slab of lead is reduced in thickness and becomes longer in length. When the desired thickness has been reached the milling stops, the lead is trimmed to the required width, and parcelled into rolls for despatch to the warehouse or site.

The milling process is a way of 'working' on the milled sheet lead. The lead is 'compressed' by the process, and its molecules are pressed closer together in a more compact mass. Hence, milled lead is more dense—that is, it weighs more per unit volume—than cast lead.

Another effect of milling is to make the metal slightly harder to work, though this is scarcely noticeable. Also, milled lead has a greater tensile strength than cast lead, and so it can be used in thinner sheets.

Standard metric sizes for lead sheet to BS 1178 are 2·40 metres wide (nominal 8ft) and up to 12 m (nominal 40ft) in length, unless otherwise specified by the customer. Lead sheet is readily available cut to any width in rolls up to 12 metres long.

Thickness may be expressed in millimetres or more conveniently by a B.S. Code number. These code numbers for strip and sheet lead correspond to the previous Imperial values of weights in pounds per square foot, as the following shows:

B S Code number	Thickness mm	in	Weights kg/m²	lbs/ft²
3	1·25	$\frac{3}{64}+$	14·18	3
4	1·80	$\frac{5}{64}-$	20·41	4
5	2·24	$\frac{3}{32}-$	25·40	5
6	2·50	$\frac{3}{32}+$	28·36	6
7	3·15	$\frac{1}{8}-$	34·73	7
8	3·55	$\frac{9}{64}-$	40·26	8

A colour code is applied for ease of identification

Code N° 3 = Green　Code N° 5 = Red　Code N° 7 = White
N° 4 = Red　N° 6 = Black　N° 8 = Orange

CHIMNEY AND ABUTMENT FLASHINGS

	Code N°	thickness (mm)
Chimney-back gutters	5 or 6	2·24 or 2·5
Chimney aprons	4 or 5	1·8 or 2·24
Ridge pieces	5 or 6	2·24 or 2·5
Cover flashings	4 or 5	1·8 or 2·24
Soakers	3 or 4	1·25 or 1·8

Commercial sizes

Commercial milled sheet lead is manufactured to BS 1178, and is 99·99% pure lead. Its commercial sizes range in length from 4·5 m to 12 m and from 2·1 m to 2·7 m width. The 'standard' size of sheet is the 2·4 m by 12 m roll, and is so called largely because it has been most frequently used in the past. Such a roll of N° 5 lead would weigh over 250 kg (5 cwt) and is not easily or quickly manhandled on the site.

Cut-to-size rolls are available on payment of a small charge at the mill. The convenience with which smaller rolls can be handled makes for an economy of time and labour which often more than repays the cost of cutting. Moreover, forethought in the ordering of roofing material can eliminate the waste which results from an oversized sheet.

Strip lead is available for flashings. It is made in rolls not less than 75 mm wide, and in thicknesses as for sheet lead. For ease of handling, the roll weights are kept to about 50 kg and so the length of the stip will depend upon its width, weight and thickness.

The material conforms to BS 1178 and offers considerable saving in time and material by reducing cutting time and wastage of lead.

Joints and fixings for sheet lead

It is well known that lead is very durable when properly fixed, and that it is extremely malleable and easily worked to shape both on and off the roof.

Joints in sheet leadwork must of course be waterproof. They must also help to secure the lead in position on the roof and at the same time allow some movement to accommodate thermal expansion. Good lead-work design and practice employs joints which meet these requirements. It is convenient to consider the various joint forms in two main groups; those which run in the direction of the roof fall or water flow, and those which run across the flow.

Joints with the flow:

Wood cored or solid rolls are used to join bays of lead along their length. They are also used where the ridge or hips of a roof are weathered in lead. Fig. 24 shows these applications.

Note carefully the shape and size of the wood core roll. Observe the undercut sides into which the bay slides, and out of which it cannot be lifted because the sides of the rolls are closer at the top than at the bottom.

Notice that only the undercloak is fixed to the wood roll. It is taken about two-thirds of the way round the roll and then close copper nailed all along the roll. Its edge is then carefully pared off with a shave-hook, so that it offers no sharp edge to the overcloak of the next bay.

Splash laps are arranged on that side of the roll which will get least wind; that is, the side away from the prevailing wind The splash lap stiffens the free edge of the bay roll, and helps to secure it to the roof. Though this extra stiffness is an advantage in fixing, it can in dusty neighbourhoods collect dust and dirt. This will absorb rain-water and encourage leakage into the roof, for water will rise between the undercloak and overcloak by capillarity and siphon wick action. If the splash lap is left off and the overcloak finished about 6 mm above the roof surface, this will be avoided.

There is no hard and fast rule about splash laps; each job must be judged according to known local conditions.

Hollow rolls are also used to join bays along their length, but only when there is no likelihood of people walking on the rolls and treading them flat. For this reason hollow rolls are, as a

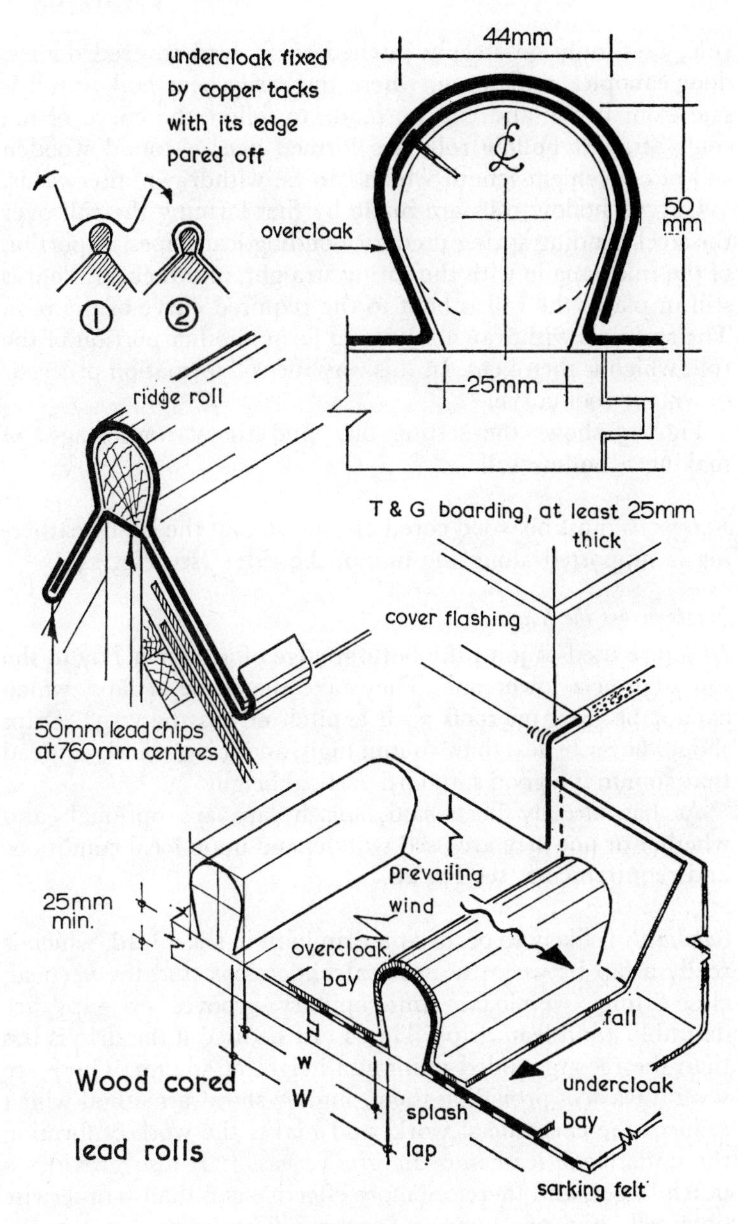
undercloak fixed
by copper tacks
with its edge
pared off
1
2
44mm
50
mm
overcloak
25mm
ridge roll
T & G boarding, at least 25mm
thick
cover flashing
50mm lead chips
at 760mm centres
prevailing
wind
25mm
min.
overcloak
bay
fall
w
W
undercloak
bay
splash
lap
sarking felt
Wood cored
lead rolls

FIG. 24

rule, used only on steeply pitched roofs, lead-covered domes, door canopies, and so on, where the easily bent hollow roll is safe from traffic and can be made to follow the curve of the roof. Straight hollow rolls are formed over a round wooden roll of convenient length which can be withdrawn afterwards.

Curved hollow rolls are made by first forming the roll over the steel bending spring used for bending lead pipes. A portion of the roll is made with the spring straight, and then, while it is still in place, the roll is bent to the required curve of the roof. The spring is withdrawn a little to form another portion of the roll, which is then bent. In this way the roll formation proceeds down the roof curve.

Fig. 25 shows the setting out, and the various stages of making a hollow roll.

Ridge rolls must be wood cored or solid so that the lead weathering is supported along the line of the ridge (see Fig. 24).

Joints across the flow:

Drips are used to joint the bottom edge of an upper bay to the top edge of a lower one. They take the place of laps, which cannot be used on roofs with a pitch of less than 15°. Drips should never be less than 40 mm high, and it is generally agreed that 50 mm is a good standard vertical height.

As has already been said, splash laps are optional, and whether or not they are used will depend upon local conditions and requirements (see Fig 26).

Capillarity is likely to occur at a drip joint in sheet lead, which is really a lap between the vertical undercloak and the vertical, close-fitting overcloak. Anti-capillary grooves or gaps are desirable at all lead drips. They must be used if the drip is less than the recommended minimum height of 50 mm. There are several ways of providing them. Fig. 26 shows a method which reduces the carpenters' work, and makes the work of forming the undercloak lead into the groove easier. It also provides a much larger, and therefore more effective gap than is otherwise obtained. Fig. 26 shows an anti-capillary groove to the lap joint of a ridge roll.

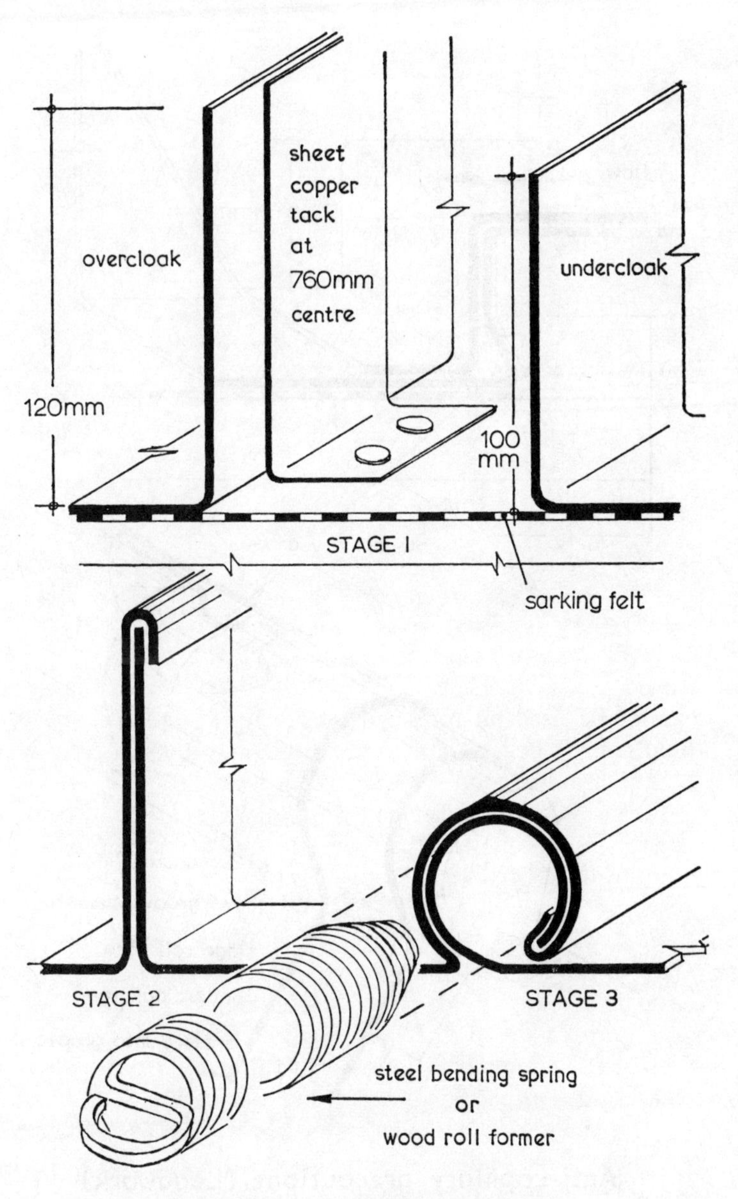

Formation of lead hollow roll

FIG. 25

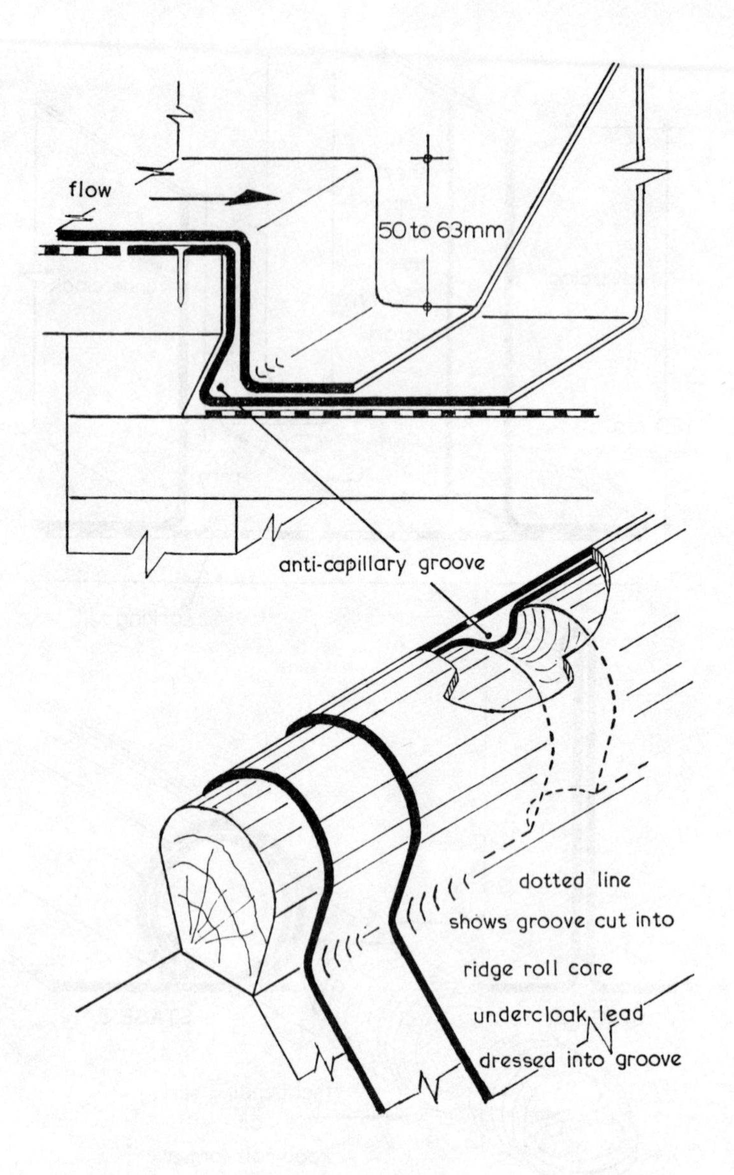

Anti-capillary precautions (Leadwork)

Fig. 26

Laps cannot be used on pitches of less than 15°. At steeper pitches they may be used, provided that proper fixing is arranged and that the sheets of lead lap over one another by at least 76 mm, measured vertically.

Welts. Leaving out the possibility of capillarity for a moment, it will be seen that rolls or drips are joints which provide a 'water barrier' equal to their vertical height. In other words, water would have to stand at least 50 mm deep on a roof before it could enter through these joints. This is unlikely to happen, unless perhaps the outlet of a cesspool which has no precautionary overflow pipe becomes blocked, but it draws attention to the reduced height and consequently reduced 'water barrier' of the welted joint. Welts are useful in many jobs, but these points must always be borne in mind:

1 Welts would not be used on flat roofs. They could be used for jointing vertically fixed sheets when they may run with or across the flow of water.
2 Welts are also used for end-jointing long lengths of weatherings; for example, in lead D.P.C.'s, in cap flashings for copings, or in cornice weatherings.
3 In the horizontal, or near horizontal position, when one is dealing for example with copings and cornices, a double lock welt used with a knowledge of its limitations may be effective with or across the flow.

Fig. 27 shows the formation of welted joints. Notice the similarity between the early stages of forming a double lock welt and forming a hollow roll.

Additional fixings for sheet leadwork

The design and placing of some lead weathering details require special fixing arrangements.

Soldered dots (Fig. 28) are used to secure vertical sheets of lead and also to provide some support for their weight, thus reducing any tendency to sag.

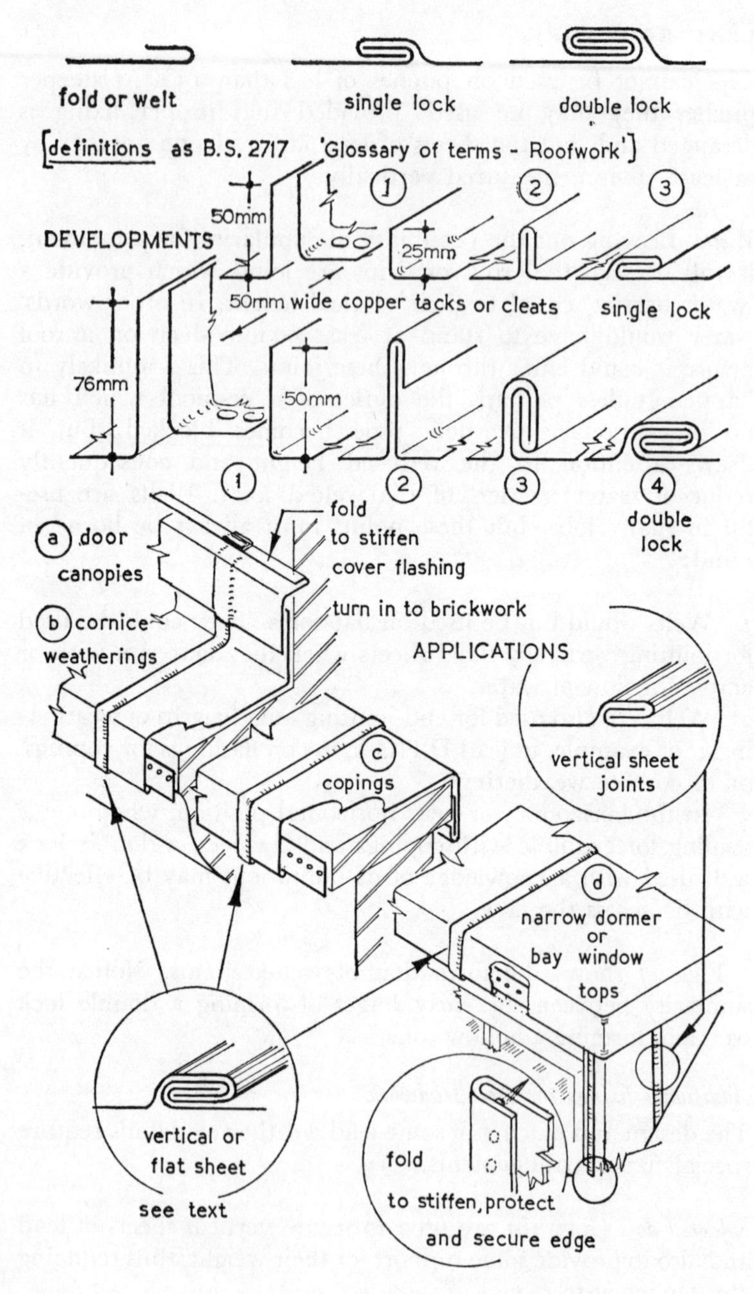

FIG. 27. Welts for Leadwork

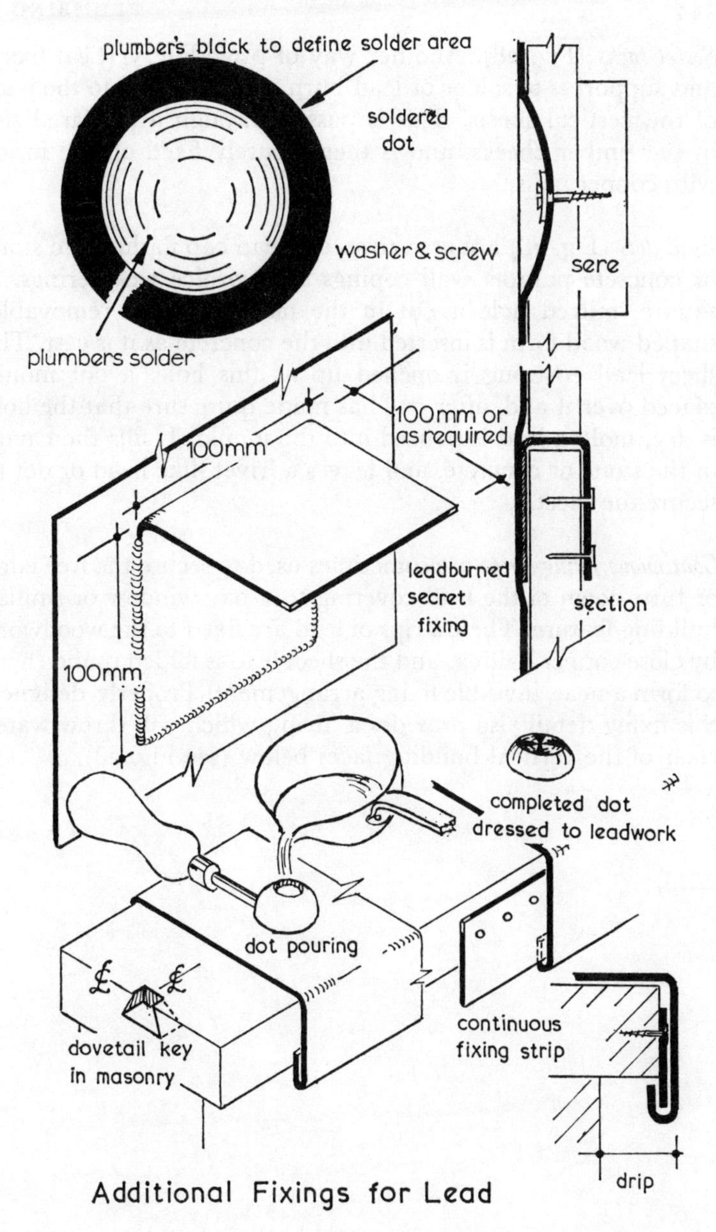

Additional Fixings for Lead

FIG. 28

Secret tacks (Fig. 28). Another way of providing vertical fixing and support is to solder or lead burn a strip of lead to the back of the vertical sheets. This is passed through a prepared slot in the timber cheeks, and is then securely fixed on the inside with copper nails.

Lead dots (Fig. 28) are used to secure lead cap flashings to stone or concrete parapet wall copings and cornice weatherings. A square, mitred hole is cut in the masonry, or a removable, shaped wood form is inserted into the concrete as it is cast. The sheet lead covering is opened up at this hole, a dot mould placed over it and, after one has made quite sure that the hole is dry, molten lead is poured into the mould. It fills the cavity in the stone or concrete, and leaves a 'rivet'-like head or dot to secure the sheet.

Continuous fixing strips are sometimes used to secure the free edge or turn down of the lead covering to a bay window or similar building feature. These strips of lead are fixed to the woodwork by close copper nailing, and the sheet lead is folded round them to form a neat, invisible fixing arrangement. Properly designed this fixing detail also provides a 'drip', which will throw water clear of the vertical building faces below (see Fig. 28).

13
Copper roofwork

In the past, copper roofwork was considered to be a bit tricky and perhaps rather beyond the plumber's ability. Most of the older sheet copper roofwork was therefore done by so-called specialists who did nothing else. Today, things are very much changed. Recent developments in the working of sheet copper have led to simple, yet efficient and craftsmanlike ways of using this material for roof flashings and weatherings. The modern plumber regards copper as a plumber's material. He is quick to learn and adapt these new and interesting skills and now he can challenge the specialists with 'anything you can do—I can do better'. Copper is long lasting, relatively light, easy to manipulate and fix, fire resistant, creep resistant, and pleasing in appearance.

Commercial sizes

Copper is obtainable either in standard-sized sheets 6 ft by 3 ft and 4 ft by 2 ft, or in strip form, in widths from 6 in to 3 ft 6 in.

Copper strip is available in any length, and is supplied in rolls. It is convenient to handle, and if ordered in widths to suit the work in hand it will be found more economical in cutting and use, and generally less costly than sheet copper.

Thickness or gauge is shown by an imperial Standard Wire Gauge number—S.W.G. for short. The bigger the S.W.G. number, the thinner the copper sheet or strip. For example 24 S.W.G. is thinner than 22 S.W.G.

1 ft^2 of 24 S.W.G. copper weighs approximately 1 lb (0·45 kg), and 22 S.W.G. copper weighs a little less than $1\frac{1}{3}$ lb/ft^2 (0·59 kg).

Gauges commonly used by plumbers for traditional roofwork; that is, for all roofwork other than patented roofing systems which you need not consider at the moment—are 26 S.W.G., 24 S.W.G. and 22 S.W.G. It is recommended that 24 S.W.G. should be used for all normal work in flat or pitched roof coverings, flashings, gutters etc. The 26 S.W.G. copper should only be used for short lengths of fully secured cover flashings.

For gutters, 22 S.W.G. is recommended, since it is rather thicker and can be used in longer lengths. Fewer drip joints are therefore needed. It is also recommended for roofs which may be severely exposed to strong winds, since its additional stiffness will help to keep the roof secure.

Metrication will result in thicknesses being expressed in mm 24 S.W.G. = 0·558 mm. 22 S.W.G. = 0·711 mm. These thicknesses will be rounded off when BS 1569 is metricated and may become 0·6 mm and 0·7 mm respectively.

Temper denotes the degree of hardness of copper sheet or strip, and for roofwork it is important that only 'dead soft temper', or fully annealed copper, should be used.

The specification or description of the material given when it is ordered should always include the gauge required, and the words 'dead soft temper copper sheet or strip, for roofing, to BS 1569'. It should also give the sizes of material required.

Substructure preparation and underlays of felt for sheet copper roofwork will all be as described in Chapter 11.

The working of sheet copper weatherings

As you know, copper is malleable to a certain extent, but it work-hardens with cold-working. Work-hardened copper can be restored to a soft state by being heated to a dull red heat and then quenched in water or allowed to cool in the air. Nevertheless, it is better always to try to form the copper with as few blows as possible. One good blow delivered at the right place is better than several taps which only tend to work-harden the material. The following points should also be remembered:

1 Accuracy in setting out is very necessary.
2 You should use sharp snips. Any accidental 'snags' along a snip-cut edge which will have to be seamed, or worked in any way, should be filed out at once. Otherwise, a tear might start at the snags as working proceeds.
3 Fixings must be adequate and secure, for although its light weight is one of copper's many advantages, this does not help to keep it in place against the force of a strong wind.
4 Clips of the material being laid must be placed at 460 mm centres along all joints in sheet copper. The clips vary from 19 mm width for batten rolls, to 25 mm for standing seams, and 50 mm for drip edges.
5 Lining plates or continuous fixing strips are used in suitable widths to secure the bottom edges of roof sheets.
6 Nails used to secure the clips must be large-headed, cut copper nails, at least 25 mm long, Iron or galvanized iron nails must not be used. If screws are used they must be of good quality brass. In all cases care must be taken to ensure that no fixing device punctures the copper roof sheeting.
7 Stiffening beads should be provided at all free or unfixed edges; for example, at the bottom edge of cover flashings. These are quite simply formed by allowing an extra 13 mm at such edges and turning this smartly through 90° on the inside or underside. It is then dressed flat over an 3 mm thick metal straight edge to make a kind of channel section which stiffens the edge and makes it less likely to lift in a strong wind.

Abutment flashings

The setting out, preparation and fixing of soakers, cover, and step and cover flashings has been described on pages 120 and 123. The only differences with copper are that the cover flashings will have stiffening beads at their free edges, and the fixings into the brick joints will be secured by rolls of narrow strips of copper called 'tags', which will be used as wedges.

Chimney-back gutters

These are quickly and easily made on the bench after the necessary measurements have been carefully taken from the chimney. The setting out and cutting of the three pieces of copper which form a chimney-back gutter are shown in Fig. 29. The procedure is as follows:

Gutter sole piece. Mark out the length D, which is equal to the width of the stack plus an allowance of 13 mm at each end. These allowances will later form the overcloak to the jointing seams, or welts.

The width of the piece is equal to the under slate allowance A, which is about 230 mm, plus the allowance B, as is necessary for the gutter sole and fillet, plus the upstand to wall allowance C, which is about 76 mm. For the 457 mm-wide stack described on page 128, the sole piece would be 483 mm long by about 457 mm wide.

The side pieces are each cut out of a piece of material about 457 mm by 380 mm, or according to the size of the chimney. They should be set out as shown in Fig. 29. Take care to mark out the seam allowances and fold the lines on opposite faces, since the two opposite-handed side pieces must bè one for the right- and one for the left-hand side of the stack. Note that the dotted line shows the actual gutter shape, size and angle to the roof. Outside this line 6 mm is allowed to form the undercloak of the welted seam joint.

The 50 mm by 50 mm triangular fillet is used in the angle of the gutter and stack wall upstand so that two 45° angles are formed instead of one 90° angle, since this arrangement is easier to work.

The seam allowance is cut neatly and sharply with radiused angles—a 16 mm radius serves to mark these curves.

To prepare the side pieces:

1 Turn stiffening beads at free outer edges.

2 Over a piece of 50 mm by 50 mm angle iron, turn the 6 mm seam allowance until it is at right angles to the sheet face.

3 Lay a straight steel edge along the fold line, turn the side cheek up 90°, and dress the 6 mm seam allowance on the under slate bit down on to the straight edge.

4 Repeat this for the opposite side cheek, again making sure that the stiffening beads and seam turns are opposite to the ones just done.

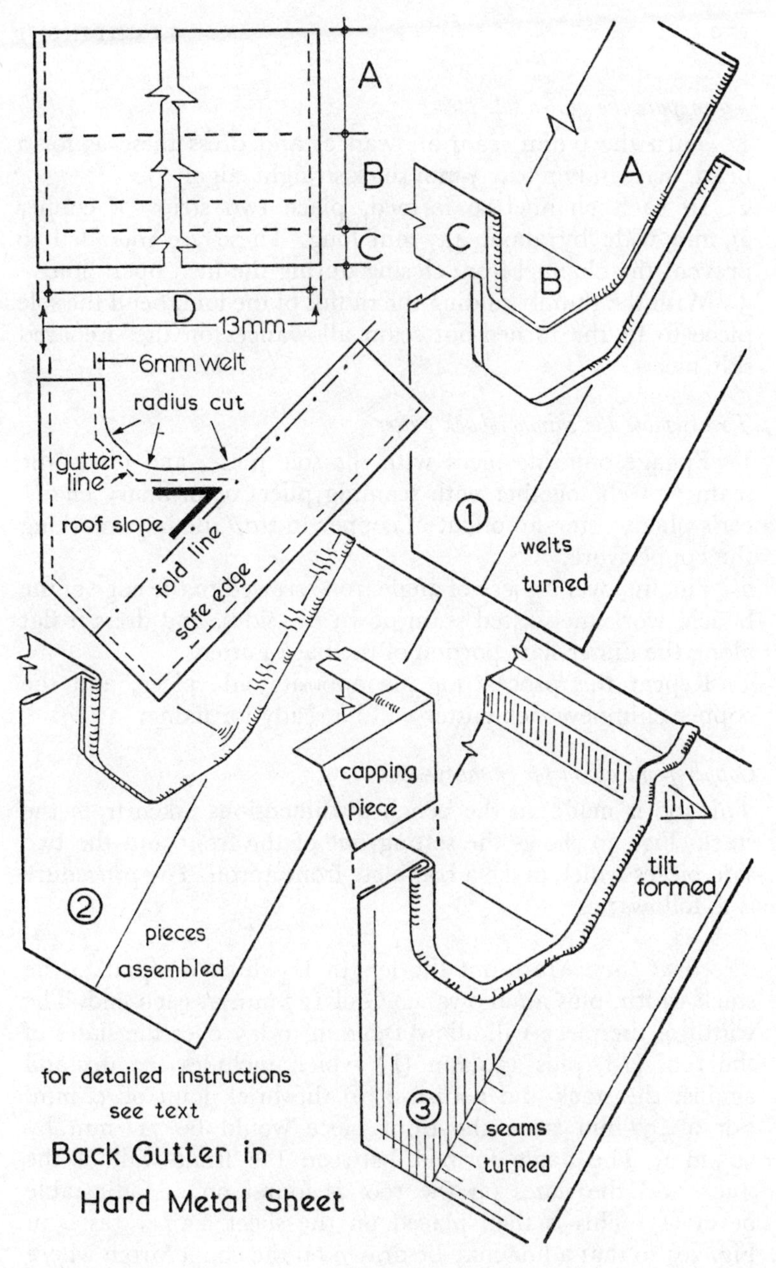
A
B
C
D
13mm
6mm welt
radius cut
gutter line
roof slope
fold line
safe edge
A
C
B
①
welts
turned
capping
piece
tilt
formed
②
pieces
assembled
③
seams
turned
for detailed instructions
see text
Back Gutter in
Hard Metal Sheet

FIG. 29

To prepare the gutter sole piece:

1 Turn the 6 mm seam allowances and dress these as for a bead, over and on to a 3-mm thick straight edge.
2 In each channel so formed, place two strips of copper 25 mm wide by about 457 mm long. These are intended to prevent the channel from closing during the next operation.
3 With the thumb guiding the radius of the fold, bend the sole piece to fit the turned-out seam allowances on the prepared side pieces.

To assemble the chimney-back gutter:

1 Engage one side piece with the sole piece, and pinch the seam or welt together with seaming pliers or ordinary engineer's pliers, using an offcut of copper to stop the jaws marking the copperwork.
2 Finally, over a piece of angle iron screwed to the edge of the bench, work the welted seam down the sides, and dress it flat along the under slate portion of the back gutter.
3 Repeat this process for the opposite side piece, and the copper chimney-back gutter will be ready for fixing.

Copper front apron for chimney-stack

This too is made on the bench to dimensions taken from the stack. Fig. 30 shows the setting out of the front and the two side pieces which make a complete front apron. The procedure is as follows:

The front piece. Mark out the length D which is equal to the stack width, plus an allowance G of 127 mm at each end. The width of the piece will allow 150 mm to lay over the slates of the roof (E), plus 150 mm (F) which includes the upstand against the stack and a turn-in to the brick joint of 32 mm. For a 457 mm stack the front piece would be 711 mm by 305 mm. The angle formed between the front wall of the stack and the slates on the roof is found on an adjustable bevel (I). This is then placed on the sheet copper (as 2 in Fig. 30) so that a line may be drawn on the copper from where the line G crosses the central fold line to the outer side edge of

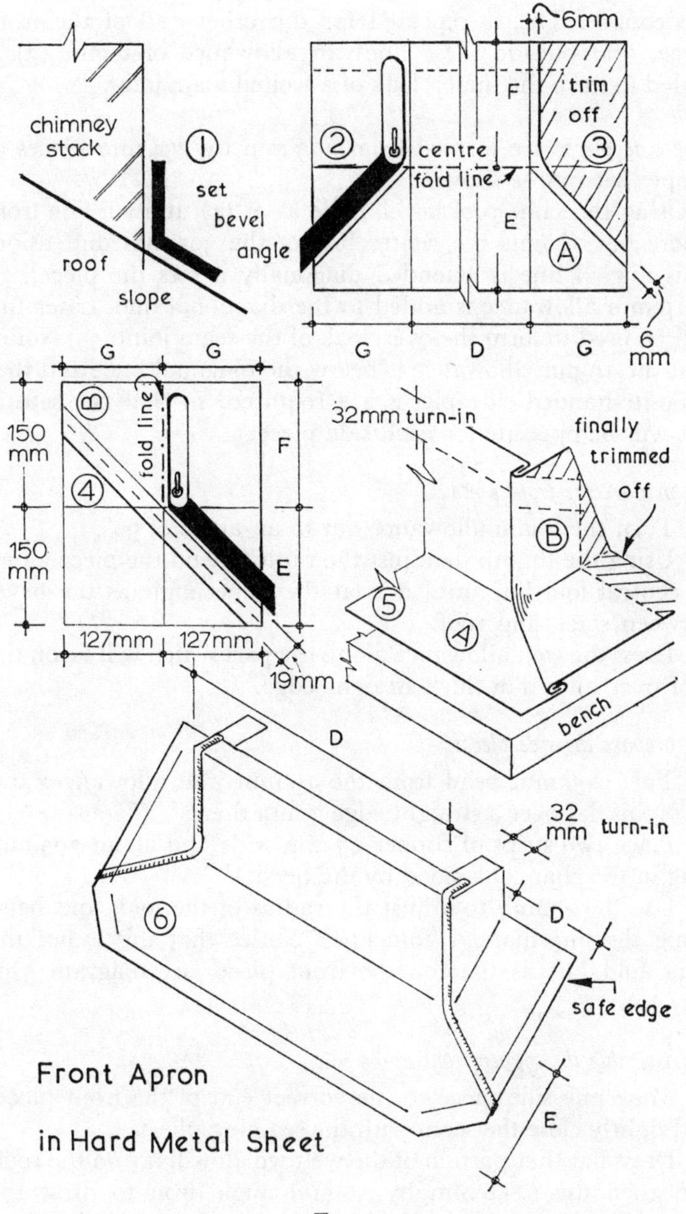
chimney
stack
1
set
bevel
angle
roof
slope
6mm
2
centre
fold line
F
trim
off
3
E
A
G
D
G
6
mm
G
G
150
mm
B
fold line
F
4
150
mm
E
127mm
127mm
19mm
32mm turn-in
finally
trimmed
off
B
5
A
bench
D
32
mm
turn-in
6
D
safe edge
E
Front Apron
in Hard Metal Sheet

FIG. 30

the copper. This is repeated for the other end of the front piece, and outside these lines an allowance of 6 mm (3) is added to form the undercloak of a welted seam joint.

The side pieces are formed from 305 mm by 254 mm pieces of copper set out as shown.

Using the same roof bevel angle as at (4), draw a line from where line G cuts the centre line of the 305 mm dimension. This angled line is extended diagonally across the piece (4). A 19 mm allowance is added to the diagnonal line. Later this will be used to form the overcloak of the seam joint. (5) Notice that the 19 mm allowance is below the diagonal line; and that opposite-handed side pieces are required, so that the setting out will be opposite for each side piece.

To prepare the front piece:

1 Turn the 6 mm allowance out to an angle of 90°.
2 Using the thumb to adjust the radius, bend the piece along its central fold line until it is at the same angle as the bevel between stack and roof.
3 Dress the welt allowance along the part which will lie on the roof over an 3 mm-thick straight edge.

To prepare the side pieces:

1 Turn a 6 mm bead from the 19 mm welt allowance, and dress this flat over a straight edge 3 mm thick.
2 Place two strips of copper 25 mm wide and about 305 mm long in the channel formed by the bead.
3 Use the thumb to adjust the radius of the fold, and bend along the line marked 'fold line'. Notice that this is not the same fold line as that on the front piece (see diagram 5 of Fig. 30).

To assemble the copper front apron:

1 Mate one side piece to the correct end of the front piece, and tightly close the seam with the seaming pliers.
2 Dress flat that portion of the welt which will lay on the roof, and then use a 50 mm by 50 mm angle iron to dress the upstand part of the welt through 90° on to the upstand face.

3 Repeat this with the other side piece.

4 Turn stiffening beads on all edges of the apron which will lie on the roof.

5 Finally, make 'dog-eared' or folded corner 32 mm down from the upstand top edge, or wherever the brick courses require. Finish off the turn-in to the brick joint, and the apron is ready for fixing in place on the chimney.

A simple, inexpensive, yet effective way of learning this setting out, preparation and assembly routine is to do it on drawing paper or stiff brown paper.

14

Aluminium roofwork

Aluminium was first produced towards the end of the nineteenth century. It is, therefore, a comparatively new material, but its properties of lightness, resistance to corrosion, and in certain forms its strength, combine to make it a very effective and much used metal. The modern plumber, quick to realise the value of these properties in a roofing material, accepts aluminium as a plumbing material.

Its properties are similar to those of copper in many ways, and so it is understandable that methods for working and jointing it have followed the same lines as those for copper.

Commercial sizes

Strip aluminium for plumber's roofwork is available in many widths, but the most convenient are the 1 ft 6 in (457 mm), and the 2 ft (610 mm), which are generally regarded as the standards. The material is supplied in rolls or coils of standard weights of 28 lb (12·6 kg) or 56 lb (25 kg). The length of the coiled strip will depend upon its gauge or thickness and, of course, its width; for example, one 56 lb coil of 20 S.W.G. aluminium 2 ft wide will cover about 110 ft² (10·25 m²). It might be interesting to compare this area with the areas that could be covered by a similar weight of other roofing materials. Aluminium is very light and easily worked. It is highly resistant to corrosion, does not creep, is fire resistant, and is quite pleasing in appearance. Gauges commonly used are 22 S.W.G. (0·7 mm) for flashings and well-secured small areas. For larger areas, for example flat or pitched roofs, 20 S.W.G. (0·9 mm) must be used.

Metrication will result in thickness being expressed in mm.

Lengths and widths will be in metres or mm. The relevant BS is not metricated at time of going to press but, in any case, the BS will quote both imperial and metric dimensions in this instance.

The purity of a metal affects its resistance to corrosion and its working properties. It is generally true to say of all metals that these properties improve as the metal becomes purer. Aluminium is made in many grades of purity. These vary from 99·99% pure aluminium through 99·8% and 99·5% to the commercial purity grade fixed at 99%. For special purposes, less pure aluminium alloys are made. The so-called 'impurities' in these are, in fact, carefully controlled quantities of other elements, which give the material further properties where these are necessary. For plumber's roofwork, however, workability and resistance to corrosion are the most important properties, and so an aluminium of high purity must be used. Commercially pure aluminium, to BS 1470 NS3, can be used in certain tempers and gauges for roofwork in areas or bays, but, for the reasons already given, the purer the material the better. Super purity 99·99% pure aluminium is recommended for all plumber's roofwork, and its extra malleability makes it very suitable for flashing and weathering details.
Temper. These materials are obtained in 'dead soft', ¼ hard, and harder tempers, but the plumber is concerned only with the soft and ¼ hard varieties. 'Dead soft' aluminium should be used for flashings, and for such weathering details as chimney flashings. Quarter hard temper aluminium is recommended for roofwork in areas where winds are strong. The greater stiffness of aluminium at this temper helps it to resist the suction effects of strong winds, which might otherwise tend to lift that very light metal. At this temper aluminium is still very easy to work, and in many cases the very slight extra stiffness is a help in manipulating and working it.

Specifications for aluminium coiled strip for roofwork would be as follows:

For flashings and weatherings—super purity aluminium strip, flashing quality, 'dead soft' temper, to BS 1470, 22 S.W.G. or 20 S.W.G., or equivalent metric thickness, as required.

For roofwork in bays—super purity aluminium strip, roofing quality, ¼ hard temper, to BS 1470, 20 S.W.G.; or, commercial quality NS3 aluminium in ¼ hard temper, 20 S.W.G. to BS 1470. But remember that the higher purity aluminiums are better for working and for resistance to corrosion.

Substructure preparation and *felt underlays* for aluminium roofwork are as described in Chapter 11.

The working of sheet aluminium

Aluminium, copper and zinc are referred to as 'hard' metals as distinct from the 'soft' metal, lead; consequently these three are all worked by much the same methods.

Work-hardening is unlikely to occur in normal applications of aluminium, and may be avoided by seeing that the blows of dresser or mallet are firmly and expertly delivered; but if it does occur at all, the sheet can be annealed by heat treatment. The annealing temperature cannot be judged against the light-coloured aluminium, and so the plumber applies heat with a blowlamp, occasionally pausing to draw a matchstick across the heated surface. When the correct temperature has been reached, the match will leave a black charcoal line on the sheet, which can then be quenched in water or allowed to cool in the air.

As with other hard metals, sharp snips must be used and snags avoided on all cut edges, or these might develop into tears during working. Tools must be clean and in no way contaminated with minute particles of other metals, particularly copper. Electrolytic corrosion could result from the thoughtless use of tools previously used on copper work.

Fixings are particularly important with aluminium as with all light roofing materials. They must be carefully arranged so that the metal is securely held on the roof in high winds. Clips of aluminium are prepared and used in the way described for copper roofwork on pages 164 and 167.

Nails for fixing should be large headed and of hard aluminium alloy to BS General Engineering Specification N6, but good quality large headed galvanised iron nails could be

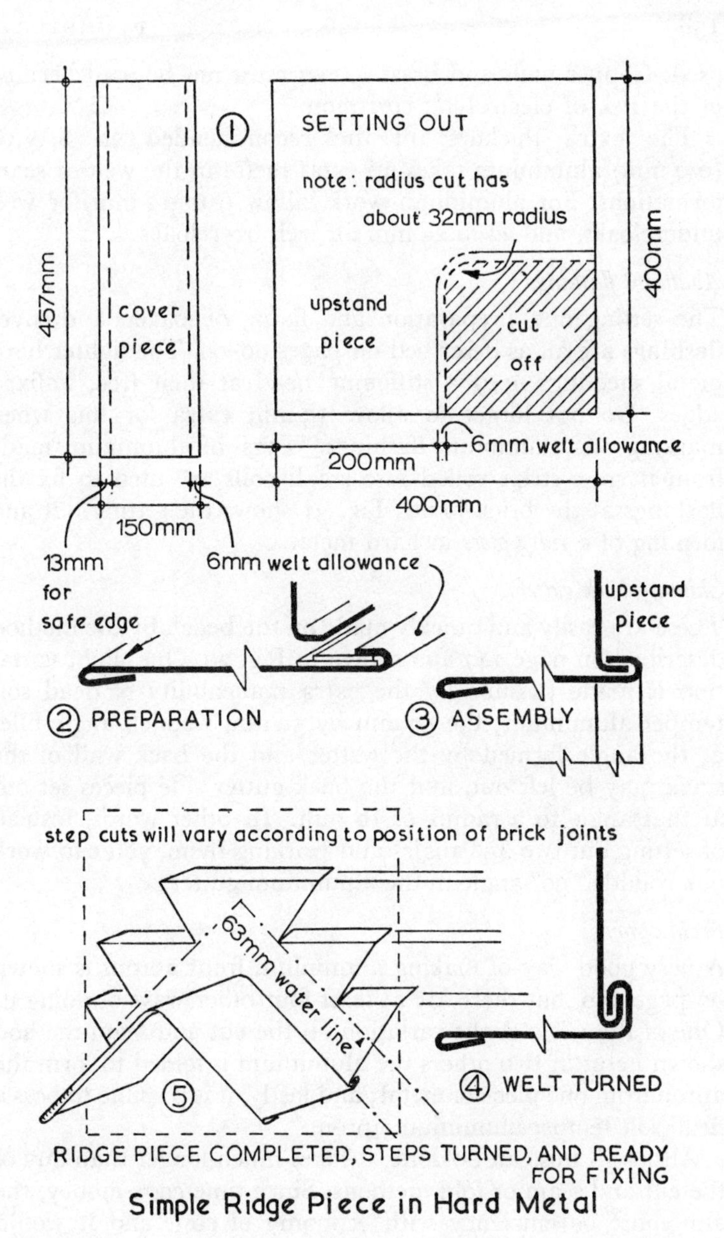

Simple Ridge Piece in Hard Metal

FIG. 31

used. Copper nails and brass screws must not be used because of the risk of electrolytic corrosion.

The extra thickness of the recommended 20 S.W.G. (0·9 mm) aluminium takes up extra metal in the welt or seam formations. For aluminium work, allow 9 to 12 mm for welt undercloaks, and 22 to 25 mm for welt overcloaks.

Abutment flashings

The setting out, preparation and fixing of soakers and cover flashings are all as described on pages 00–00 The lighter hard metal sheetings need a stiffening bead at their free, unfixed edges Do not forget to allow 13 mm extra for this when marking out aluminium flashings. 'Tags' of aluminium made from narrow strips rolled into small coils are used to fix the flashings at the brick joints. Fig. 31 shows the setting out and forming of a *ridge piece* in hard metal.

Chimney-back gutters

These are easily and quickly made on the bench by the method described on page 147 and shown in Fig. 29. One slight variation is made possible by the extra malleability of dead soft temper aluminium; the 50 mm by 50 mm wooden angle fillet at the angle formed by the gutter and the back wall of the stack may be left out, and the back gutter side pieces set out at that angle to a radius of 16 mm. In other words, instead of setting out two 45° angles and working them, you can work one rounded 90° angle in the aluminium gutter.

Front aprons

A very good way of making aluminium front aprons is shown on page 150, but there are at least four other ways of doing it. One of these is a slight variation on the cut and seam method shown here; in two others the aluminium is folded to form the apron from one piece of metal; and lastly, it is possible to boss a dead soft temper aluminium apron.

Although this can be done, it takes much longer than any of the cut and seam or fold methods. Since time costs money, the aim must be efficiency with economy of cost, and it would therefore be better to use one of the quicker ways and keep all bossed work to a minimum.

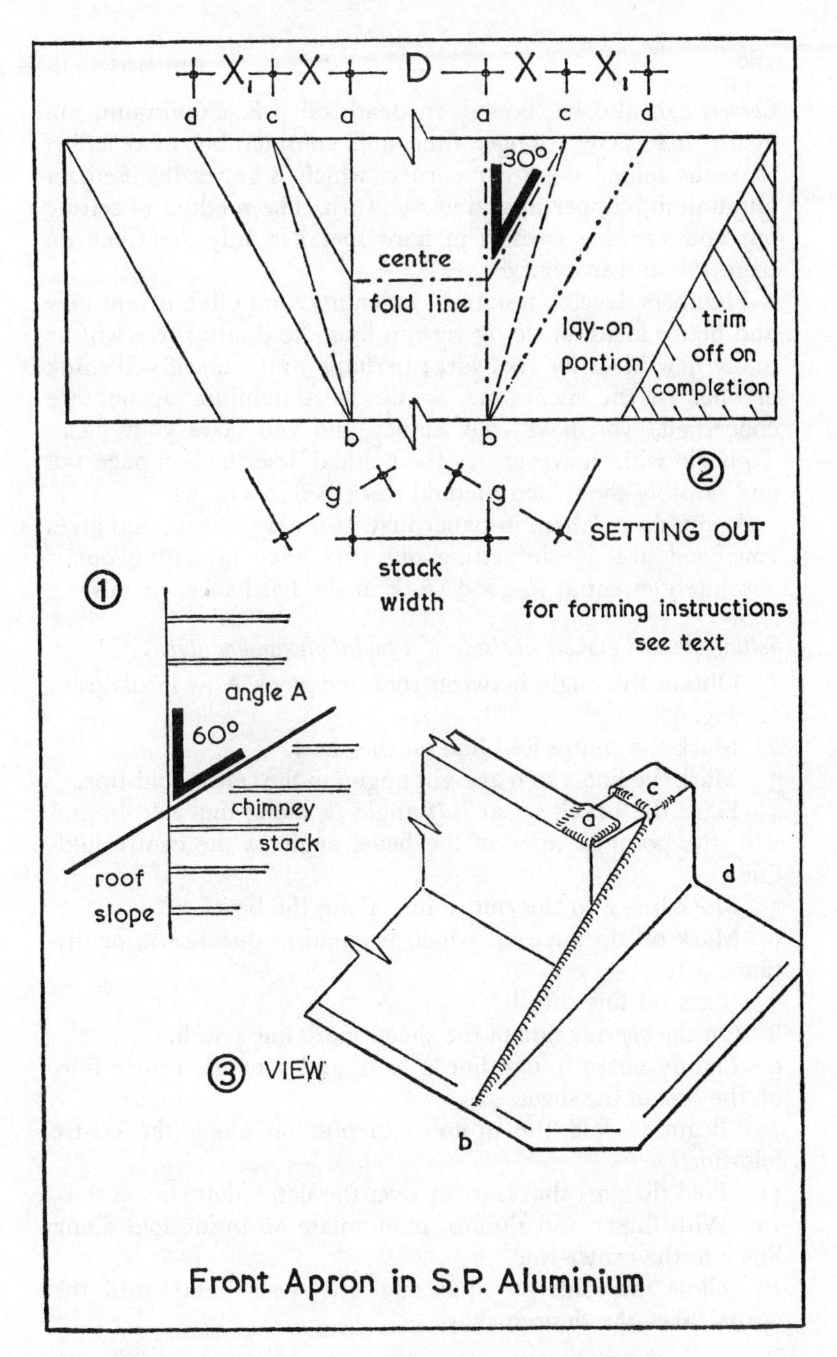
X1 X D X X1
d c a a c d
30°
centre
fold line
lay-on
portion
trim
off on
completion
b b
g g
② SETTING OUT
stack
width
①
for forming instructions
see text
angle A
60°
chimney
stack
roof
slope
a
c
d
③ VIEW
b
Front Apron in S.P. Aluminium

FIG. 32

Corners can also be 'bossed' in dead soft S.P. aluminium, but again these take a longer time and considerably more effort than the folded dog's-ear corner, which is generally used for aluminium, copper and zinc roofwork. The method of setting out and working corners in hard metal is fully described on page 179 and in Fig. 38.

Plumbers develop a natural ingenuity, and often invent new and better ideas for doing certain jobs. No doubt there will be many new ideas for roofwork; perhaps you yourself will think of one. In the meantime, so far as aluminium aprons are concerned, 'you pays your money and you takes your pick'. To begin with, however, try the method described on page 158 and possibly the folded method described in Fig. 32.

Again, do try them in paper first. It is easy, cheap, and gives you good practice in setting out. Accuracy in setting out is absolutely essential to good work in the hard sheet metals.

Setting out and making one side of a folded aluminium apron:

1 Obtain the angle between roof and stack A as in diagram (1) Fig. 32.

2 Mark the centre fold-line on the sheet.

3 Mark the line a to b at right angles to the centre fold-line.

4 Place the bevel, set at *half* angle A, along line a to b, and with the point or apex of the bevel angle at the centre fold-line.

5 Mark line c to the centre-line along the bevel.

6 Mark off distance x_1, which is equal to distance x, or distance a to c.

7 Mark off line d to b.

8 On the *opposite* side of the sheet, mark line c to b.

9 Lightly chase wedge line c to b, and a to the centre line, on the *back* of the sheet.

10 Begin to fold the upstand to position along the centre fold-line.

11 Fold the part that is to lay over the slates along line d to b.

12 With finger and thumb, manipulate an inside fold along line c to the centre-line.

13 Close the folds by squeezing with your hand until the apron takes the desired shape.

14 Finally, dress the folds close, but not necessarily dead close; trim off, turn the edge stiffening beads, and prepare the turn-in to the brickwork.

Note. Both ends of the apron should be worked at the same time, following this sequence of operations.

15

Joints and fixings for copper and aluminium roofwork

As has already been said, there is some similarity in the working of copper and aluminium for abutment flashings, chimney weatherings and so on. These materials are also worked and fixed in much the same way when applied in large areas on flat or pitched roofs. The following descriptions will therefore generally apply to both metals, but any special fixing requirements are given under the heading 'Fixings' on page 164. You should note especially the type of nails to be used, the sizes of the fixing clips or cleats, and the correct spacing for these.

Joints with the flow

The solid or *wood cored roll* may be used on flat or pitched roofs, but it must be used where people are likely to walk on the roof. The simplest, easiest to work, and therefore probably the most common roll form, is the batten roll (Fig. 33).

Sometimes the wood rolls are undercut slightly, but not for the same reason as are rolls for sheet leadwork. In the case of copper and aluminium, the purpose of the undercut is simply to provide an expansion space.

These metals are secured entirely by special clips, which are fixed to the substructure or beneath the rolls, and are folded or seamed in with the sheeting joints to give an indirect but secure fixing. The importance of these fixing clips cannot be over-emphasized. They must be the right size and spaced properly in order to hold these light metals against strong winds. Fig. 33 shows the simple steps in the formation of a batten roll.

The standing seam could perhaps be called the hard metal

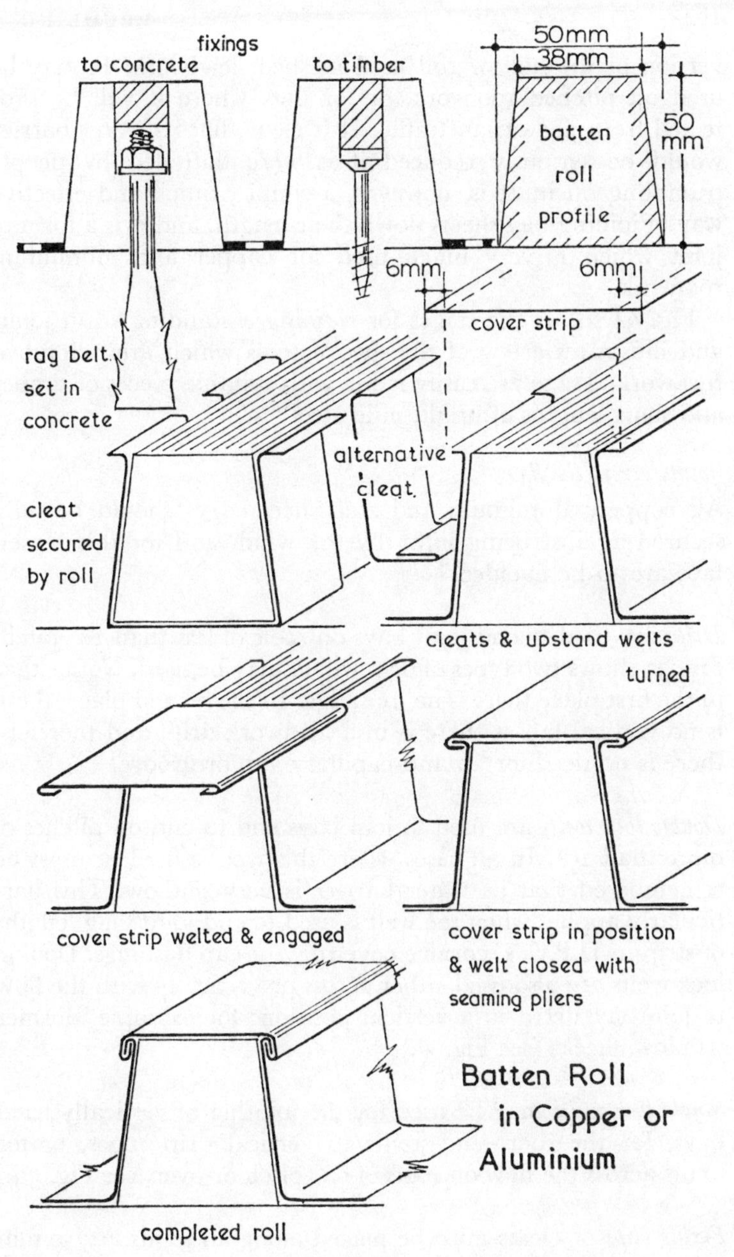
fixings
to concrete
to timber
50mm
38mm
50
mm
batten
roll
profile
6mm
6mm
cover strip
rag belt
set in
concrete
alternative
cleat
cleat
secured
by roll
cleats & upstand welts
turned
cover strip welted & engaged
cover strip in position
& welt closed with
seaming pliers
Batten Roll
in Copper or
Aluminium
completed roll

FIG. 33

version of the hollow roll used in sheet leadwork. It may be used on pitched roofwork, or for flats where it will be protected from pedestrian traffic. It is clear that its water barrier would be seriously reduced if it were flattened by people trampling on it. It is, however, a simple, quick and effective way of joining bay sheets down their length, and it is a form of joint which is very much used for copper and aluminium roofwork.

Fig. 34 shows the stages for forming a standing seam joint, and illustrates a few of the special tools which are helpful in this work. These are easily made with suitable pieces of timber and short lengths of bright mild steel bar.

Joints across the flow

All copper, aluminium and zinc sheet edges should be fully secured against being lifted by the wind, and for this reason laps are to be avoided.

Drips are used to end joint bays on roofs of less than 10° pitch. Fig. 35 shows two types of drip which may be used. Notice that in the first place there is no free edge. In the second place, there is no vertical lap as there is in a leadwork drip, and therefore there is no need for an anti-capillary gap or groove.

Double lock welts are used to join bays end to end on pitches of more than 10°. In all cases where this welt is used, it must be remembered that its water barrier is very shallow. This particularly applies when the welt is used to end joint long lengths of strip in D.P.C.'s, cornice coverings, or cap flashings. Double lock welts are also used either across or vertically with the flow to join bays fixed in a vertical position; for example, dormer window cheeks (see Fig. 35).

Single lock welts may be used for the jointing of vertically fixed bays, for the short end joints to vertical drip edges, or for joints across the flow on roofs of 60° pitch or over (see Fig. 35).

Fixing clips or cleats must be placed along all joints at 350 mm to 460 mm intervals, centre to centre.

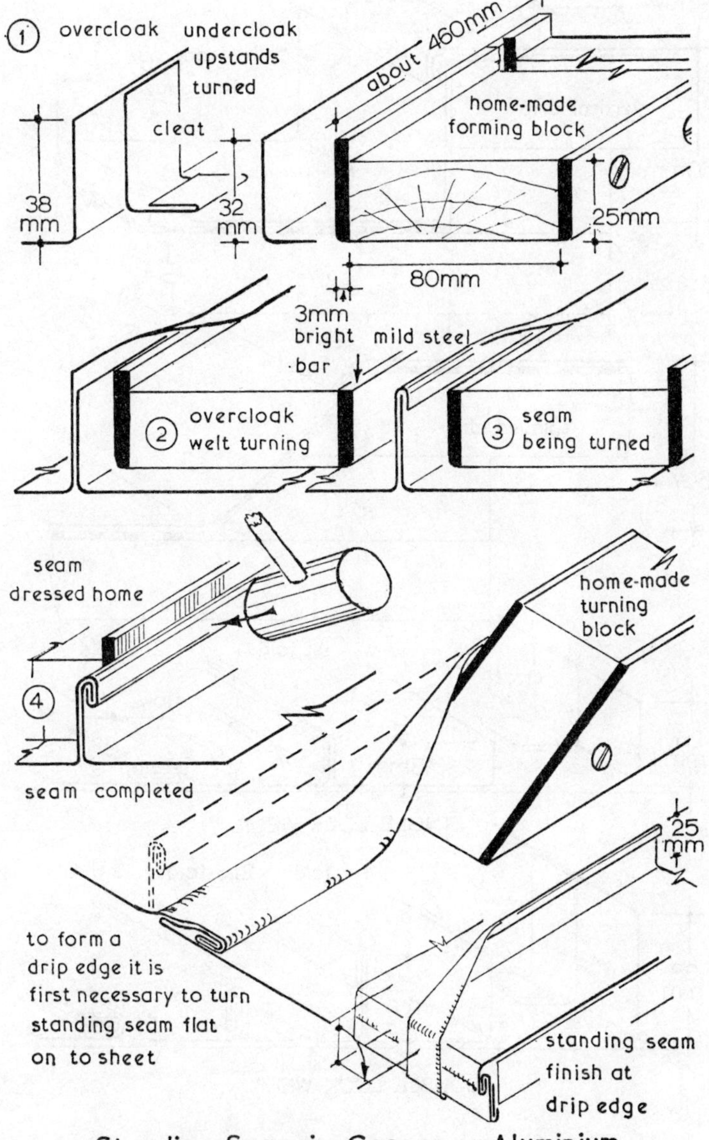

Standing Seam in Copper or Aluminium

FIG. 34

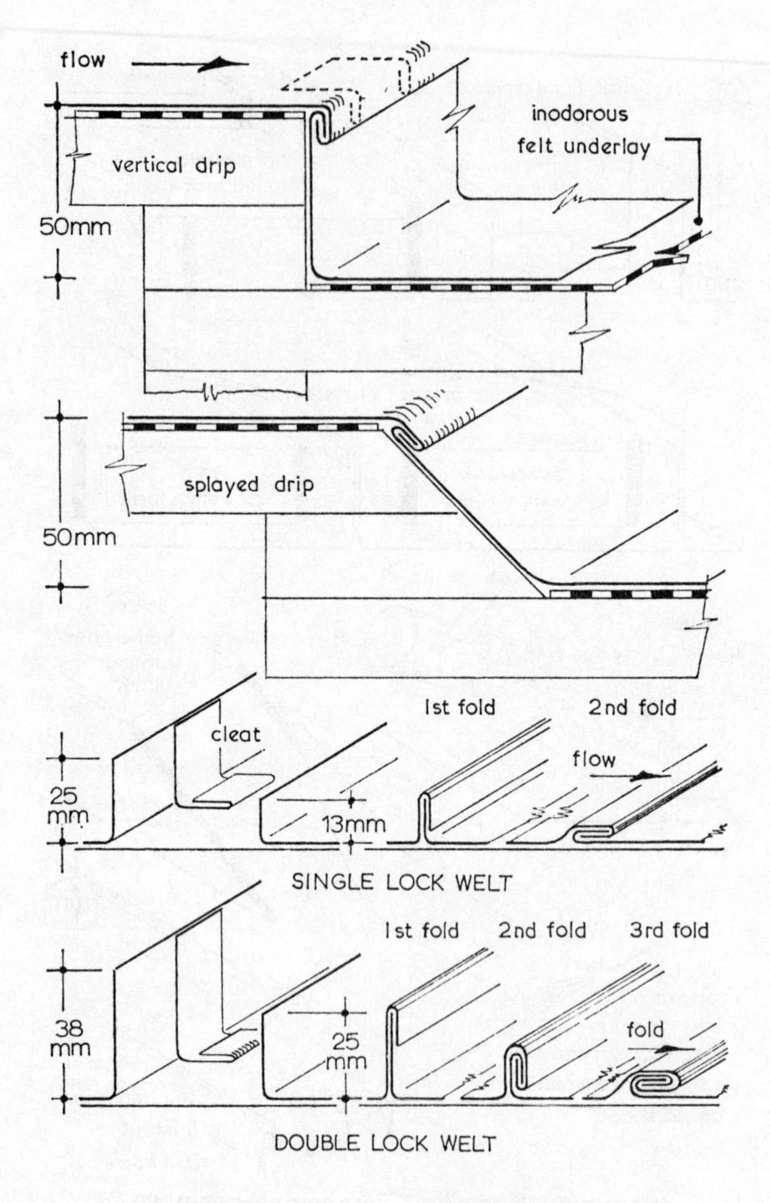

Drips and Welts in Aluminium & Copper

FIG. 35

Additional fixings for sheet copper and aluminium

The minimum fall for a flat roof covered with either copper or aluminium should be 50 mm in 3000 mm. (1 in 60). The area of a bay should not be more than 1 to 1·5m² in order that the effects of thermal expansion should be reduced and to allow frequent fixing at the joints. The shape and dimensions of the bays will depend upon the shape of the roof and the width of the copper or aluminium strip used. Try to work the bay shapes so that they can be cut economically from standard-sized rolls or coils of metal. Allowance must be made for roll upstands, drip upstands, and seaming on the bay. The amount of material left decides the width between the rolls or standing seams, and the length of bay between drips, or drip and drip edge of the roof. For example, 610 mm-wide copper strip might be used with batten rolls 40 mm high for which upstands of 40 mm plus a seam allowance of 6 mm at each side must be made. The width of the bay, after these allowances had been deducted, would be 518 mm and the rolls would be fixed accordingly. The spacing of the bay end joints would be decided in the same way.

In the case of gutter bays, there should be not more than 3 m between drips. These should be of the square type, and should preferably be 63 mm high.

Gutters do take a wide variety of shape in cross section, and since some of these shapes give stiffness they may be used in longer lengths. Advice on gutter designs should be got from a specialist. The Lead, Copper, Zinc and Aluminium Development Associations provide such specialist advice quite free of charge. Their addresses may be found in the reference section of any library.

Dormer window cheeks and similar triangular areas may have long free edges at the roof abutment. It may be desirable to cut the bays vertically, so as to introduce the need for a welted joint, along which additional fixings can be inserted.

Continuous fixing strips, sometimes called lining plates, are always close nailed to the vertical drip edges of a roof, cornice, or cap

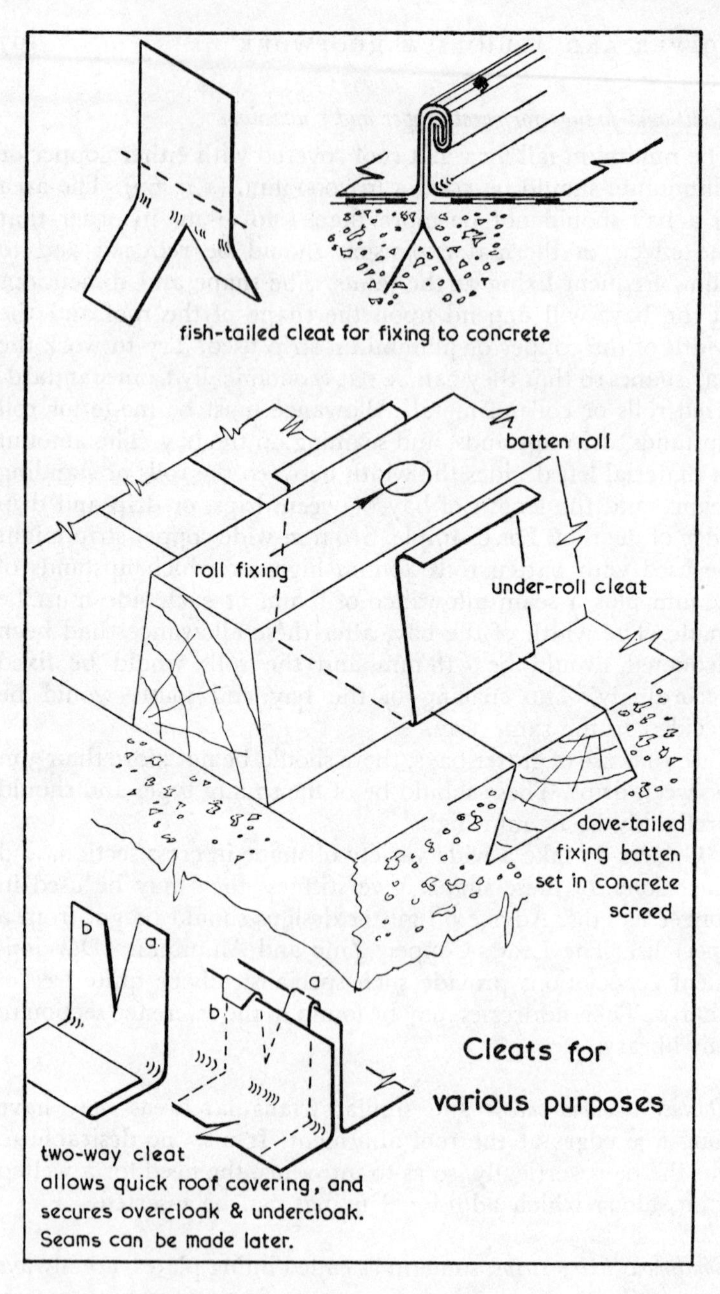
fish-tailed cleat for fixing to concrete
batten roll
roll fixing
under-roll cleat
dove-tailed
fixing batten
set in concrete
screed
b
a
a
b
Cleats for
various purposes
two-way cleat
allows quick roof covering, and
secures overcloak & undercloak.
Seams can be made later.

FIG. 36

flashing, and the free edge of the copper or aluminium sheet is single welted round these (see Fig. 34, page 165).

Fixings to concrete substructures may be fishtailed cleats embedded in the fine concrete screed which is laid over the flat structural roof slab, and trowelled smooth and to falls. Alternatively, dovetailed battens might be embedded in the surface screed and clips fixed to these just as they are for ordinary timber deckings. Figs. 33 and 36 show examples of these fixings.

Linseed oil painted along welt edges before the welt is made will repel water and help to prevent possible capillarity at such joints.

Bituminous paint on all surfaces of aluminium work in contact with new concrete will help to prevent corrosion, but if felt underlays are properly used this precaution should seldom be necessary.

16

Zinc roofwork

Zinc is a plumber's metal. Apart from being extensively used in hot dip galvanising to provide a protective coating for mild steel tubes, cylinders and cisterns, it is regularly used in the form of sheets for roof weathering.

Zinc is quite light in the gauges specified for roofing. It is relatively cheap and offers a reasonably good resistance to atmospheric corrosion. A roof laid in accordance with recommended practice may be expected to give a useful and trouble-free life of forty years, and, quite probably, much longer. Examples of zinc roofing 100 years old are not uncommon, a notable one being the spire of Wellington College Chapel in Berkshire, which was weathered with zinc sheets in 1868 and has needed no maintenance since.

Working techniques

Zinc is a hard metal, and the methods of working it are the same as those for the other hard sheet metals.

Zinc is said to have a 'grain'. This runs lengthwise down the sheet just as grain in timber runs down the length of the piece. The grain does not affect the normal working of zinc, but great care should be taken when one is forming folds along the grain; such folds should not be too sharp. A rounded fold of radius at least twice the thickness of the metal will be satisfactory. The marking out of zinc sheets is best done with a blunt centre punch or similar tool. A pencil mark does not show up well, and a sharp tool such as a scriber will cut into the zinc. This does not matter on a line which is to be cut later on with the snips, but if the set outline for a fold is cut it could seriously weaken the metal at this point. The sharp scriber is not therefore recommended for the purpose.

Commercial quality zinc to BS 849 is chiefly used for the specially developed system of zinc roofing known as roll-cap roofing, a system so well devised some sixty years ago that it has not been possible to improve upon it.

Ordinary commercial quality zinc sheets work more easily when slightly warm. In cold weather it is therefore an advantage to store the sheets in a warm place. In really severe weather a sheet may be warmed by passing over it a few times a blowlamp burning with a not too powerful flame. A high temperature is not necessary, and indeed should be avoided since this can permanently upset the crystal structure of the metal, and make it brittle. The zinc need only be heated to about or a little above hand temperature.

Some roofing details in this quality zinc involve cutting and soldering. Soldering in any form is not considered to be good practice in roofwork, and should be avoided if it is possible to do so by using a different design or fixing method.

Special soft temper zinc is now available. It is chiefly used for covering roofs by the standing seam method, described for copper and aluminium on page 162 and in Fig. 34. This material can also be used on simple seamed details. Ridge pieces, chimney-back gutters and chimney front aprons are easily and quickly made in it by the method described on pages 147 and 150, and illustrated in Figs. 29 and 30. Simple ridge pieces can also be made in soft temper zinc by the method described on page 151 and as shown in Fig. 31.

Here are some points to be noted when one is making these details in this material:

1 Allow 6 mm seams as with copper.
2 The second turn of seam will be found to move much more easily if it is lightly warmed by having a blowlamp flame, or a gas jet, passed over it once or twice.
3 Zinc is much stiffer than copper or aluminium. In other words, it has more spring back. Once the material has been

handled for a while and the extent of this springiness gauged, due allowances can be made for it.

4 The zinc does not work-harden to an appreciable extent, and there is no need for annealing.

Soldering on sheet zinc, where this is unavoidable, is quite easy. Care must be taken to see that the soldering iron is not too hot because of the relatively low melting point of the zinc (see Table 8, page 112). Solder should be of the tin-rich tinman's variety, with a proportion of 50%, or better still 60%, of tin to lead. It should be quite free of antimony, since this would tend to produce brittle joints. The flux generally used for this work is a solution of zinc chloride commonly known as 'killed spirits'. The plumber often makes it by dissolving zinc scraps in hydrochloric acid, taking care not to breathe in the fumes which are given off during the process. Zinc ammonium chloride solution is an improvement on killed spirits, and is made simply by adding sal-ammoniac to the killed spirits. Many proprietary brands of flux obtainable at the ironmonger's are of a similar composition and it is really much easier to buy it than to make it.

All soldered joints should be well swilled with water when they are finished so as to remove all traces of flux, since this is corrosive.

Copper bit soldering can be used for soldering some zinc weathering details and will be dealt with in the next chapter.

Zinc sizes, gauges, and temper

Commercial quality zinc and the special soft temper zinc are both supplied in standard sheets 7 ft (2·1 m) or 8 ft (2·4 m) by 3 ft (0·9 m). Zinc gauge is a measurement of thickness peculiar to zinc. As the zinc gauge number increases, so does the thickness of sheet. S.W.G., the sheet measurement used for other hard metals, works in the opposite direction. Zinc manufactured to British Standard bears its zinc gauge number, and below this appears its nearest S.W.G. size. The table below is helpful for comparing various thicknesses.

In *Metrication* zinc sheet thicknesses will be expressed in mm as shown below. Zinc gauge Nos. S.W.G. thickness standards will become obsolete. Lengths and width measurements will be rounded off in metres. For example, 8 ft = 2·4 m but may become 2·5 m on metrication of the BS 849 when eventually published.

Zinc gauge number	*Nearest S.W.G.*	*Weight of zinc* oz/ft²	kg/m²	*Thickness in mm*
12	23	15	4·5	0·6
13	22	16·75	5·0	0·7
14	21	18·5	5·6	0·8

The recommended zinc gauges for roofwork are as follows:

Soakers (12 Z.G.) 0·6 mm thickness

All other work in (14 Z.G.) 0·8 mm thickness.

17

Working techniques in soft and hard metal roofwork

In comparison with iron, copper, zinc and roofing-quality aluminium, lead is relatively soft. This property of softness together with certain others—notably its high degree of malleability—allows lead to be worked in ways which cannot be extensively used with other metals.

Thus the commonly used plumber's metals can be conveniently classified as the soft metals; that is, lead sheet and pipe—and the hard metals; that is, copper, zinc, and aluminium sheet. Today one talks of soft metal techniques of working—for example, the bossing of lead, or of the hard metal techniques which involve cutting and seaming of the materials to form similar details in the less malleable hard metals.

Soft metal techniques in roofwork

The ease with which sheet lead can be worked to complicated shapes under the skilled application of tools is one of its big advantages. Lead can be made to 'flow' from one part of the sheet to another. In this way surplus lead can be bossed outward and away from where it is not needed, or bossed inward to provide additional lead where it is required.

Setting out and bossing a simple corner in sheet lead is the first step towards gaining the skill necessary for bossing. The best way to learn how to do it is to have a skilled plumber demonstrate and explain his aims as the work proceeds, but the following notes will be helpful if you have some lead and wish to practise on your own.

Fig. 37 shows the setting out necessary. The working procedure will direct you so that the job goes smoothly. With

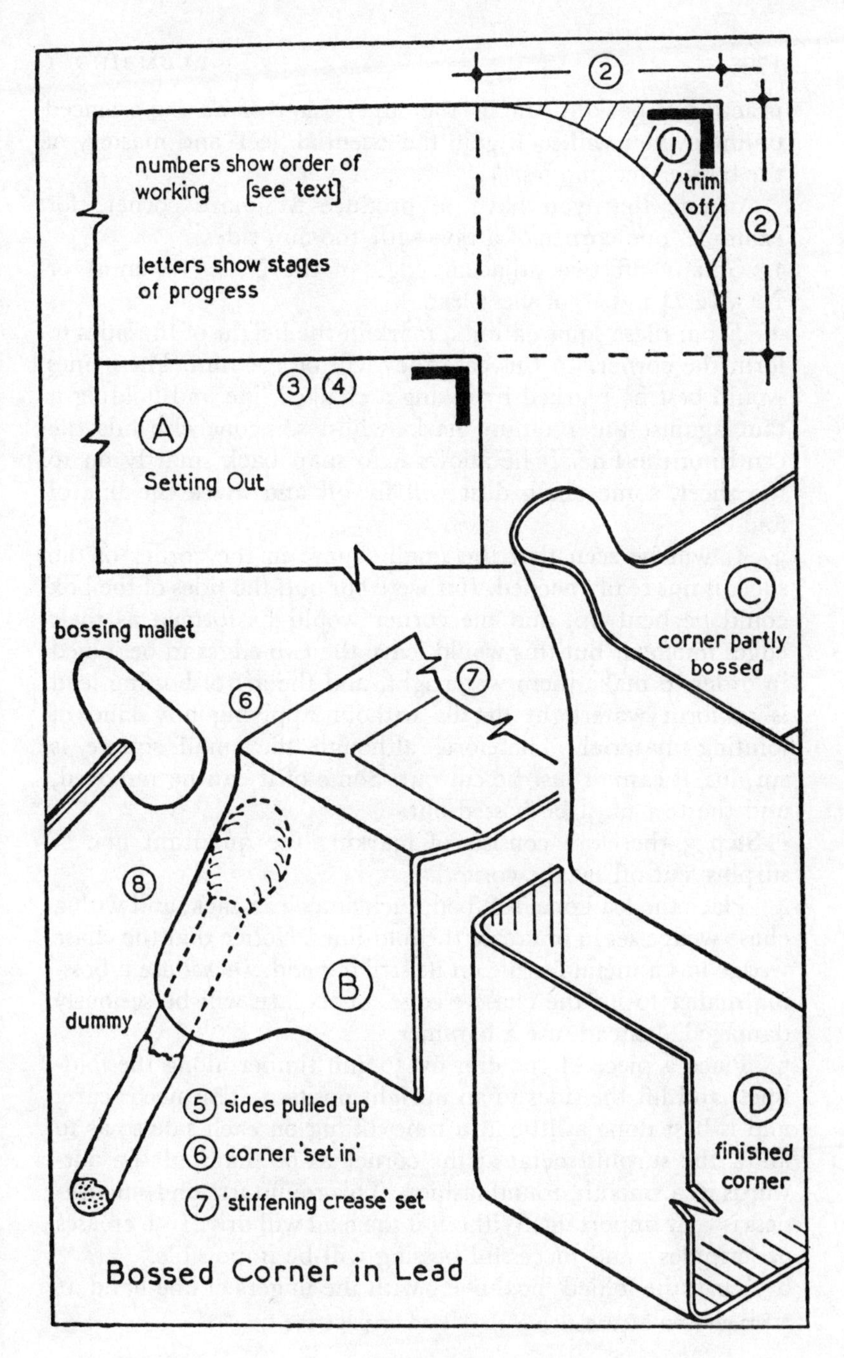
2
1
trim off
2
numbers show order of working [see text]
letters show stages of progress
3 4
A
Setting Out
C
corner partly bossed
bossing mallet
6
7
8
B
dummy
5 sides pulled up
6 corner 'set in'
7 'stiffening crease' set
D
finished corner
Bossed Corner in Lead

FIG. 37

practice, especially under the supervision of an experienced plumber, you will soon gain the essential 'feel' and mastery of the bossing techniques.

Assume that you have to produce a square corner; for example, one corner of a box with 100 mm sides:

1 Square off two adjacent edges of the N° 4 (1·8 mm) or N° 5 (2·24 mm)* of sheet lead.

2 From these squared lines, mark in the height of the sides to form the corner. In this case they will be 100 mm. These lines would best be marked by taking a chalked line and holding it taut against the 100 mm marks whilst someone else lifts the centre of the line. If he allows it to snap back smartly on to the sheet, some chalk dust will fly off and mark the line of fold.

3 It will be seen that the small square in the corner of the sheet is not really needed. If it were cut out, the sides of the box could be bent up, and the corner would be formed as their edges touched. But this would leave the two edges to be joined in order to make them watertight, and the art of bossing lead is to form watertight details without applying any kind of jointing material. Therefore, although the small square is surplus, it cannot just be cut out. Some of it can be removed, and the rest must be bossed out.

Step 3 therefore consists of marking the quadrant line of surplus 'cut off' at the corner.

4 Place the lead on a soft bed, such as a clean sack, and with a chase wedge set in or crease the fold-lines. Notice that the chase wedge has a metal ferrule on its striking end. *Do not* use a bossing mallet to hit the chase wedge, or its face will be seriously damaged. Instead, use a hammer.

5 Place a piece of 100 mm by 50 mm timber along the fold-lines, and lift the sides to an upright position. This needs care, and is best done a little at a time, lifting on each side so as to allow the surplus metal at the corner to be manipulated outwards in a smooth, round fashion. This roundness and smoothness is very important. Without it the lead will drive into creases or 'cripples', and successful bossing will be impossible.

6 Turn the folded 'box' over, with the fingers of one hand at

* See note on Metrication of sheet lead on page 134.

the inside bottom of the corner to get the feel of where you will be hitting. Use a 'dresser' to 'set in' the corner to about 13 mm up its height. This clearly defines the corner, helps to stiffen it against the bossing action, and also helps to check on the size of the piece of bossed leadwork. Nothing is more annoying—or more wasteful in time, energy and material—than to boss or work any plumber's metal to shape only to find that it is either too big or too small for the job it was made for.

7 While the 'box' is still on its back, set in the 'creasing' lines or 'stiffening creases' about 50 mm in from the folded edges. These creases will help the lead to keep its shape during the bossing process.

8 Turn the box to a convenient position. This is very important, for everyone can work better if they are comfortable. Place the head of the dummy tool inside and against the bottom of the corner, and then with the bossing mallet direct blows systematically round the corner, gradually working out towards the free edge. The dummy is really a form of portable anvil, and between it and the mallet the lead is bossed, 'squeezed', or 'flowed' to where it is wanted—in this case out and away to the free edge where, as it collects, it can be cut off.

9 Finally, the corner is neatly dressed square, but not too square in the angle. A slightly rounded corner is better able to accommodate the effects of thermal expansion and contraction than is a sharp corner.

Tools required:

1 Steel try squares.

2 Rule. This would preferably be a steel rule since it is more durable and more suited to plumber's work than a boxwood rule. The rule would be divided in millimetres.

3 Chalk line. This is best kept on a wooden reel or bobbin.

4 Snips. These need to be about 310 mm long and of good quality. Snips are normally made for right-handed workers, who hold the snips in their right hand whilst the left holds the material being cut off the sheet. Sometimes left-handed workers using right-handed snips have to cut the sheet off the surplus, and this is so awkward that they unfortunately

struggle against nature to use their right hand. Otherwise, they have to force the snips to cut the metal, using them incorrectly in their left hand. All this effort is quite unnecessary because 'left-handed' snips are made, and although some inexperienced tool salesman might think you were pulling his leg in asking for them, they can be got and they are a boon to the left-handed plumber.

5 Dresser. One with a 176 mm face length is quite large enough.

6 Chase wedge. One with a 76 mm wide blade is quite a useful size to begin with.

7 Dummy. This has a 'hard lead' head shaped rather like an egg, which is mounted on a malacca cane handle about 310 mm long.

8 Bossing mallet. These are made in a variety of sizes, and are measured according to the diameter of the larger hemispherical face. A 38 mm diameter mallet will be found to move the lead just as effectively, and with less effort, than the heavier, larger mallets.

9 An 'eye' and a 'feel' for quality of workmanship. These cannot be bought at the counter of a tool shop. They are nevertheless essential tools of the trade. Without them the best tools will not produce quite such good work.

Hard metal techniques in roofwork

Soft temper copper and aluminium strip or sheet, and special soft temper zinc sheet are malleable to a degree, but not to such an extent that they can be bossed as lead can. All these hard metals tend to work-harden in use (see page 146). An awareness of this is of the first importance in developing a hard sheet metal working technique. All unnecessary blows or working on the metal must be avoided. Very soon one learns to direct blows just where they are needed, when they are needed—and more.

Accuracy in measurement and setting out is equally important, for whereas the soft metals may be adjusted to some extent after a detail has been worked, no such adjustment is possible with hard metal. A clearly marked steel rule, a good try square, sharp snips, the knowledge and ability to set out angles, and a 'feel' for the hard metals are all essential to success in this most modern type of plumber's roofwork.

Corners in hard metal roofwork are formed by the dog's-ear method, in which the surplus metal is folded and turned back upon itself.

Fig. 38 shows the setting out necessary, and the following sequence of operations should be helpful:

1 Square two adjacent edges of the sheet.
2 Mark on the heights of the sides. This may be done in pencil.
3 Optional. It will be found that if the diagonal line a–b in the surplus metal square is lightly chased, the folding process will be much easier since the metal tends to follow the line of the chase. A bricklayer's brick bolster that has been blunted is useful as a chase wedge for this purpose.
4 Using a block of wood as a support, pull up the sides. Pull both to an angle of about 45° first, and then manipulate the surplus corner metal into a dog's-ear.
5 Having made sure that the fold is in the correct place, pinch it close with the seaming pliers. Finally, fold and dress it close to one outer side of the corner to complete the detail.

Practise this detail in stiff paper. It provides good experience at little cost and in a very short time. With experience thus gained, making corners in copper, zinc or aluminium will be no trouble at all.

Tools required:
1 Steel try square
2 Steel rule
3 Snips
4 Seaming pliers
5 Dresser

Corners in Nuralite are formed in very similar manner to that described for hard metals except that:
1 When Nuralite is heated it softens and becomes flexible. In this state it can be manipulated by hand to the forms shown at A, B and finally C in Fig. 38.
2 Simple wood blocks used to help form and press the hot Nuralite to shape, give a neater finish and protects the hands from getting too hot when manipulating the heated material.

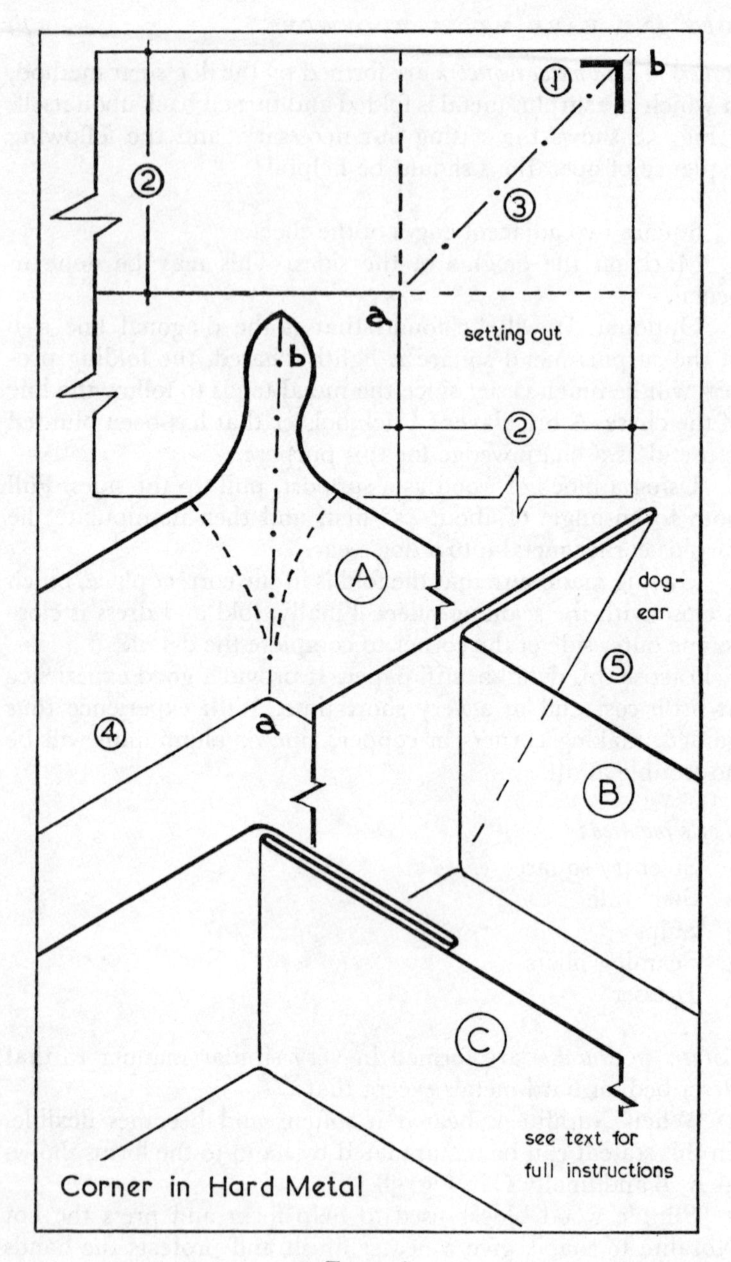
1
b
2
3
a
setting out
b
2
A
dog-
ear
5
4
a
B
C
see text for
full instructions
Corner in Hard Metal

Fig. 38

18

Non-metallic flexible weathering

The constant demands for speedier forms of construction have resulted in the development of many new materials and improved methods of using old ones.

One notable development which finds increasing use in the hands of the skilled plumber is durable and quickly applied material for roof covering and weathering details. Manufactured from asbestos fibre and bitumen, both of which have been recovered from Egyptian tombs, ample evidence of the long life of the constituents of this new material, it is known as non-metallic flexible weathering to distinguish it from the more traditional plumber's metals and from the pliable rolled roofing felts.

Nuralite, as this material is known, is sold in flat sheets measuring 1438 mm × 914 mm × approximately 2 mm thick (8 ft × 3 ft × $\frac{3}{32}$ in). British Standard Code of Practice 143, Part 8 covers recommended practice for roof weatherings in this material.

Properties

It may be described as a bitumen bonded asbestos sheet which is completely water-proof. It is resistant to abrasion, stiff but flexible at air temperatures. It becomes pliable and easily manipulated to intricate forms when heated and, once formed and allowed to cool, the formation is permanent unless reheated.

Although initially black in colour, Nuralite quickly takes on a pleasing grey appearance on exposure to the weather.

It is highly resistant to atmospheric corrosion, even in

industrial or coastal areas, and, as it is non-metallic, it cannot rust nor is it subject to electrolytic decomposition.

Nuralite is the simplest of all plumbers' materials to cut and fabricate. It may be cut with snips or by making a deeply scored line and then bending to break along the line of score.

Its thermal plastic property is the essence of its easy application. Using simple techniques, details such as complete sets of chimney flashings as illustrated in Fig. 49 are quickly made with a minimum of manual labour and considerable saving in time.

The forming of an internal corner, for example, follows exactly along the lines described and illustrated for hard-metals in Chapter 17 and Fig. 38, the only difference being that the Nuralite sheet is warmed to a pliable state by the application of heat from a propane gas torch which every plumber uses today.

The ideal Nuralite forming temperature is 182°C at which temperature the material can be accurately and deftly formed to the desired shape.

The method of manufacture of Nuralite allows simple forms of jointing by de-laminating the edge of one sheet edge, applying a weld block of bitumen which is heated and the softened bitumen rubbed about 19 mm wide along both sides of the other sheet edge to be inserted in the opened laminate. This done, and the sheets mated together, heat is applied along the line of joint, the laminae pressed close on to the insert sheet and further heat is applied to fuse the sheet material and the 'tinned' edges into what can only be described as a weld, since it is a fusion of both materials joined to form a really sound joint. (See Figs. 41 and 42).

The term de-lamination, as applied to this simple jointing technique, suggests that the material is made in the form of layers, whereas it is a homogeneous material but lends itself to careful parting along its heated edges.

The manufacture of Nuralite is briefly described as follows. Selected asbestos fibre obtained from Africa and Russia is dry mixed in careful proportion with pelleted bitumen, that is, bitumen crushed into tiny particles of uniform size.

This mix is then placed into a container and water added to make a consistency resembling thin porridge. Thus the raw materials are prepared for the manufacturing process.

The wet mix is fed at regulated speed on to a porous conveyor belt the same width as the finished Nuralite sheet. The conveyor belt, about 3 metres long at its top travel, passes over a series of rolls which keep it level and flat. At one point in the travel of the belt it passes over a perforated roll which has air continuously sucked from its inside and as the wet mix which has been deposited on the belt passes over this roll, much of its moisture content is sucked out. In this way the mixture, wetted to make it fluid initially, becomes drier and stiffer as the belt continuously rotates and more wet mix is applied to build up to the requisite thickness of the finished sheet.

From this it will be seen that the laid down material is not, strictly speaking, in layers but in one well knit mass.

When the thickness build up is correct, the machine operator deftly cuts across the formed sheet, now resembling damp stiff cardboard, and the sheet is transferred to another conveyor belt for the finishing process.

The limp sheets of asbestos fibre and granulated bitumen pass along the conveyor belt and through a temperature controlled, heated tunnel. Here the bitumen melts and diffuses throughout every fibre of the sheet.

When the sheets emerge at the end of the tunnel they are sufficiently cool to handle and are now in the familiar stiff but flexible black Nuralite sheets as used on site. A final trimming to exact length finishes the product which is carefully parcelled up for convenience of handling.

The finished sheets weigh approximately 2·5 kg/m² ($\frac{1}{2}$ lb/ft²), are 1438 mm long, 914 mm wide, and approximately 2 mm thick.

One surface of Nuralite is smoother than the other. Although the material is equally water-proof whichever way applied, it is, of course, better to use the smooth side to the weather because this sheds the water better. The *thermal expansion* of Nuralite is less than that of zinc and lead and similar to that of copper (see Table 7, page 92). Its coefficient of expansion is 0·000019 and so no trouble is likely from the effects of thermal movement. Even in high summer Nuralite will not become tacky on account of the high softening point of the material.

Application may be almost exactly as for fully supported roof weathering in traditional plumbers' metals. Concrete roof slabs will need to be finished with a trowelled, smooth screeded surface. Boarded roof decks will need the same care in preparation and laying as would be given if they were to receive lead, copper, aluminium or roofing. (See Chapter 11 for notes on the preparation of timber deckings for roofs.)

Nuralite is particularly suitable for flashings and all types of weathering details for domestic, public, and industrial buildings covered with slates, tiles, corrugated sheeting and a variety of other roofing materials.

The following are some typical uses:

Fully supported lining for gutters
Ridge and hip capping
Chimney flashings
Abutment wall flashings and hanging step flashings
Chimney flashings
Flashing to rooflights
Dormer roofs and checks

It may also be used for damp proof courses, gutters over window and door openings in cavity walls.

Jointing

Apart from the de-laminated joint previously described, a simple form of roll cap jointing may be used in small flat areas such as dormer window tops. For large flat roof areas the '*Twinrib*' method of roofing is preferred. In this case, specially manufactured base strips are nailed to the prepared roof deck spaced so that a full sized sheet of Nuralite just lies within strips and laps them up to the centre stop. Before laying the roof sheet *Nuralite Adhesive No. 10* is applied, 3 mm thick to the roof deck in 125 mm squares spaced 250 mm in from one narrow edge and at 175 mm centre to centre down the sheet.

Across the narrow dimension of the sheet the adhesive areas are spaced 150 mm in from one long edge and then at 305 mm centre to centre. This gives 18 adhesion areas in all and the purpose of these is to retain the sheet close to the roof and so restrain lifting due to high winds.

The adhesive applied, the roof sheet is placed into position and, when adjacent sheets have been so laid, the capping strip pre-formed and pre-coated with bitumen like its partner base strip, is located over the junction of the sheets.

A 'soaking' heat is applied all along the line of joint with a propane gas torch. This fuses the bitument coating of the base strip, the Nuralite sheets and the capping strip, and the joint is completed by 'ironing' it tight to ensure a completely homogeneous bond.

This form of roofing is applicable in large flat areas as flat as 1 in 100 up to 40° pitch.

Tools required for working Nuralite are simple. Most of them will be found in the plumber's tool kit. Other specialised items such as sealing irons used to press into place and generally finish joints in a craftsmanlike way, may be purchased from the Nuralite Company, Whitehall Place, Gravesend, Kent, or may be made by the plumber to suit his needs (see Fig. 58A, detail 1–3).

The following list forms a guide to the equipment necessary.

1 A bench which should have a mild steel angle let in flush with the bench top and side edge. The edge of the angle should have its sharpness removed so that its radius is about 1·5 mm that is, just not a 'knife-edge'.

If a groove 16 mm wide by 16 mm deep is cut along the length of the top of the bench and about 76 mm in from the bench edge, this will be found most useful, used in conjunction with an 8 mm diameter mild steel rod, to form stiffening beads on details like abutment flashings, hanging step flashings and the front edges of chimney aprons. To form such a bead the detail is laid face down over the groove in the bench and positioned so that the proposed line of bead is over the groove. The sheet is then heated along this line and the 8 mm rod

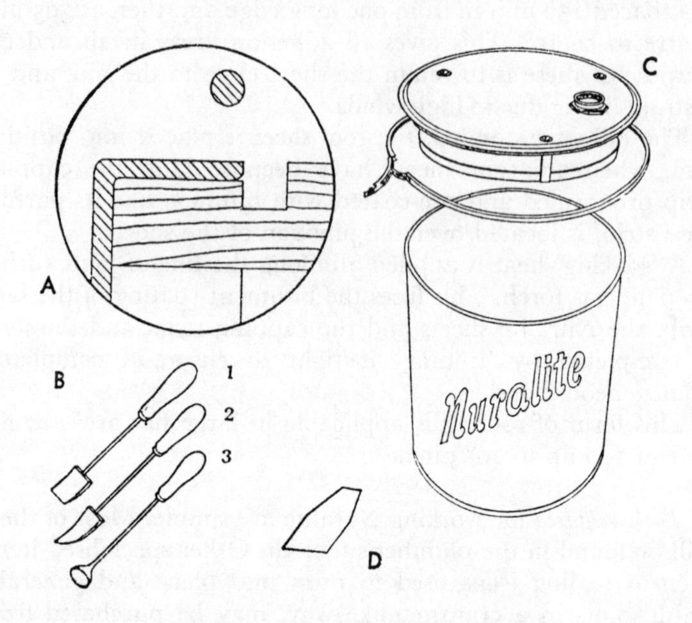

Fig. 58A (see page 211—Fig. 58B).

pressed down upon it and so taking the now pliable sheet into the groove to form a half round bead.

2 Straight edges, steel.

3 Knives for cutting and de-laminating. These need to be with thin, straight blade.

4 Tinsmith's snips for shaped cuts which cannot be conveniently done with the cutting knife.

5 Propane gas torch for making joints and heating Nuralite prior to bending or forming. For laying sheets on site work a torch having gas consumption in the order of 3·5 kg per hour (124 oz/h) is recommended to ensure a good heat 'soak'. For smaller details fabricated in the workshop, a smaller head will prove adequate and, indeed, a hand torch will do here just as well as the torch with flexible hoses as would be preferred for *in situ* work.

6 A selection of planed timber blocks will be found useful in the job of shaping and forming the hot Nuralite.
7 Sealing irons. These are shaped flat pieces of mild steel welded to a 10 mm mild steel handle provided with a file handle for comfort and convenience of working. The irons are heated as required and are used to compress and generally finish joints.

For 'Twinrib' roofing, a gas fired boiler is required for heating the jointing compound which is used in place of the bitumen weld block for bench made details. Brushes, trowels and scrapers will also be required.

The cold moulding technique is a more recent innovation. It forms a convenient alternative to heating the sheet to make it pliable for forming.

Purpose made containers to hold a rolled sheet of Nuralite are obtainable from the Nuralite Company.

A quantity of softening agent is poured into a special compartment of the container holding the sheet to be softened and in such a way that only the vapour arising from the liquid softening agent contacts the sheet.

The container thus prepared is sealed air tight and left for 24 to 72 hours, depending upon seasonal temperature and the area of sheet being softened (see page 211).

The advantage of this method lies in the absence of heat so that flashings and the like can be quickly carried out using the minimum of tools. Unless one is pre-fabricating many items over a long period, the time it takes for the vapour softening process to work may prove less advantageous than the quicker heating method.

Cold moulded Nuralite is jointed by a special cold jointing compound sold as *Nuralite Jointing Compound No. 20.* A lap joint may be used when a minimum of 76 mm lap is required, or delaminated joints may be made when a minimum 25 mm depth of insert is required.

The following diagrams and related descriptions illustrate some of the working techniques and methods of setting out and applying some of the more commonly used Nuralite weathering details.

Cutting Nuralite

1 *Straight Cuts.* Score deeply with the knife, and break directly over the metal edge of the bench (see A and B of Fig. 39).

2 *Irregular Shapes.* Use the tinsmith's snips exactly as for metal (see C and D of Fig. 39).

Bending Nuralite

Warm to form is the invariable rule, which must be strictly obeyed.

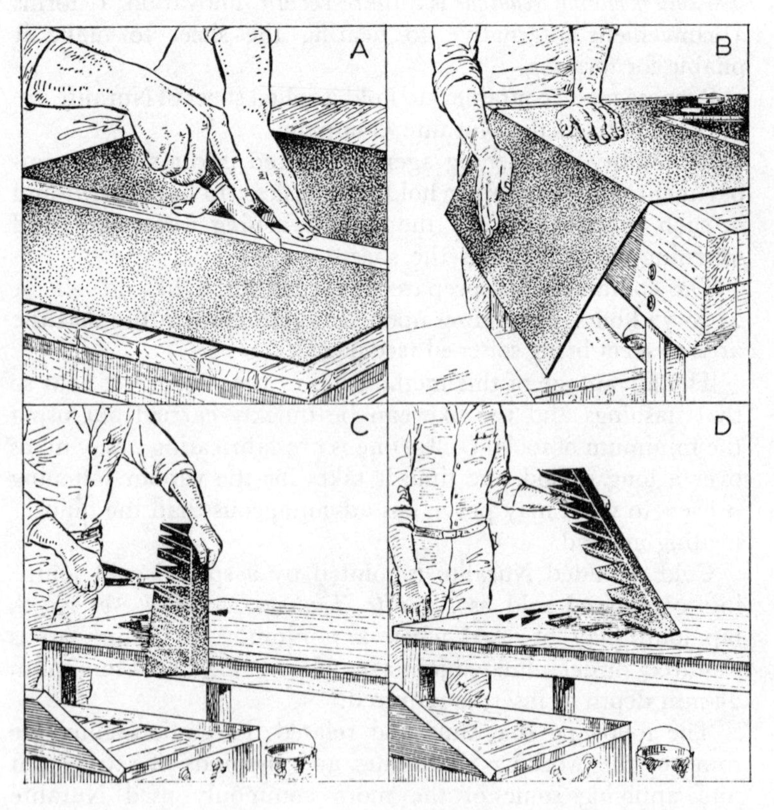

FIG. 39

1 Simple bends up to 90°. Mark out with engineer's chalk. Place line squarely over the metal edge of the bench. Hold sheet in position with one of the straight edges, parallel with and about 50 mm (2 in) from the bench edge. Apply the blow lamp flame up and down the bend line to obtain an even rise in temperature, (see A of Fig. 40), until the sheet begins to lose its rigidity. The exact degree of softening will soon become obvious with experience. When the material is sufficiently softened, bend it down slowly and firmly with the second straight edge, to the angle required (see B of Fig. 40). The bend now formed should be allowed to cool, and will be permanent.

2 Bends up to 180° as for welts. The minimum recommended radius for 180° bends is the thickness of the material. To form an acute bend, first proceed as in 1 above. Having obtained a 90° angle, turn sheet over, and align angle with edge of bench. Rewarm line of bend (see C of Fig. 40). Place straight edge of suitable thickness inside and close up to angle of bend, press upstand gently over with other straight edge (see D of Fig. 40), allow to cool.

3 To form a bead. Place sheet to span bead groove, with edge parallel to and approximately 8 mm ($\frac{5}{16}$ in) beyond groove, (see E of Fig. 40). Heat evenly with blow lamp, as before described. Place the 8 mm ($\frac{5}{16}$ in) rod directly over the groove and press down gently and evenly from end to end. Finish finally by pressing straight edge down on top of rod, thereby sharpening the return angles (see F of Fig. 40). Remove from groove and allow to cool.

Joints for Nuralite

The laminations of one piece have to be parted to receive the thickness of the other piece to a depth of 25 mm.

The following procedure must be adopted—Warm the edge of the first piece with the blow lamp to the normal working temperature (see A of Fig. 41), insert one of the knives at the corner, between the laminations, as near as possible to the centre of the sheet thickness (see B of Fig. 41). Take the other knife in

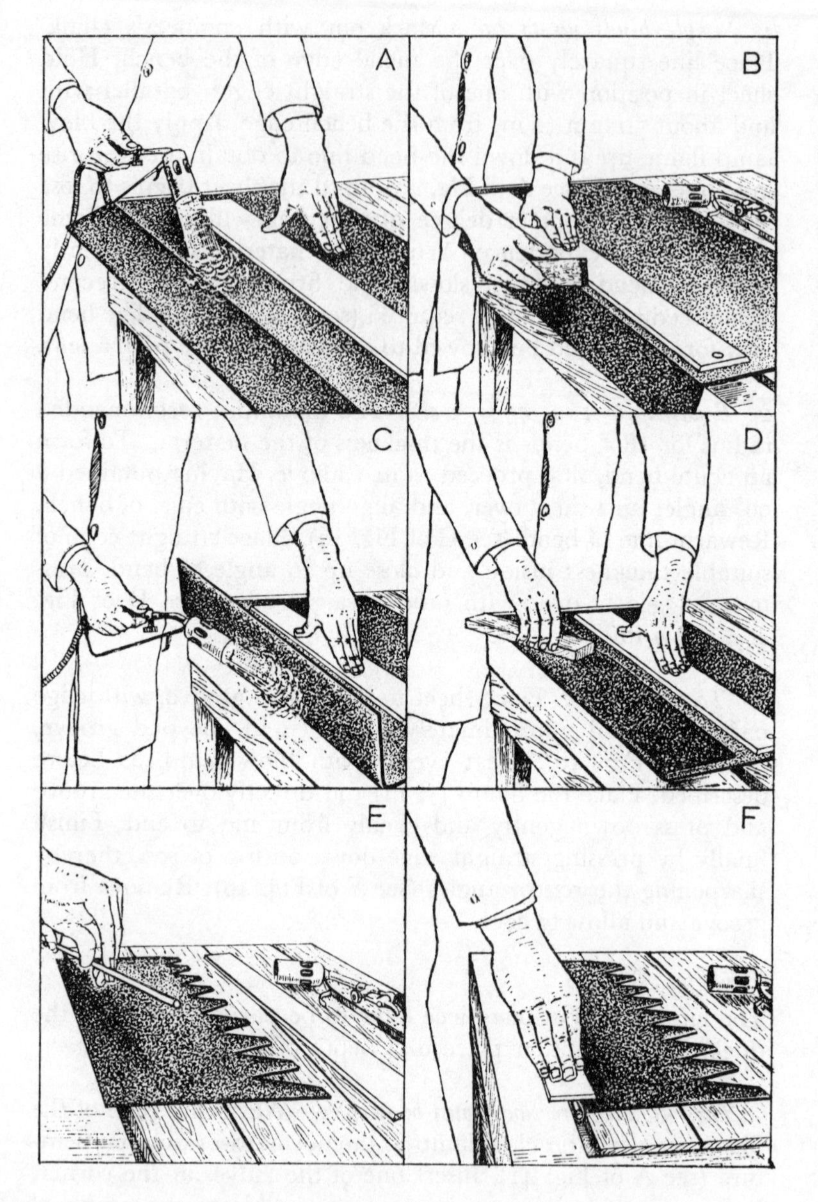

FIG. 40

the other hand and part the intended joint, (see C of Fig. 41). The art of carrying out this operation is soon acquired.

Coat the joining edge of the second piece of Nuralite on both sides with the Fibrous Welding Plastic, supplied for the purpose, which before application must be heated to an easily workable consistency (see D of Fig. 41). The coating should be 1·5 mm ($\frac{1}{16}$ in) thick and 25 mm (1 in) wide. Heat the edge again, causing the bitumen in the welding plastic to amalgamate with the bitumen in the sheet. While this edge is still warm, quickly heat the open laminations of the other edge. Insert the coated edge (see E of Fig. 41). Close the laminations and press the whole firmly together with the straight edge. Use No. 1 Sealing Iron which must have, meanwhile, been heated to smooth the joint and seal the edge as shewn (see F of Fig. 41).

With a little experience, a neat and impervious joint will be achieved.

If much jointing is undertaken at one time, the plastic block should be heated in a tin or other suitable vessel, and taken out as required with the knife. For limited jointing, the end of the block can be softened in the blow lamp flame and the application made directly by rubbing this softened end where required on the Nuralite.

No. 3 or No. 4 lap joint

This joint should only be used in a position where water will shed over the lap. To form a lap joint both pieces of Nuralite to be joined must be precoated on the surfaces being brought together with Nuralite Jointing Compound to a minimum width of 76 mm (3 in). This Compound must be heated before application to a temperature not exceeding 180°C in a propane gas fired boiler. The surfaces to be joined must be dry and dirt free. The following procedure is to be employed.

Precoating

Both pieces to be joined are heated over the area where the joint is to be made with a large capacity propane gas torch until the bitumen component of the Nuralite has a moist appearance. Hot liquid Nuralite Jointing Compound is then

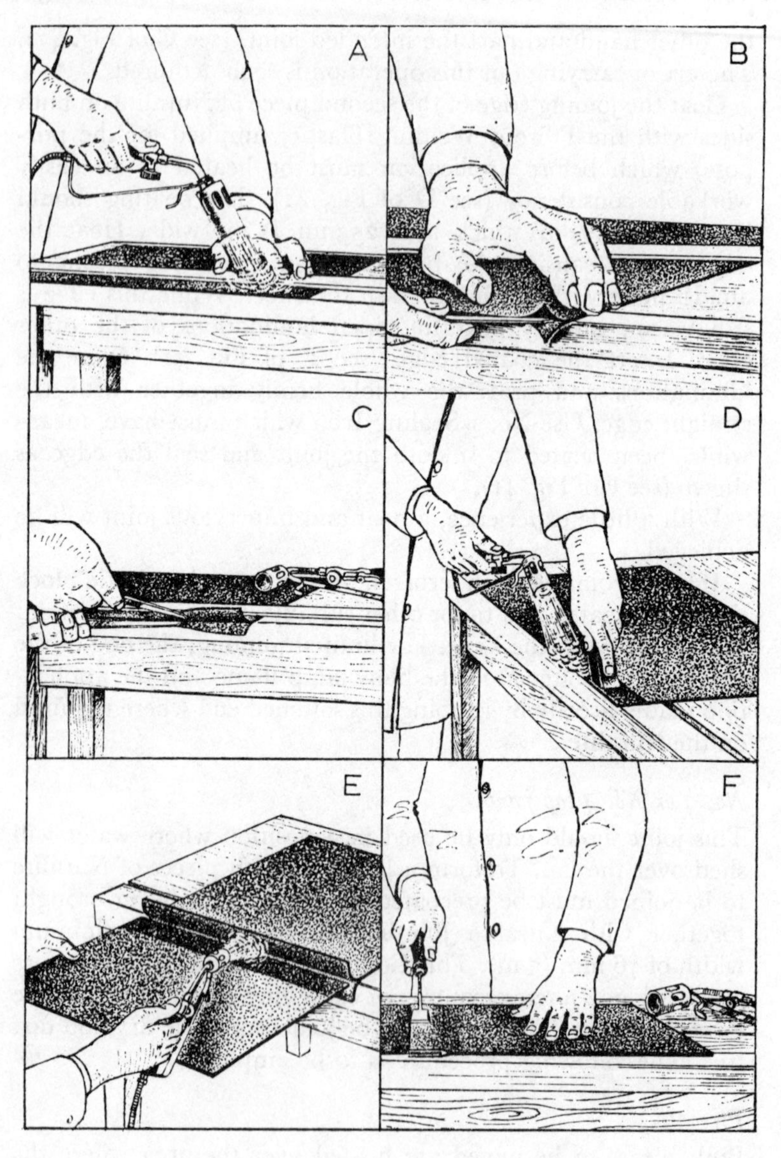

FIG. 41

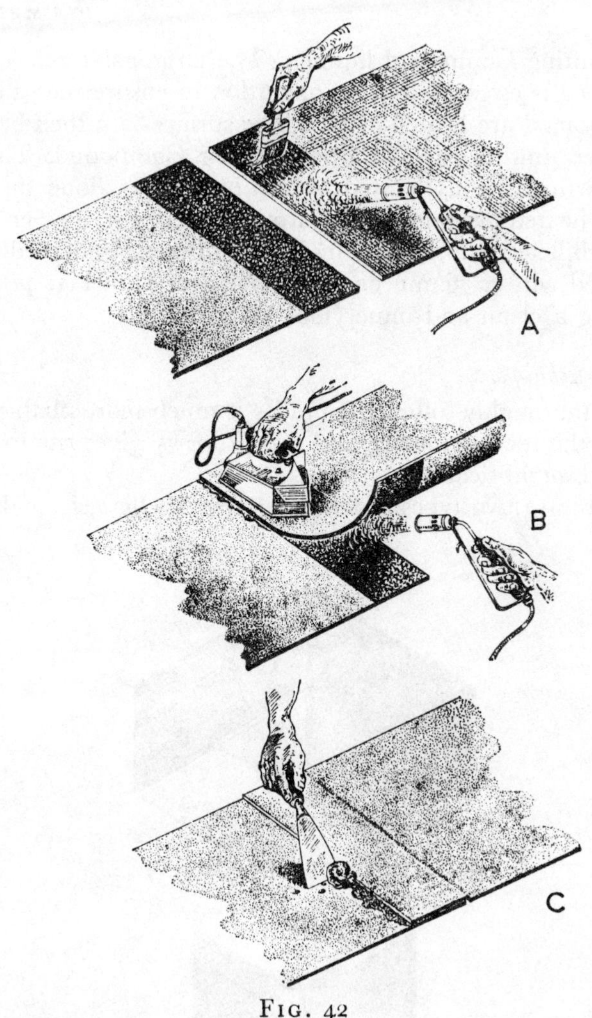

FIG. 42

brushed on to both surfaces (see A of Fig. 42), so that it becomes merged with the bitumen in the Nuralite. The precoated surfaces can then be left to cool and are thus prepared for the joint to be made.

Making the joint

The two precoated surfaces are heated simultaneously until

the Jointing Compound liquifies. The large gas torch already described is essential for this operation to ensure that all areas being joined are heated evenly. The surfaces are then brought together uniting the Nuralite Jointing Compound. The lap joint is finally consolidated by ironing firmly along the joint with a heated propane gas flat iron (see B of Fig. 42), ensuring the complete exclusion of air. Excess bitumen squeezed out is removed with a 76 mm paint scraper, (see C of Fig. 42), thus forming a clean and impervious lap joint.

Folding technique

When thoroughly softened Nuralite is much more pliable than any of the roofing metals, and this property gives rise to other methods of fabricating.

There are two types of fold, the first is where, in making a

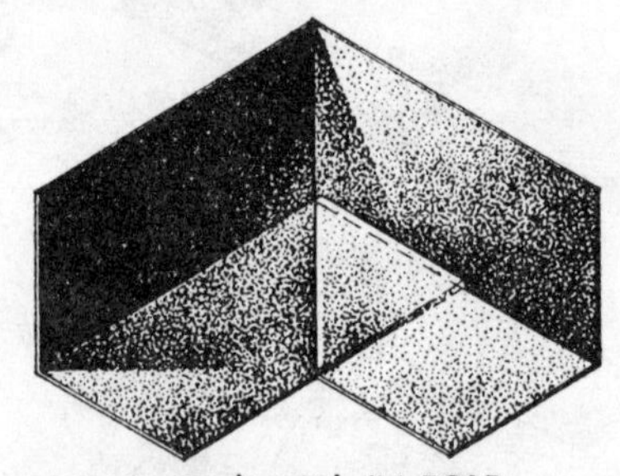

MATERIAL IS 'LOST' IN FOLD . . . .

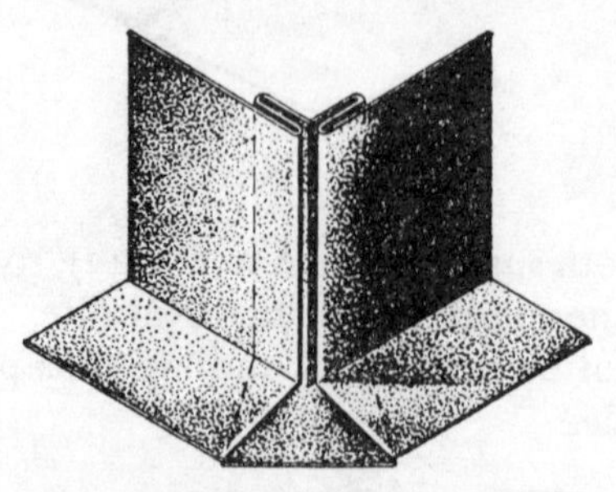

. . . . AND GAINED IN GUSSET

FIG. 43

bend, there is some excess material which—in the traditional way—must be cut out and rejoined by delamination. In the folding method this is neatly folded away. The second occurs when, in making a form, some extra material is required. Traditionally this is done by inserting a gusset. In the folding method the extra material is obtained by making a fold and then drawing the material required from it. This is shown in Fig. 43.

Setting out for folding

Where material is required to be 'lost', the maximum amount

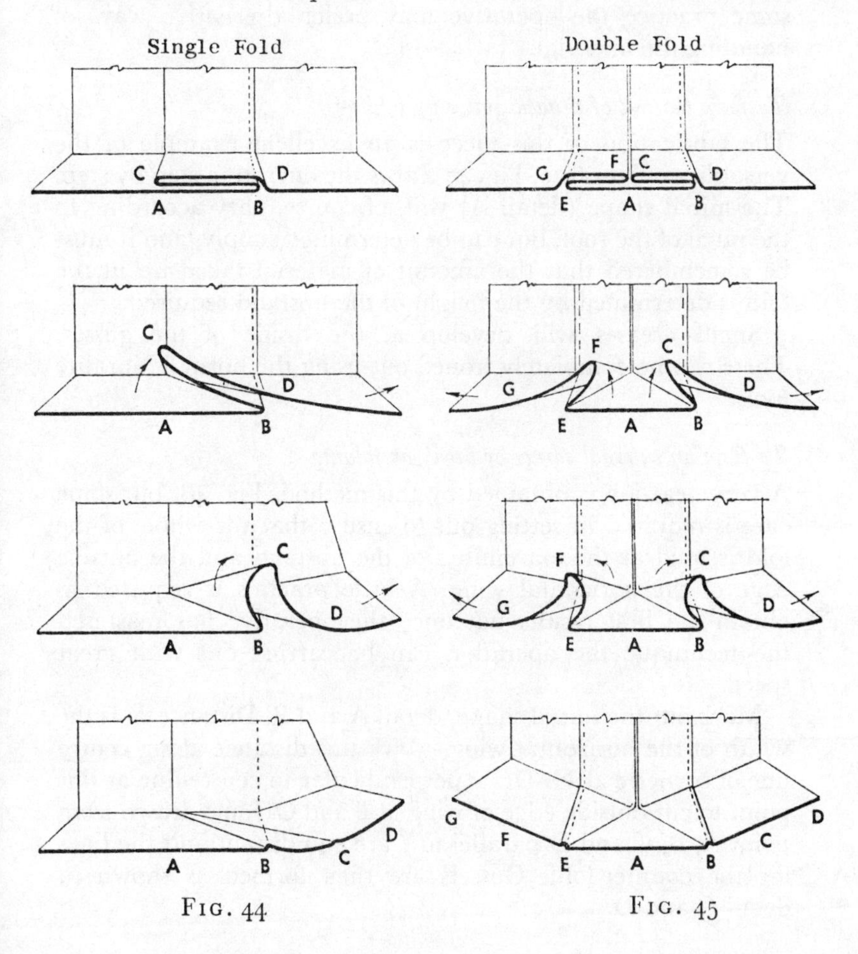

FIG. 44

FIG. 45

is first gathered in the fold. Where material is required to be gained, as in the case of a gusset, the total girth of the fold (Figs. 44 and 45) is obviously the greatest distance across the gusset. Only a little experience is required to set out, knowing the angle required and the height of the upstand. Setting out is simply dimensional and not geometrical.

To form a Gusset by folding

This operation is best carried out using the counter-fold. The illustration (Fig. 46) shows how the work is done. After some practice the operative may prefer alternative ways of handling the material.

To form the end of a back gutter by folding

The fabrication of this piece is an excellent example of the versatility of Nuralite. Fig. 47 shows the operation step by step. The initial shape (detail A) will, of course, vary according to the pitch of the roof, but can be determined simply, and it must be remembered that the amount of material taken up in the fold is determined by the height of the upstand required.

Small creases will develop at the inside of the gusset. These can be completely ironed out using the normal Nuralite tools.

To form an external corner or break by folding

A very neat job is obtained by this method, Fig. 48, but some care is required in setting out to ensure that the whole of the fold is used at the extremities of the upstand and the outside edge of the horizontal wing. A little practice is required to obtain the best results but once the operative has mastered the technique the operation can be carried out with great speed.

Mark out the initial shape, detail A and B. Distance X is the width of the horizontal wing. Mark this distance along centre line of corner ¢ at O. Draw perpendicular to centre line at this point to cut outside edge of wing at B and C. Lines drawn from point A, B, C and D parallel to ¢ are equidistant and the lines for the counter-fold. Gussets are then formed as shown in detail C and D.

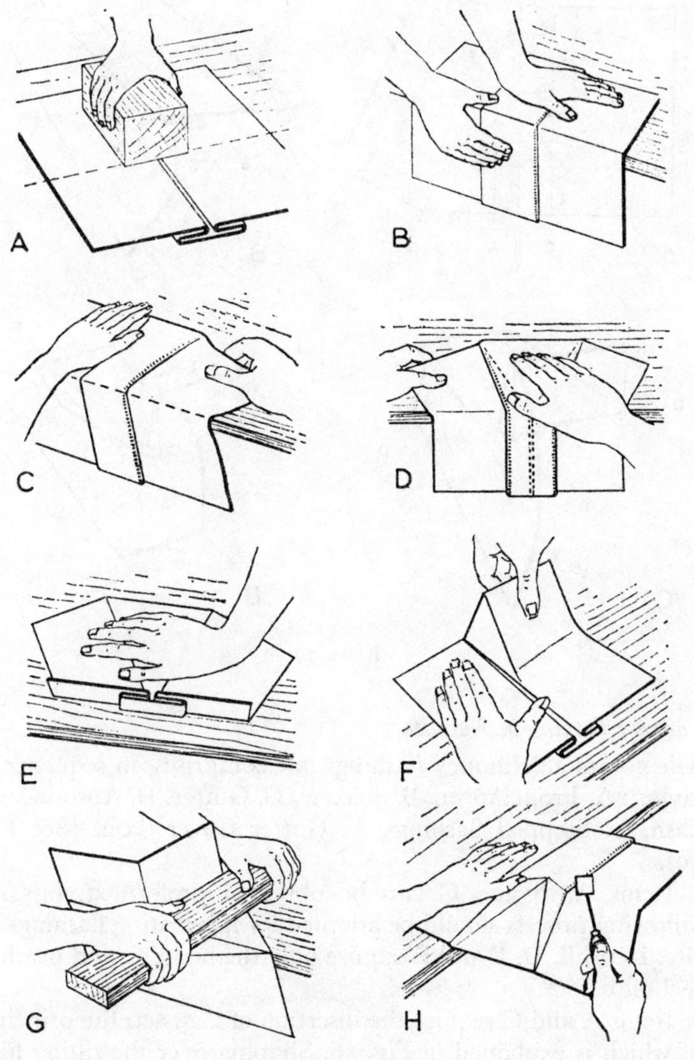

FIG. 46

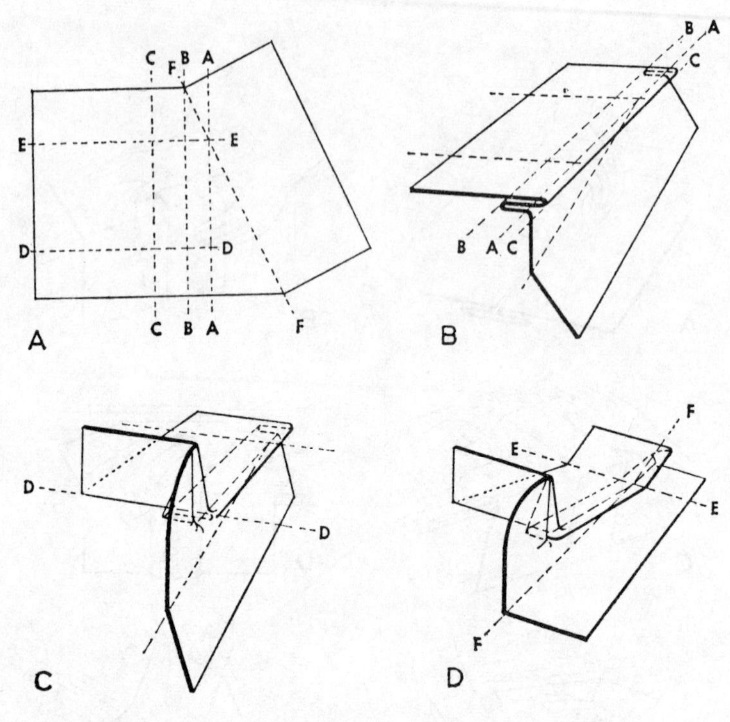

Fig. 47

Chimney flashing in Nuralite

The complete chimney flashing unit comprises, in sequence of laying: A. Front Apron, B. Soakers, C. Gutter, D. Apron Cover flash, E. Stepped flashings, F. Gutter Cover flash. (See Fig. 49).

Items A, B and C can be obtained preformed, but the following process should be adopted for fabricating flashings on site. Items B, D, E and F require only the bending and beading technique.

Items A and C require the insertion of a gusset, the principle of which is explained in Fig. 56. Shaping over the tilting fillet at the gutter is easily effected by forming the area of Nuralite over the fillet to the required contour when in its mouldable state.

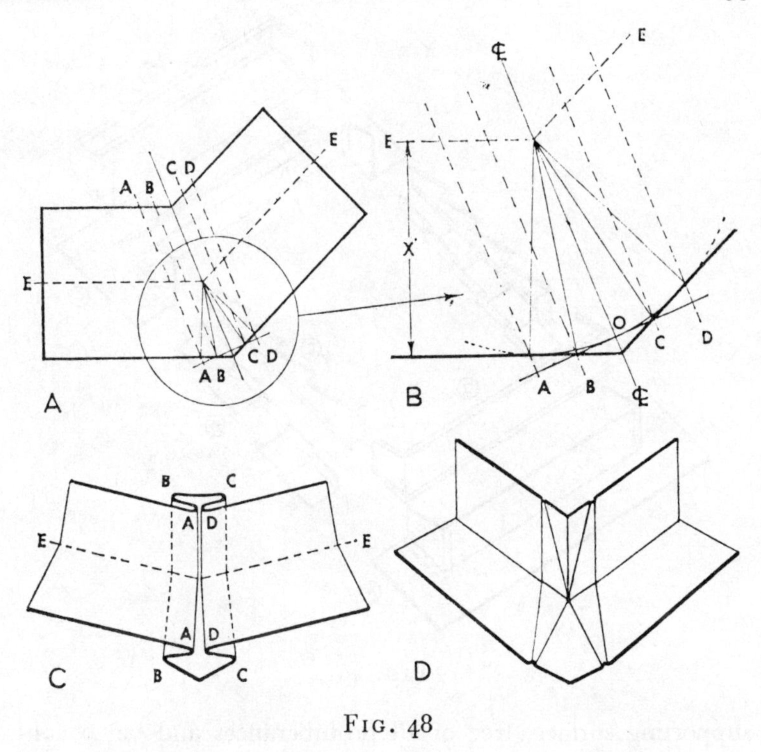

FIG. 48

Roll cap roofing in Nuralite

Whilst the Twinrib system, as later described, is now generally used for areas of near-flat or low pitched roofing there are occasions where, for aesthetic reasons, the traditional roll cap system may be preferred.

For roll cap application, the decking should be fully supporting. As Nuralite is completely non-metallic, troublesome electrolysis is not encountered. It can, therefore, be used in juxtaposition to any of the metal roofs, to which it can be satisfactorily jointed. Lead, copper and zinc roofs can, of course, be discharged on to Nuralite without corrosive effect.

Decking

Roof decking must be capable of sustaining a nail or screw fixing to retain wood rolls and present a smooth, firm, fully

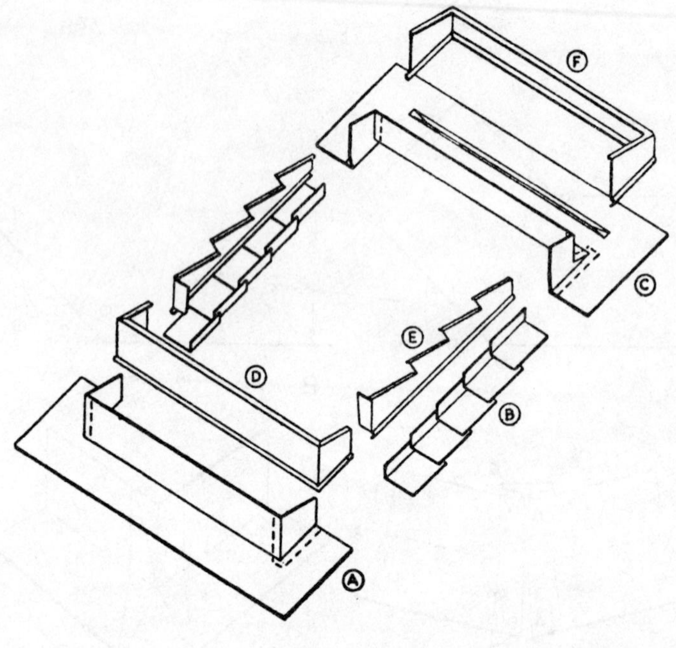

FIG. 49

supporting surface, free of all protuberances and be so constructed as to provide a minimum of 1 in 60. It may consist of any of the following—

Timber. This should be well seasoned, 25 mm (1 in) thick, with all uneven edges and rough surfaces dressed off and nail heads well punched in. Tongued and grooved boarding is preferred, but close boarding is satisfactory provided that it presents an even and firm surface.

Chipboard. A type resistant to excessive atmospheric moisture movement should be used and fixed in accordance with the particular manufacturer's instructions, care being taken to see that all joints are flush and firm.

Plywood: A type resistant to excessive atmospheric moisture movement should be used and fixed in accordance with the

Plywood Manufacturer's Association recommendations, care being taken to see that all joints are flush and firm.

Compressed Straw Slabs: A type with a shower-proofed surface is preferred. It must be fixed in accordance with the manufacturer's instructions and all site cut edges sealed with self-adhesive tape supplied by the manufacturer. All exposed edges to be trimmed with wood fillets.

Woodwool: This decking must have a suitable cement and sand screed topping, thus providing a firm, smooth surface.

Concrete: Must have a smooth surface preferably with a suitable cement and sand screed topping.

Drips

Where a roof has fall of less than 1 in 8, drips must be provided to allow for expansion and facilitate end weathering of the sheet. These drips should be in the form of vertical steps 50 mm (2 in) deep, spaced to coincide with the 2·4 m (8 ft) long sheet after due allowance has been made for the amount of material taken up in forming the welted edge on the lower end of the sheet and the upstand at the higher end.

Wood rolls

Should measure approximately 44 mm (1¾ in) high and 32 mm (1¼ in) wide. They should be spaced 838 mm (2 ft 9 in) apart in order that the turned up sides of sheets should finish just below the tops of the rolls. The best Nuralite practice at drips is for the upper rolls to over-run the lower roll by 44 mm (1¾ in) and be chamfered back from this point at 60°.

Gutters

Where roof drainage is effected by Box, Valley or Parapet Wall gutters, built into the roof structure, they can be made with the same falls and drips as recommended for the roll cap system.

Sequence of application

Nuralite sheets are laid starting at the eaves and working progressively backwards towards the highest roof level. This procedure ensures correct weathering sequence at drips. Where built-in gutters occur, their lining with Nuralite must, of course, be undertaken first, commencing at the gutter outlet and proceeding to the highest level as for flat roof.

Treatment at eaves

First attach anchor strip to the roof decking with 19 mm large headed galvanised clout nails (see Fig. 50A). Clip welted edge of pre-formed sheet to anchor strip and place tray into position between wood rolls (see Fig. 50B). Clip roll cap stop end over welted edge of Nuralite (see Fig. 50C). Lower roll cap over roll (see Fig. 50D).

Treatment at drips

Having laid the first course of sheets with the dog eared corners tucked behind the wood rolls, fix anchor strips at the drips in

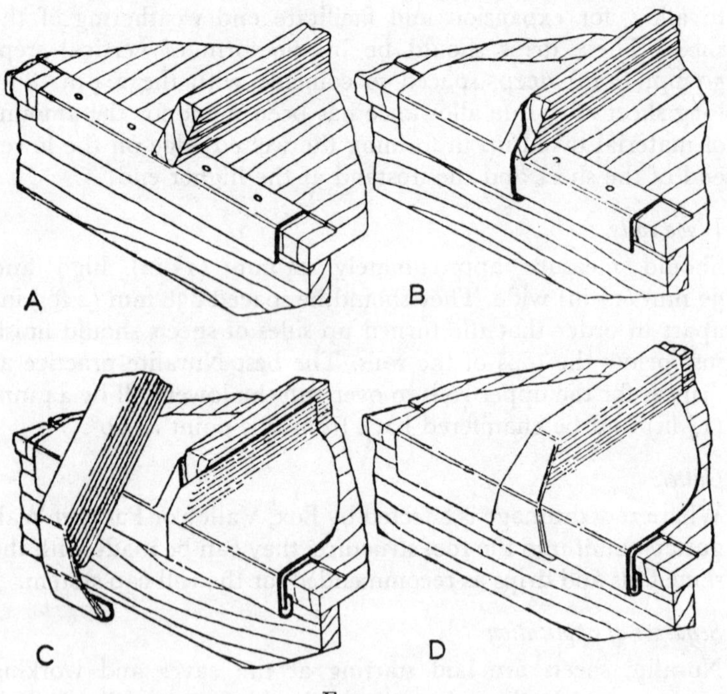

FIG. 50

the same manner as at eaves (Fig. 51A) (Roll cap omitted for clarity). Now fasten the wood rolls into position for the next course, over-running and thus securing, the lower roll cap (Fig. 51B). Clip welted edge of the next course to anchor strips as at eaves; these sheets must have a gusset inserted where upstand is cut, to allow welted edge to be bent down and return against roll (Fig. 51C). Form offset in roll by the folding technique or inserting gussets as necessary. Apply welding plastic to secure end at lap (Fig. 51D).

Treatment at back wall

Having placed the Nuralite sheets in position with dog eared corners tucked behind the wood roll (Fig. 52A and B) fix the roll cap having pre-formed saddle piece (Fig. 52C). Chase the nearest appropriate joint to a depth of 25 mm and wedge into position a suitable Nuralite cover flash, having a beaded edge. Fill joint and point in the normal manner (Fig. 52D).

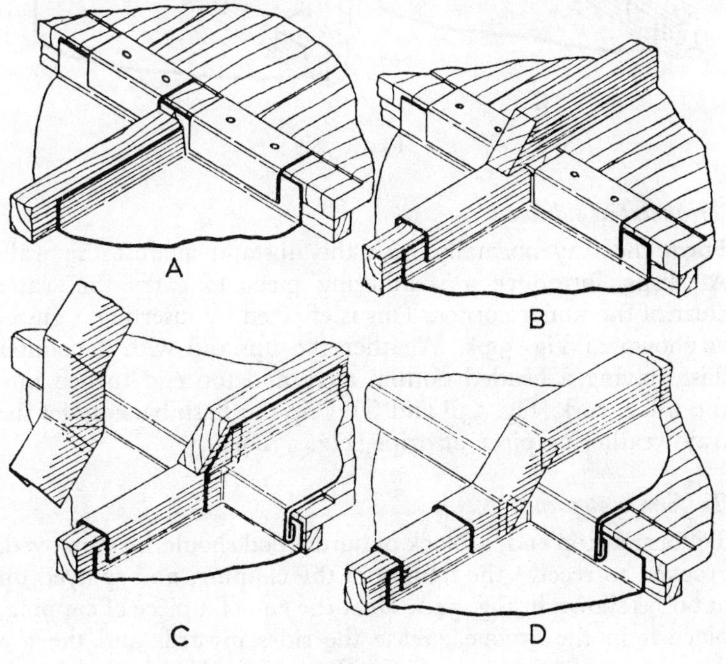

FIG. 51

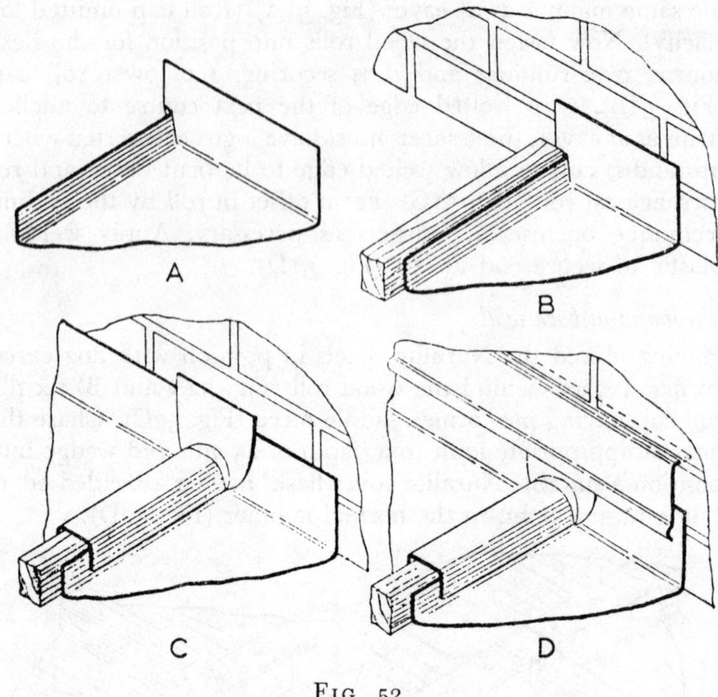

Fig. 52

Treatment at side wall

Form the tray normally with the upstand against the wall. At drips, introduce a weathering piece to carry the water clear of the under corner. This is effected by inserting a gusset as shown in Fig. 53A. Weather the upstand with a counter flash having a beaded bottom edge and top end turned into the brick work (Fig. 53B and C). Clip this flash back under the tray weathering piece at drips (Fig. 53D).

To form a stop end

To form a stop end, a block of hard wood should be employed, grooved to receive the flanges of the capping, and splayed off at 60° as shown in Fig. 54A. Heat the end of a piece of capping, place it in the groove, crease the sides inwards and the top downwards, as shown in Fig. 54B and C). The further bends

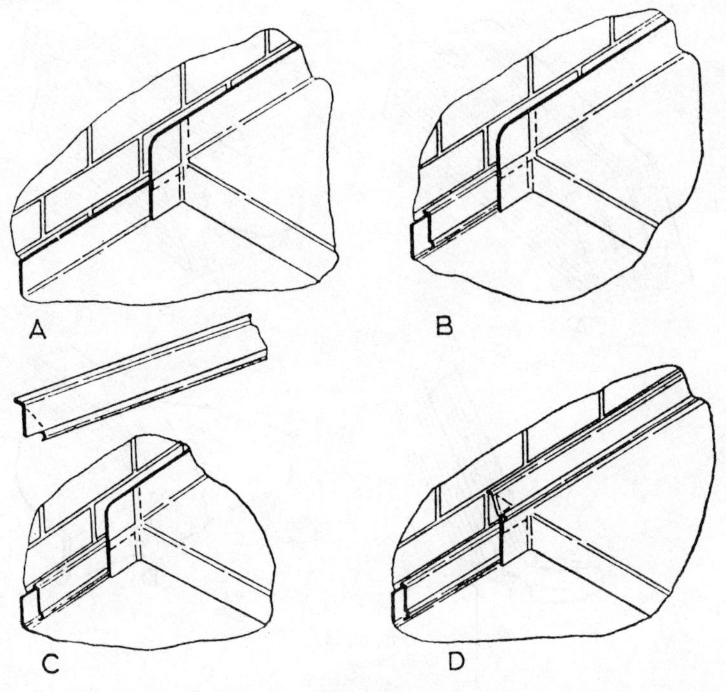

FIG. 53

to clip round the welt should be made after the piece is removed from the block (Fig. 54D).

To form a saddle piece

Heat one end of a normal length of roll cap and flatten out (Fig. 55A). Form a double fold of about 13 mm wide across the flat surface by pulling the flattened end first forward (Fig. 55B) and then backwards. Turn up at right angles in line with the outer edge of the fold (Fig. 55C). Complete the saddle by pressing down at side of the capping against a wood roll (Fig. 55D).

To insert a gusset

Fig. 56A. Use the standard joining technique in this case. De-laminate the parent material (Fig. 56B). Treat the gusset

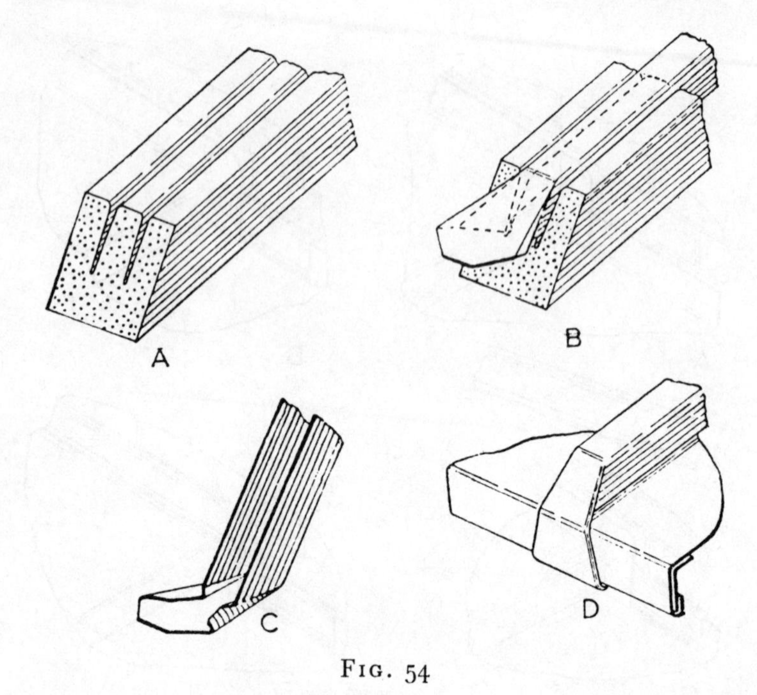

FIG. 54

with the heated welding plastic both sides to the depth of the intended joint (Fig. 56C). Insert and seal and make slightly with the hot sealing iron (Fig. 56D).

Pipe flashing

The Anglemaster variable plate is a prefabricated detail available from the Nuralite Company. It can be simply adjusted to suit any angle of pitched roof up to 45°.

Preparation of an Anglemaster

Heat enlarged portion of the spigot base and when sufficiently pliable, push the spigot down vertically, thus forming a uniform collar and then immediately pull back to the approximate angle of the roof slope, ensuring that the spigot is kept vertical when viewed from the front or back (Fig. 57). If desired, the flashing can then be allowed to cool for later use.

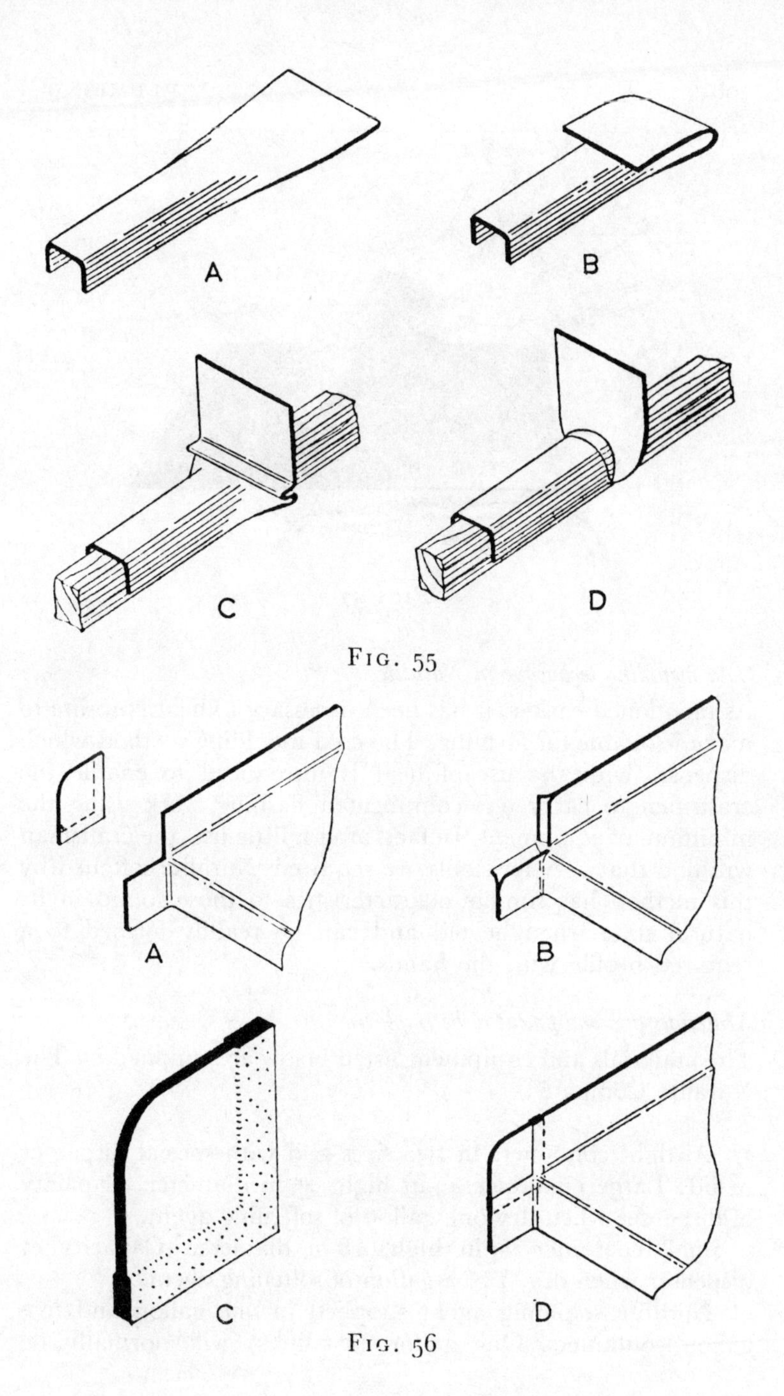

FIG. 55

FIG. 56

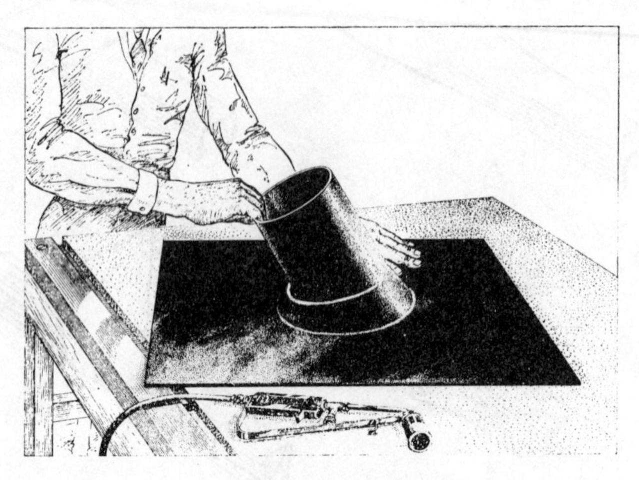

FIG. 57

Cold moulding technique in Nuralite

As mentioned earlier, it has been necessary to heat Nuralite to make it pliable for forming. The cold moulding method which dispenses with the use of heat is introduced to enable the craftsmen to carry out complicated flashing work using the minimum of equipment. In fact, after a little use, the craftsman will find that very few tools are required. Nuralite softened by this method has similar characteristics to those found in its natural state when heated and can be readily formed to a required profile with the hands.

Materials and equipment to be employed

The materials and equipment listed below are supplied by The Nuralite Company.

1 Airtight containers in two sizes and with special dispenser in lid: Large container 42 in high, 23 in diameter. Capacity of dispenser when dry one gallon of softening agent.

Small container 24 in high, 18 in diameter. Capacity of dispenser when dry, $\frac{3}{4}$ of a gallon of softening agent.

2 Nuralite softening agent supplied in one gallon and five gallon containers. One gallon (4·5 litres) will normally be

sufficient to soften 12 of the 8 ft 0 in × 3 ft 0 in (2·4 m × 0·9 m) standard sheets.

3 Nuralite Jointing Compound No. 20 supplied in one pint (0·57 litres) and one gallon (4·5 litres) containers. The latter contains sufficient compound for approximately 125 ft (38 m) run of 3 in (76 mm) wide jointing.

Preparation of Nuralite

1 Position airtight container on level surface to ensure that softening agent does not spill from dispenser.

2 Roll up each 8 ft 0 in × 3 ft 0 in (2·4 m × 0·9 m) standard sheet or piece cut to required size and place *loosely* inside each other in container so that the vapour can penetrate readily among them. Care must be taken not to fracture the Nuralite when rolling.

The Nuralite Twinrib roofing system is a method of flat roofing exploiting the ability of Nuralite to be joined impermeably with special compounds. It uses a sandwich form of jointing simulating the delaminated joint, and has been designed to meet the need for an alternative to the roll cap system, having all the latter's advantages but dispensing with the complicated roof structure with rolls and drips needed for the roll cap method.

The Twinrib system is intended to be applied by the skilled plumber and uses the standard 8 ft 0 in × 3 ft 0 in (2·4 m × 0·9 m) sheet.

This system is particularly suited to contemporary designs where the trend is towards unobstructed areas of near flat roof. Its range of application is, however, from falls of 1 in 80, up to a maximum of 40° pitch.

The following deals fully with the detailed method of application.

Fabrication of Nuralite to suit twinrib system of roofing

It is necessary to carry out the three basic operations—cutting, bending and joining—to enable the craftsman to apply Nuralite using the Twinrib method. Cutting and bending should be carried out as shown in Figs. 39 and 40.

The No. 3 or No. 4 lap joint (Fig. 42) is used for jointing the sheeting and the delaminated joint for fabricating accessories where it is impracticable to form a 76 mm lap joint. The No. 3 joint is used for roofs from minimum pitch up to 25° and employs Nuralite Compound No. 3. The No. 4 joint is used from 25° to the maximum of 40° and employs Nuralite compound No. 4.

The twinrib joint employs the same principles as described for lap jointing. Details are given in the text and Fig. 60.

Tools

Described below is a complete set of tools and equipment for applications referred to in this twinrib system.

1 A bench which should have a mild steel angle edge with arris radiused to 2 mm. The first board should have a groove 12 mm wide × 12 mm deep and a 6 mm diameter rod for forming beads, detail A (Fig. 58b).
2 Straight edges.
3 Engineer's chalk.
4 Knives for cutting and delaminating.
5 Tinsmith's snips for irregular cutting.
6 Propane gas torch for making joints and heating Nuralite. The torch should have a possible consumption of 124 oz per hour for preparing and laying sheets. A torch with a smaller head is more suitable for preparing flashings on the bench.
7 Wooden blocks, lubricated to avoid sticking to hot Nuralite.
8 Nuralite sealing irons (1, 2 and 3), detail B (Fig. 58b).
9 A propane gas-fired boiler for heating Nuralite Jointing Compound No. 3 or No. 4.
10 A suitable brush for applying hot No. 3 or No. 4 Jointing Compound.
11 Propane gas flat iron with wind shield for consolidating joints.
12 76 mm paint scraper for removing excess bitumen when finishing the joint.
13 A small trowel for applying Nuralite No. 10 Adhesive.
14 An airtight drum with a dispenser in lid to contain Nuralite softening agent, detail C (Fig. 58b).

15 Nuralite Adhesive No. 20 applicator, detail D (Fig. 58b).

Note: Items 8, 14 and 15 are obtainable from the Nuralite Company.

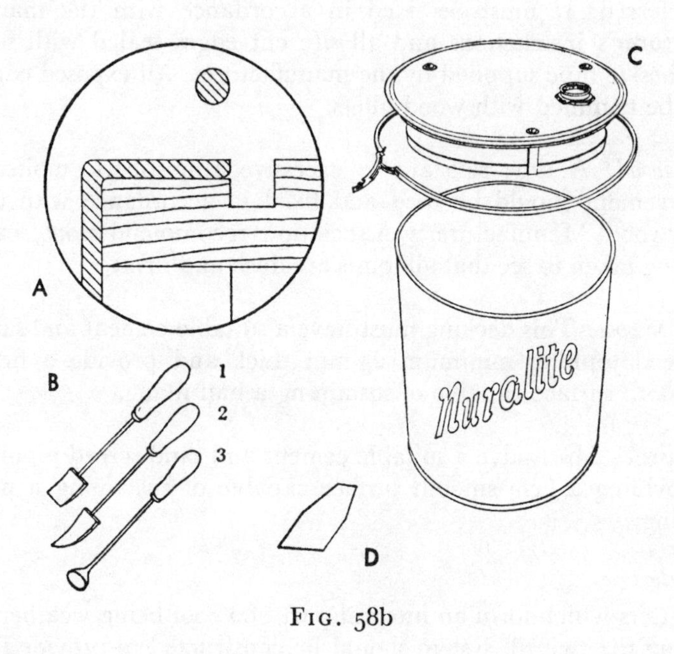

FIG. 58b

Decking

Roof decking must be capable of sustaining a nail or screw fixing and presenting a smooth, firm, fully supporting surface, free of all protuberances and be so constructed as to provide a minimum fall of 1 in 80. It may consist of any of the following:

Timber: This should be well seasoned, minimum 25 mm thick with all uneven edges and rough surfaces dressed off and nail heads well punched in. Tongued and grooved boarding is preferred, but close boarding is satisfactory provided that it presents an even and firm surface.

Chipboard: A type resistant to excessive atmospheric moisture movement should be used and fixed in accordance with the particular manufacturer's instructions, care being taken to see that all joints are flush and firm.

Compressed straw slabs: A type with a shower-proofed surface is preferred. It must be fixed in accordance with the manufacturer's instructions and all site cut edges sealed with self-adhesive tape supplied by the manufacturer. All exposed edges to be trimmed with wood fillets.

Plywood: A type resistant to excessive atmospheric moisture movement should be used and fixed in accordance with the Plywood Manufacturer's Association recommendations, care being taken to see that all joints are flush and firm.

Woodwool: This decking must have a suitable cement and sand screed topping minimum 25 mm thick and provide a firm, smooth surface capable of sustaining a nail fixing.

Concrete: Must have a suitable cement and sand screed topping providing a firm smooth surface capable of sustaining a nail fixing.

Gutters

Gutters which form an integral part of a roof being weathered using the twinrib system should be constructed to provide full support for the material, be of a minimum width of 150 mm, and have a fall of at least 1 in 160.

Linings should be fabricated from the standard Nuralite sheet and laid lengthwise, each length being lap jointed to the next (minimum lap 76 mm). Laying should be commenced at outlets and, provision made for correct weathering at structural expansion joints.

Materials to be employed

All the materials listed below to be used in conjunction with laying a twinrib roof are supplied by the Nuralite Company.

1 Standard Nuralite sheets 8 ft 0 in × 3 ft 0 in (2·4 m × 0·9 m) supplied flat.

2 Pairs of prefabricated Nuralite twinribs 2·4 m long, 76 mm girth; supplied in cardboard cartons containing 12 or 24 pairs. These comprise:

(a) base strip precoated topside having longitudinal central rib ¼ in (6 mm) high ¾ in (5 mm) wide.

(b) cover strip precoated underside having central bead to locate over base strip rib.

3 Nuralite jointing compound No. 3 or 4 supplied in bags each containing sufficient for approximately 150 ft (45 m) run of 3 ft (76 mm) wide jointing.

4 Nuralite adhesive No. 10 which should be protected from temperatures below 1 °C supplied in 1 gallon (4·5 litres) or 5 gallon (22·5 litres) airtight containers. One gallon is sufficient for approximately 25 square yards of sheeting. Or 4·5 litres is enough for approximately 22 m².

5 Nuralite masonry nails with washers; supplied in boxes each containing sufficient for approximately twelve pairs twinribs and the adjoining Nuralite sheeting.

Mechanical fixings

These should always have a ½ in (13 mm) minimum diameter head, or a washer of similar size introduced, must be stable and incapable of removal from the decking by a direct pull of 40 lb (20 kg).

The following are fixings which have proved to be satisfactory under normal conditions:

Galvanised clout nails for timber, chipboard or plywood.

A large headed patent nail supplied by the manufacturers for compressed straw slabs.

A special hardened masonry nail (type to be determined on site) with a washer for cement and sand screed.

Sequence of application

Should there be a gutter at the lowest part of the roof this is firstly lined with Nuralite then the roof decking is marked out to accept the twinribs at 3 ft 0½ in (0·9 m) centres running in the direction of the fall. The base strips are then securely mechanically fixed to the decking. The Nuralite sheets are subsequently precoated where the joint is to be made and laid between the base strips commencing at the lowest part of the roof, thus ensuring correct weathering at joints. The twinribs joints are capped with a cover strip.

Treatment at gutters

When a gutter is an integral part of the twinrib roof, the lining is cut from 8 ft 0 in × 3 ft 0 in (2·4 × 0·9 m) Nuralite sheets to the required girth, heated and formed to shape and precoated with Nuralite No. 3 or No. 4 Jointing Compound on all surfaces to be jointed.

Having determined that the temperature is suitable, immediately prior to laying each individual length apply Nuralite Adhesive No. 10 ⅛ in (3 mm) thick with a trowel to the gutter sole. The areas of adhesive should be 3 in (76 mm) wide, 3 ft (0·9 m) long and 12 in (305 mm) apart in gutters up to 12 in (305 mm) wide, and in 4½ in (114 mm) minimum squares in two or more rows in accordance with the application, see Fig. 62, if the sole exceeds 12 in (305 mm) in width. The lining is then laid, (Fig. 59A–C), starting from the outlet, each length being lap jointed to the next and secured mechanically at maximum 6 in (150 mm) centres with the lap joint. The 3 in (76 mm) portion dressed on to the roof surfaces is fixed mechanically at 9 in (228 mm) centres and 1½ in (38 mm) from the edge. The lining shown in Fig. 59A is fixed mechanically to the fascia and the drip turned down as at eaves. Stop ends are formed using Standard Nuralite Techniques, e.g. Fig. 59D and E, the end of the lining being bent up and ears thus formed, turned behind the upstand end. Bends, outlets, expansion rolls, drips and rainwater sumps are also formed using the standard Nuralite techniques.

Treatment at valleys

Strips of the required girth 8 ft 0 in (2·4 m) long are cut from the standard Nuralite sheet and precoated on all surfaces to be joined. Valleys are then lined as in Fig. 59F, No. 10 Adhesive being applied as for gutters. The dotted lines indicate where the bottom edge of the sheeting will terminate to form a channel. Laying is commenced at the lowest point, and each length lap jointed to the next and secured to the decking with

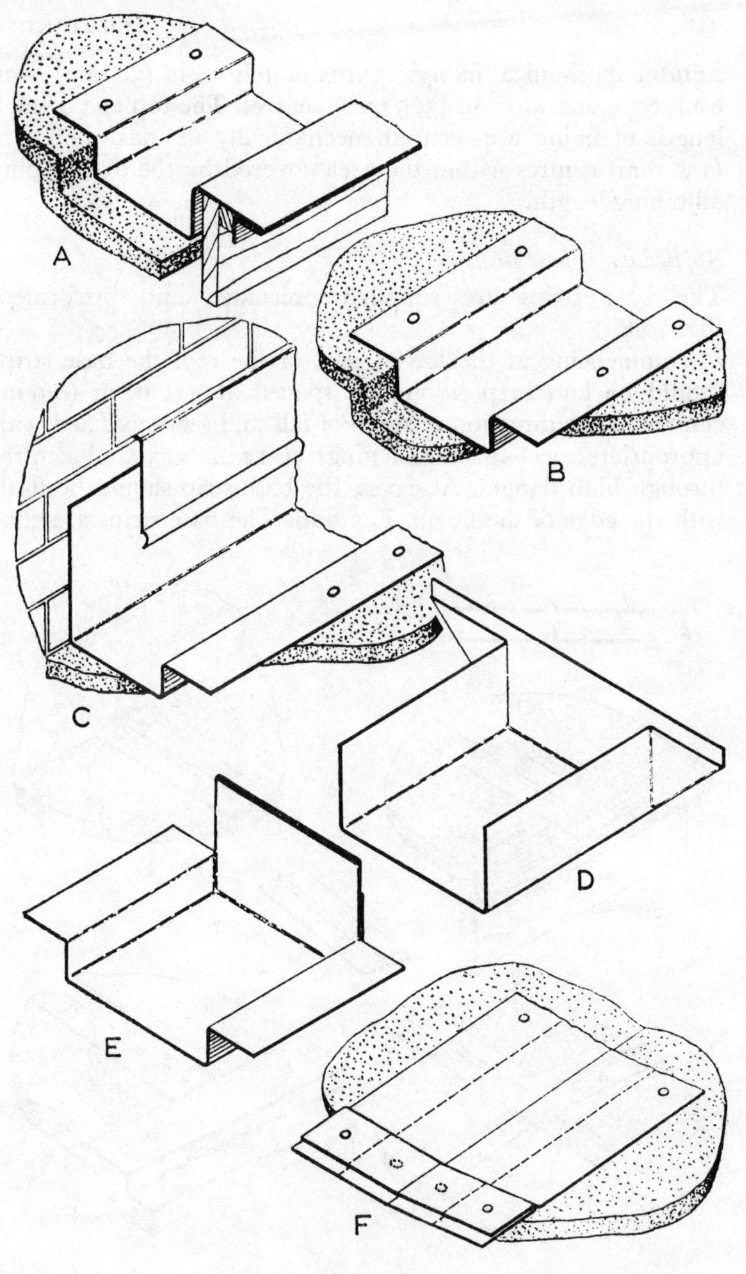

FIG. 59

suitable mechanical fixings approximately $1\frac{1}{2}$ in (38 mm) from each edge and at 12 in (305 mm) centres. The top end of each length of lining are secured mechanically at maximum 6 in (150 mm) centres within the area covered by the overlapping adjoining length.

Application of base strips

The base strips are supplied precoated and preformed, Fig. 60A.

Commencing at the lowest part of the roof the base strips should be laid in position and spaced at 3 ft $0\frac{1}{2}$ in (0·9 m) centres in the direction of the roof fall and securely fixed with appropriate mechanical fastenings at 12 in (305 mm) centres through both flanges. At eaves, the base strip should be flush with the edge of the fascia, Fig. 60B. The base strips are con-

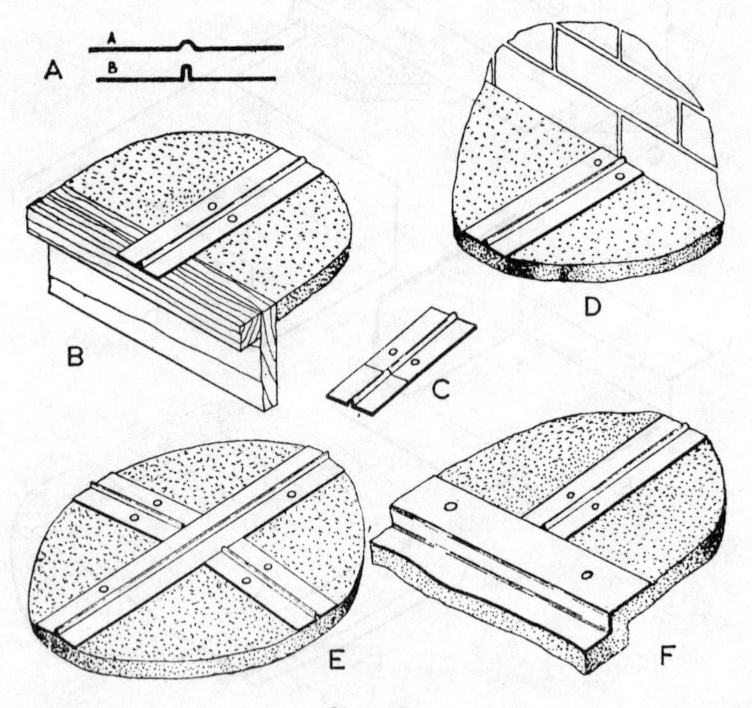

FIG. 60

tinued up the roof by butt jointing the ends of each length to the preceding lengths Fig. 60C, being mechanically fixed as previously described. They are also butt jointed to base strips at ridges and hips, valleys and gutter linings, and abutments at the top of a roof slope, Fig. 60D to F.

At side abutments 3 in (76 mm) base strips are employed cut from standard 8 ft 0 in × 3 ft 0 in (2·4 m × 0·9 m) sheets. These are pushed tightly against the abutment, mechanically fixed at 12 in (305 mm) centres and precoated.

Twinrib base strips are laid at ridges and hips with the rib at the apex of the ridge or hip intersection and fixed mechanically as before.

Treatment at eaves and verges

Edge of the Nuralite eaves and verge sheets are formed to the profile shown in Fig. 61A and firmly secured to the fascia with suitable mechanical fixings at 6 in (150 mm) centres. Care must be taken to ensure that the upper surface of the horizontal

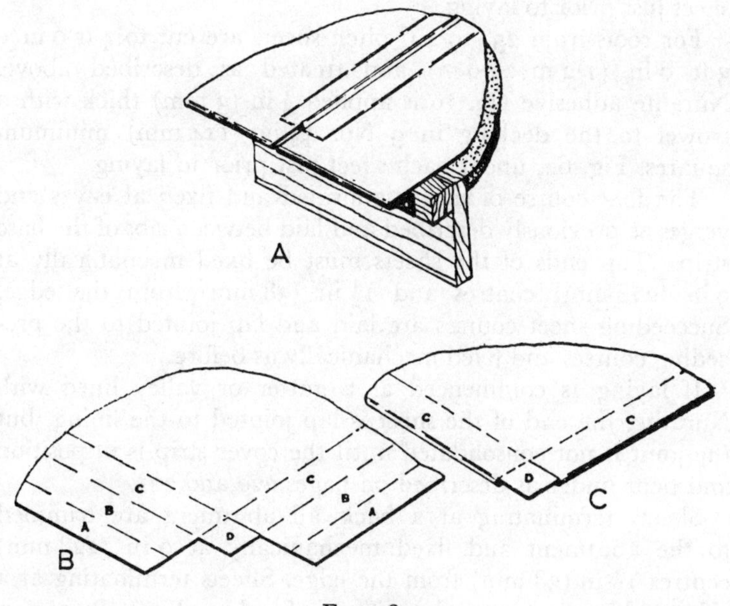

FIG. 61

leg of the 'L' shaped portion which is turned under is flush with the top surface of the decking and that the vertical leg is approximately $\frac{1}{4}$ in (6 mm) away from the front of the fascia, this will enable the drip to be bent down satisfactorily. The part to be turned down should be sufficiently wide to form a drip at least $1\frac{1}{2}$ in (38 mm) deep. To form a corner, the sheet is cut and formed as Figs. 61B and 61C parts A being nailed to the fascia, and parts B, C and D to remain projecting.

Applying sheets

For roofs with a fall of $1\frac{1}{2}$ in in 10 ft 0 in (1 in 80) up to 25° pitch Nuralite 8 ft 0 in × 3 ft 0 in (2·4 m × 0·9 m) sheets are firstly precoated on the underside edges to a width of $1\frac{1}{2}$ in (38 mm) where they are to be in contact with the twinrib base strips and 3 in (76 mm) at side abutment base strips, and on all other surfaces where lap joints are to occur. Having determined that the temperature is suitable Nuralite adhesive No. 10 is applied $\frac{1}{8}$ in (3 mm) thick with a trowel to the decking in 18 No. $4\frac{1}{2}$ in (114 mm) minimum squares, see Fig. 62, under each sheet just prior to laying.

For roofs from 25° to 40° pitch sheets are cut to 4 ft 0 in × 3 ft 0 in (1·2 m × 0·9 m) and treated as described above. Nuralite adhesive No. 10 is applied $\frac{1}{8}$ in (3 mm) thick with a trowel to the decking in 9 No. $4\frac{1}{2}$ in (114 mm) minimum squares, Fig. 62, under each sheet just prior to laying.

The first course of sheets is formed and fixed at eaves and verges as previously described and laid between ribs of the base strips. Top ends of the sheets must be fixed mechanically at 9 in (228 mm) centres and $1\frac{1}{2}$ in (38 mm) from the edge. Succeeding sheet courses are laid and lap jointed to the preceding courses and fixed mechanically as before.

If laying is commenced at a gutter or valley lined with Nuralite, the end of the sheet is lap jointed to the lining, but the joint is not consolidated until the cover strip is in position and bent under as described on pages 220 and 221.

Sheets terminating at a backwall abutment are trimmed to the abutment and fixed mechanically at 9 in (228 mm) centres $1\frac{1}{2}$ in (38 mm) from the edge. Sheets terminating at a ridge or hip at a twinrib junction are fixed mechanically at 9 in

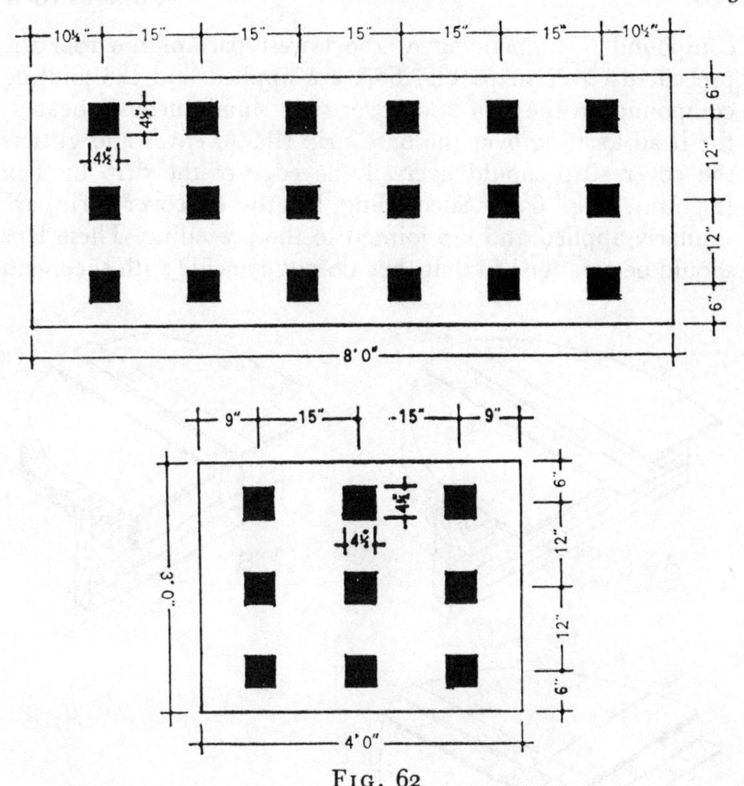

FIG. 62

(228 mm) centres ¾ in (19 mm) from the edge and through the base strip.

When meeting a side abutment the sheets are cut to fit and precoated on bottom surfaces at side edges where to be joined to the side abutment base strips and twinrib base strips. All joints between sheets and base strips are now heated thoroughly with a large capacity propane gas torch and consolidated with a hot iron. It is desirable at this stage to define the joggle arising from the sheet stepping over the edge of the base strip. Lap joints at sheet ends are consolidated in the same manner.

Application of cover strip

All sheet surfaces to receive the cover strip and other flashings are thoroughly heated and coated with hot Nuralite jointing compound. Commencing at the lowest part of the roof the precoated cover strips Fig. 60A are applied and the jointing compound on the roof and cover strip simultaneously heated, the bead locating over the base strip rib. At eaves and gutters the cover strip should oversail the edge of the drip by 1 in (25 mm), Fig. 63A. Succeeding lengths of cover strip are similarly applied and lap jointed to the preceding. These laps should be arranged so that they do not coincide with sheet end

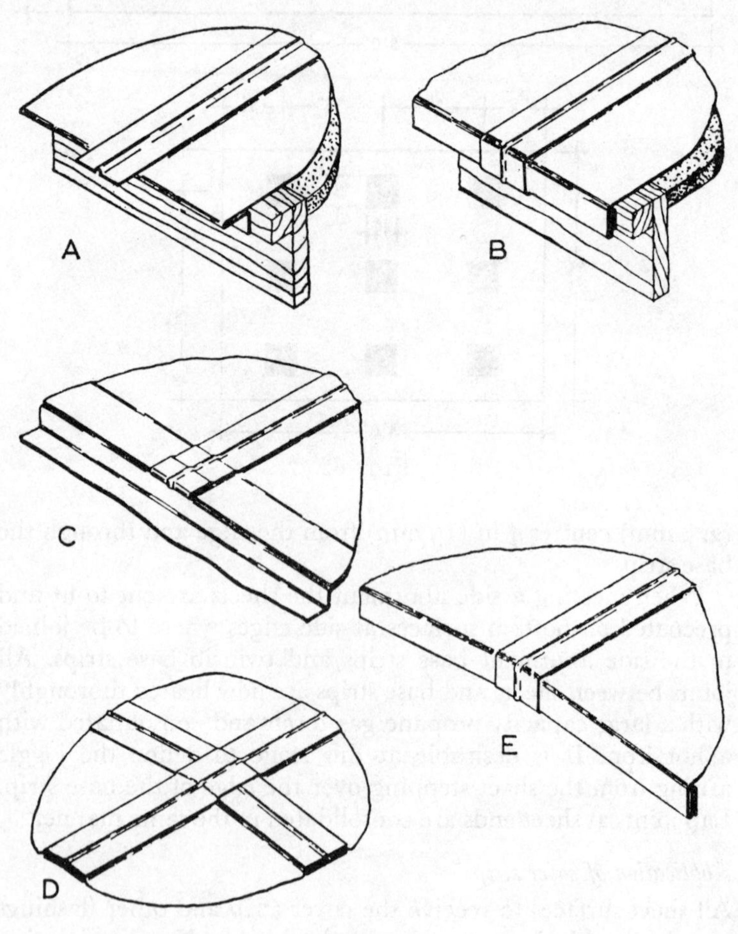

Fig. 63

laps. At abutments, ridges and hips the cover strip is continued to the extremity.

Cover strips are now thoroughly heated and consolidated with a hot iron. At eaves, the oversailing end of the cover strip is heated and hooked under the drip. Finally the drip and cover strip ends are heated and turned down, Fig. 63B.

Where sheeting is commenced at gutters or valleys the oversailing end of the cover strip is precoated on the topside, heated and hooked under the sheeting and the lap joint between the sheeting end and lining completed by heating and consolidating with the hot iron, Fig. 63C.

The cover strip completing ridges and hips is now applied as before described, particular care being taken to ensure a snug fit over intersections, Fig. 63D. Termination at abutments is weathered with a separate upstand. At verges the cover strip is heated and hooked round the drip, Fig. 63F.

All joints are consolidated with a hot iron, excess bitumen being removed with the 3 in (76 mm) paint scraper.

Treatment at upstands

Strips 8 ft (2·4 m) long cut from a Nuralite sheet to the required girth are formed into an 'L', the upstand leg being 4 in (100 mm) minimum and the base 3 in (76 mm) minimum. The base and ends of the flashings thus formed must be precoated where to be jointed. Each length of flashing is then placed in position, Fig. 64A, being jointed to the sheeting and the next length of flashing. Particular care should be taken to see that the base is heated and formed to fit neatly over the ends of twinribs. External and internal angles are weathered by using the standard Nuralite techniques. Fair ends are formed in a similar manner to external corners and thus weather the end of the drip, Fig. 64B.

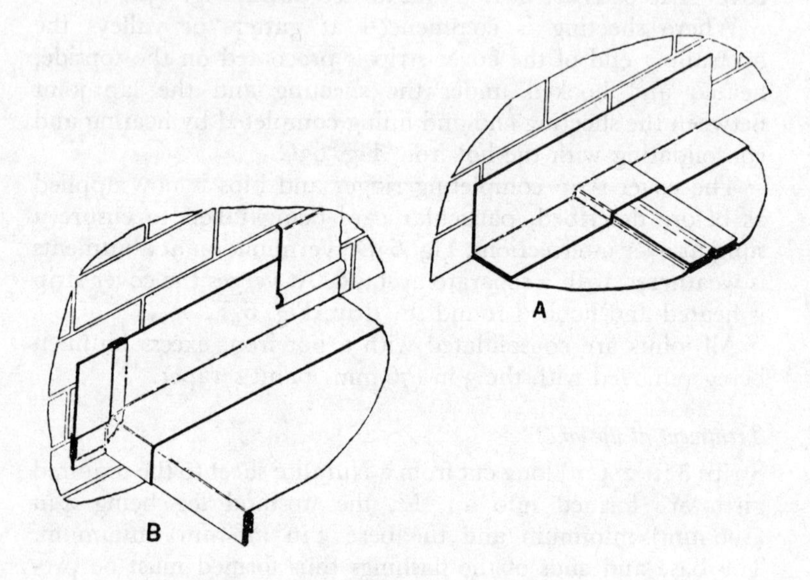

FIG. 64

19
Soft soldering

Copper bit soldering has already been mentioned as a jointing method for some zinc weathering details. Nowadays copper tube joints are generally made with solder capillary fittings, where the solder is melted in a blowlamp flame, and runs into the joint. Copper bit soldering tends to be neglected and this is a pity, for properly used it is a skill that can still be put to very good use.

What is soldering? Very briefly, it may be described as the joining of two metals by another, the solder, which has a lower melting temperature than either of the others. This is necessary because the two metals to be joined are not meant to melt and fuse, as happens in welding. In soft soldering, as copper bit work is called, the solder melts, flows between, and actually combines with the surfaces of the metals to be joined.

You will remember that when oxygen comes in contact with metals it reacts with the metal to form an oxide, which is non-metallic. You may have noticed that cleaned copper tube changes colour when it is heated. This change is caused by the rapid combination of atmospheric oxygen with the hot metal, which forms a non-metallic skin of copper oxide. This is a chemical change (see page 74). Metals can be soldered but non-metals cannot, so that before soldering or welding is begun the area to be treated must be cleaned free of the oxide. With lead or iron, heat will be enough to melt away the oxide, revealing the pure metal underneath. The oxide will float on the surface of the molten weld pool, so that if correct methods are used these metals can be cleaned and welded quite easily. The oxides of aluminium have a higher melting point than the metallic aluminium underneath, so that it is rather more difficult to weld this metal. However, by enveloping the electric arc in a 'shield' of inert gas—argon—the oxygen can be kept away from the metal and the weld made. This is called the argon-arc process.

A similar method is adopted for the welding of plastics materials such as polythene and P.V.C., both of which will readily oxidise when heated. In this case nitrogen is used. The gas, at a pressure of 14 to 21 kN/m² (2 to 3 lbf/in²), is heated in special torches to about 300°C, at which temperature it will fuse the plastic surfaces (see also page 59).

With copper, brass and zinc, however, heat by itself is not enough to remove the oxide, and special reducing agents, called fluxes, are needed. When applied together with heat, these will bring about a local reduction of the oxide, and float it away to reveal the true metal underneath. Since heat encourages oxidation, it is also important that the flux prevents the oxygen in the atmosphere from coming in contact with the metal surface. Moreover, flux helps the solder to flow over surfaces, so that it 'wets' or tins them easily.

It is necessary to clean the surfaces to be soldered. Physically, this can be done with a file, wire brush or sandpaper. Chemically, normal oxide films can be removed with special, very active, reducing fluxes. Care must be taken if these are used since they are very corrosive.

Adequate heat is essential if the solder is to penetrate the metal surfaces properly. Obviously, the heat must be great enough not only to melt the solder but also to raise the temperature of the metals being joined so that they will readily accept the 'wetting' or tinning action of the solder, and allow it to form a combination, or bond, with them.

Four simple rules will help you to do good soft solder work. They are:

1 Clean surfaces to be joined.
2 Use the correct flux.
3 Use adequate heat.
4 Use the correct solder.

The soldering iron

The copper bit, or soldering iron, is a wooden-handled iron shaft with a copper 'bit' on the end. Copper is used because of its high heat conductivity which allows it to transfer its heat quickly to the soldered joint. It also heats up very rapidly. Its capacity for heat; that is, the amount of heat it can transfer to the joint in a certain time, depends upon the size of the 'bit'.

Copper bits are sized according to the weight of the copper used, and, generally, the bigger the copper bit the easier and better the soldering work will be. A 1 kg (2 lb) bit is none too large for most jobs of soldering in plumbing.

When using the heated copper bit on the work, take care to apply it so that as much heat as possible can flow in a short time. Avoid using only the pointed end of the bit but, rather, apply one of the large, tinned, flat faces to the work (see Fig. 65).

Tinning the bit is done by heating it to the correct temperature, filing its four flat faces clean, rubbing them into powdered sal ammoniac, and at the same time touching the surfaces with solder. As for the correct temperature, red-hot is too hot. The tinsmith of old used to judge the heat by holding the bit some inches from his face—a practice which is not recommended as a safe one. Judgement of heat comes from experience. A lively 'sizzle' of the flux when the heated copper bit is inserted is a fairly good indication that the temperature is correct. If the bit is too cold the solder will not melt or run, and if it gets too hot its tinning will burn off, leaving a dirty oxide film. The oxide is a poor conductor of heat, and will not attract the solder, so that soldering will be very difficult, if possible at all.

Solders

A wide range of solders for soft soldering are available. Mixtures or alloys of metals, chiefly of lead, tin and small amounts of antimony, or must lead and tin, are generally used, depending on the type of soldering that is to be done. Table 9 lists some of the more common plumber's solders, together with their proper fluxes and uses.

These solders might contain very small amounts of iron, arsenic and copper. If the quantities are kept below the limits laid down in BS 219, the solder will prove satisfactory.

Small amounts of aluminium, cadmium, or zinc might also be present, and these could affect the working properties of the

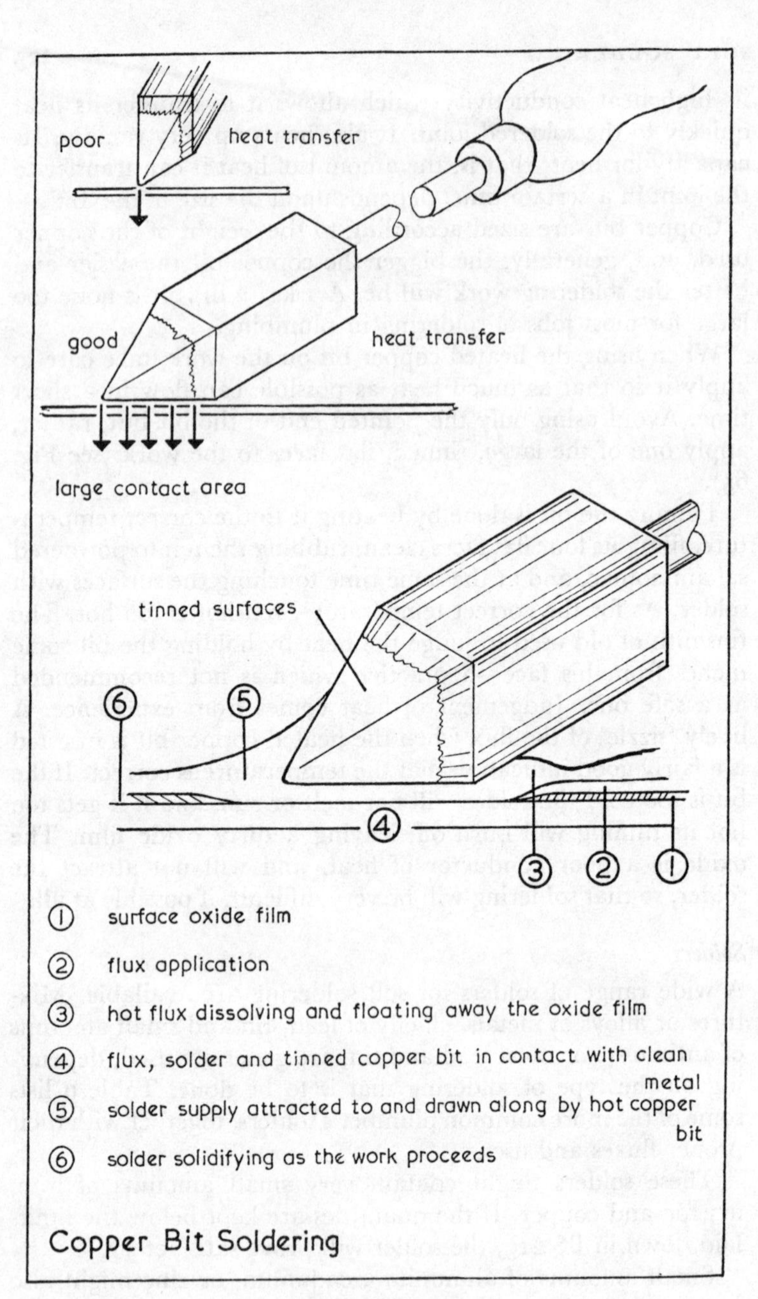
poor
heat transfer
good
heat transfer
large contact area
tinned surfaces
6
5
4
3
2
1
① surface oxide film
② flux application
③ hot flux dissolving and floating away the oxide film
④ flux, solder and tinned copper bit in contact with clean metal
⑤ solder supply attracted to and drawn along by hot copper bit
⑥ solder solidifying as the work proceeds
Copper Bit Soldering

FIG. 65

TABLE 9 THE COMPOSITION AND USES OF SOLDERS *See also BS 219, Non-antimonial solders*

BS 219	*% Composition*			*Melting temperatures °C*		*Typical uses*
Grade	*Tin*	*Antimony*	*Lead*	*solid*	*liquid*	
				'plastic' range		
A	65	0·6	The remainder	183—185		Electrical and radio work
K	60	0·5		183—188		
F	50	0·5		183—212		Copper bit soldering Tinning unions, etc.
G	40	0·4		183—234		Blowpipe soldering, for example, in gas fitting work
J	30	0·3		183—255		Plumber's wiped joints on lead pipework
Antimonial Solders						
B	50	3·0		185—204		As for Grade F above
M	45	2·7		185—215		As for Grades F and B
C	40	2·4		185—227		As for Grade G
L	32	1·8		185—243		As for Grade J
D	30	1·7		185—248		As for Grades J and L

Note: In plumbing, Grade F is commonly used for copper bit work, and Grade D for wiped joints.

solders, making them faulty. Such impurities would not be acceptable in solder made to the BS 219 specification.

Paste fluxes

Powdered resin is difficult to apply to soldering work, especially to work in awkward positions. Liquid fluxes are more easily applied, but they tend to run away from the joint area. Paste fluxes avoid both these difficulties and for this reason are more commonly used in plumber's work.

Resin dissolved in alcohol provides a non-corrosive flux for radio and electrical work. The addition of small amounts of oleic acid to this produces a more active, corrosive flux for general soldering work. Oleic acid is an organic acid obtained from animal fats. Other paste fluxes are formed from zinc chloride and ammonium chloride in solution with petroleum jelly. Alternatively, tallow or lanolin, a fatty substance obtained from wool, may be used as the base to carry the more active flux ingredients.

All fluxes are corrosive to a degree, since some acidity in the flux is very necessary to the chemical reduction; that is, to the dissolving and removal of the metallic oxides which the flux aims to bring about in the soldering processes.

'Safe' fluxes are those which are not aggressively corrosive in use or in after-effect. Tallow and resin are generally regarded as safe fluxes because although they are sufficiently acidic at soldering temperatures, they do not corrode most metals at ordinary room temperature.

TABLE 10 THE COMPOSITION AND USES OF FLUXES FOR SOFT SOLDERING

Flux	*Uses*	*Notes*
Resin	Tinning brass and copper	Resin is a gum-like substance obtained from pine tree bark. It is used in powder form, and is sprinkled on surfaces to be tinned. It is mildly corrosive at soldering temperature, and non-corrosive when cold
Tallow	Soldering on lead	Tallow is obtained from the fat of animals, particularly cattle and sheep. These organic fats contain glycerin, which makes them mildly acidic.

TABLE 10—*cont.*

Flux	*Uses*	*Notes*
Zinc-chloride (killed spirits)	All forms of copper bit soldering *except* electrical work	Produced by dissolving zinc in hydrochloric acid. Dangerous hydrogen gas fumes are given off during this process. The flux is actively corrosive and must be completely removed by washing the soldered article to remove all traces of flux.
Zinc-ammonium chloride (killed spirits plus 10% ammonium)		This flux is 'active' at a lower temperature than killed spirits; this is helpful in soldering metals with a relatively low melting point, such as zinc. It is as corrosive as zinc chloride, and similar treatment of the articles is needed after soldering. Most of the proprietary liquid fluxes are of this composition.

20

Alloys

An element, you will remember (page 72), is a substance made up of atoms all of one kind. Lead, copper, zinc and aluminium are all elements; each is composed entirely of atoms of lead, copper, zinc or aluminium.

It is possible to melt down and mix together copper and zinc, but the mixture, when set, would not be an element because it would contain atoms of both copper and zinc. Such a mixture of metals is called an alloy and in this case, if the proportions were right, by alloying copper and zinc together brass would be made. The properties of the mixed metal would be different from those of either the copper or the zinc which were added together to make it.

An alloy is sometimes described as a metal composed of two or more metals. This is of course quite true, but an alloy can be made by alloying a metallic element with a non-metallic element. A better way of describing an alloy, therefore, is as follows: 'An alloy is a metallic substance made by mixing two or more elements, at least one of which is a metal.'

You have seen that copper and zinc when blended or mixed in the right proportions will make brass. Lead and tin, two more metallic elements, can be blended to make solder, which, incidentally, has a lower melting point than either the lead or tin of which it is composed. This is but one of many strange changes of physical properties which occur when alloys are formed (see 'Mixtures', page 75).

Steel, used for making tools and so on, is an alloy. It is a mixture of iron and carefully controlled amounts of carbon. Carbon is an element but it is not a metal, so that steel is one example of an alloy composed of metallic and non-metallic elements.

TABLE 11 ALLOYS USED IN PLUMBER'S WORK

Lead Alloys

Composition %							Uses
	Antimony	*Tin*	*Cadmium*	*Silver*	*Copper*	*Lead*	
Silver-copper-lead BS 1085	—	—	—	0·005	0·005	The remainder	Has greater tensile strength than BS 602 lead. Pipes of BS 1085 may have thinner walls than BS 602 for given working pressures
Ternary alloy No. 2 BS 603	—	1·5	0·3	—	—	The remainder	Similar to BS 1085 but with greater resistance fatigue; i.e. to 'work tiredness'. Very suitable for use under roads subject to heavy traffic vibrations
Antimonial lead or 'Regulus metal'.	12	—	—	—	—	The remainder	Produces a 'hard' lead capable of being turned to shape in a lathe. Can also be screw threaded. Useful for making cocks, etc., for acid works.
Soft Solders BS 219	See Table 9, page 227						See Table 9

Copper Alloys

	Tin	*Zinc*	*Lead*	*Copper*	
63/35 Brass 'basis brass'	—	The remainder	0·1	65	A ductile alloy useful for cold pressings in sheet or strip form
60/40 Brass 'yellow metal' or 'Muntz metal'	—	The remainder	—	60	For hot stampings and for casting brass fittings
Hot forging brass	—	The remainder	2·5	58	For hot pressings and hot pressed pipe fitting manufacture. The lead included in this alloy makes for easier thread cutting and machining.

TABLE 11—*Cont.*

Composition %					*Uses*
	Tin	*Zinc*	*Lead*	*Copper*	
70/30 Brass	1·5	The remainder	—	70	For extruded tube for towel rails, etc.
Phosphor-bronze	—	The remainder	—	94	Bourdon gauge tubes
Gunmetal (bronze)	10	The remainder		88	Highly resistant to corrosion. Used for pipe fittings to be buried underground

Iron Alloys

	Carbon	*Chromium*	*Nickel*	*Magnesium*	*Iron*	
Cast iron	2—4	—	—	—	The remainder	Rainwater pipes and fittings, boilers, etc.
Mild steel	up to 0·5	—	—	—	The remainder	Water service pipes and fittings
Tool steels	1·2	Possibly small amounts See page 233			The remainder	Saws, hammers, snips, etc. Carbon content varies according to purpose of the tool. Generally tools to withstand shock, such as hammer heads, may have only 0·6 %carbon whereas snips may have up to 1% carbon
'Staybrite' steel	0·12	18	11	0·24	The remainder	Cold pressings, Kitchen sinks, etc.
Invar	0·2	—	36	—	The remainder	Has very low thermal expansion coefficient. Almost invariable in length under changing temperatures, hence the name 'Invar'. Used in bi-metal thermostatic devices

Table 11 on pages 231–2 shows the composition of the important alloys used in plumber's work, together with the purposes for which they may be used. Some alloys may be further improved by the addition of very small and carefully controlled quantities of other metals, which will invest the alloy with their own particular properties:

Chromium added to steel increases its resistance to corrosion.
Magnesium increases the toughness of steel.
Silicon added to steel destroys its magnetic properties.
Nickel, like chromium, increases steel's resistance to corrosion.
Vanadium added to steel makes it more resistant to damage by shock.

For example, you may have a spanner or a pair of pliers of chrome-vanadium steel. Such tools will not only resist corrosion; they will also withstand sudden wrenching shocks without fracturing.

21

Sanitary appliances: general

Sanitary appliances can conveniently be divided into two groups:

1 Soil appliances, designed for the collection and discharge of excretory matter. A W.C. pan is a typical example.
2 Waste appliances, for the collection and discharge of water used in washing and cooking. A lavatory basin would be used in a bathroom, and a sink in the kitchen.

Most houses have a soil fitment in the form of a W.C. pan, and also a kitchen sink. Many have a lavatory basin for washing and also a kitchen sink. Many have a lavatory basin for washing, but it is perhaps surprising to discover the relatively large number of older houses that do not yet have baths.

Design of sanitary appliances

Sanitary appliances are of many types, designs and materials, but the dimensions that are important in their fixing—height, width, pipe connection positions and so on—are all made to British Standards. Thus appliances are reasonably interchangeable, and this is very helpful in the replacement of damaged appliances and in the prefabrication of pipework.

Besides having suitable strength and durability, sanitary appliances are expected to have a pleasing appearance, and above all to be hygienic. To this end, good design aims at simplicity of line. Crevices which might harbour dirt or harmful germs are avoided.

The appliance must be as far as possible self-cleansing. All inside surfaces of a W.C. pan, for example, should be effectively washed by its flushing water. All lavatory basins, baths and sinks should have a natural, self-draining action, and all appliances should have smooth, easy, clean lines and surfaces.

Materials for sanitary appliances

The basic requirements of materials used for sanitary appliances are that they should be impervious to water; that is, that they should not absorb water or allow it to pass through them; and that they should be strong. Advances in the development of materials and design, coupled with a growing demand for better-looking and more easily cleaned appliances, have resulted in a wealth of new ideas for sanitary ware.

Pressed steel, either stainless or vitreous enamelled, is now very much used for sink and drainer units, lavatory basins and even baths.

They are also being made in *plastic materials*, notably *perspex*.

Cast-iron baths are common, but although lavatory basins have been made in this material they do not appear to have become very popular. On the other hand, kitchen sinks with drainers in cast iron do seem to be gaining in favour especially for low-cost rented dwellings, where equipment must be able to stand up to hard wear.

Ceramic ware is the up-to-date term for pottery ware or fittings made from moulded and 'fired' clays. For many years sinks, lavatory basins and W.C. pans have been made in one form or another of this material. At the moment, indeed, it is the only material in which W.C. pans are made. But the rapid development in this field may soon bring W.C. pans in other materials.

Ceramic ware, however, has all the qualities necessary for sanitary equipment. It lends itself to easy moulding into hygienically simple outlines. It is of ample strength, and its smooth surfaces are easily cleaned, giving no hold to dirt or germs.

Fireclay is a form of ceramic ware noted for its robustness. It does absorb water but it can easily be made impervious by use of the well-known glaze, applied during the manufacturing process.

W.C. pans and lavatory basins are made of *vitrified china-ware* 'Vitrified' means glasslike, and it is well known that glass is impervious to water. White or coloured glaze is therefore only used on these fittings for the sake of appearance, and to make them easy to clean.

Fireclay Sinks: BS 1206 and BS/MOE 1/7

These waste appliances in white glazed fireclay are solid, strong and reliable. They cannot corrode, are easily kept clean, do not discolour, and unless badly misused they are virtually everlasting. Sinks are intended simply for use in the kitchen, and for this purpose their appearance can hardly be improved on. They will combine easily with the built-in fitments, cupboards, and so on, which have become a popular feature of the modern kitchen.

The *Belfast fireclay sink* is 254 mm (10 in) deep, and the commonly used size is 610 mm (24 in) by 457 mm (18 in). It has an overflow of the weir type in order to make it easy to clean. This integral overflow discharges into the sink's waste outlet, as is shown in Fig. 66. The sink outlet fitting, which has a standard diameter of 38 mm, is usually made of chromium-plated brass. It may be of the pattern specially devised for use with this type of sink, or it may be a fitting of the slotted waste outlet type, similar to that used for lavatory basins. Fig. 67 shows sketches of both kinds of waste outlet fitting, and the path of overflow and waste water flows can be clearly seen.

A kitchen sink is seldom left unattended with the taps running, and so it is most unlikely that an overflow would occur. It is for this reason that the combination of the overflow discharge with the waste discharge is allowed, although the Water Undertaking's bye-laws require that overflows should discharge in an easily seen place, and *not* into a gutter, waste pipe or drain.

A suitable waste outlet plug secured by a chromium-plated chain completes the sink's equipment, apart from its waste pipe and trap which are dealt with on page 253.

Belfast sinks have been, and still are, popular for home

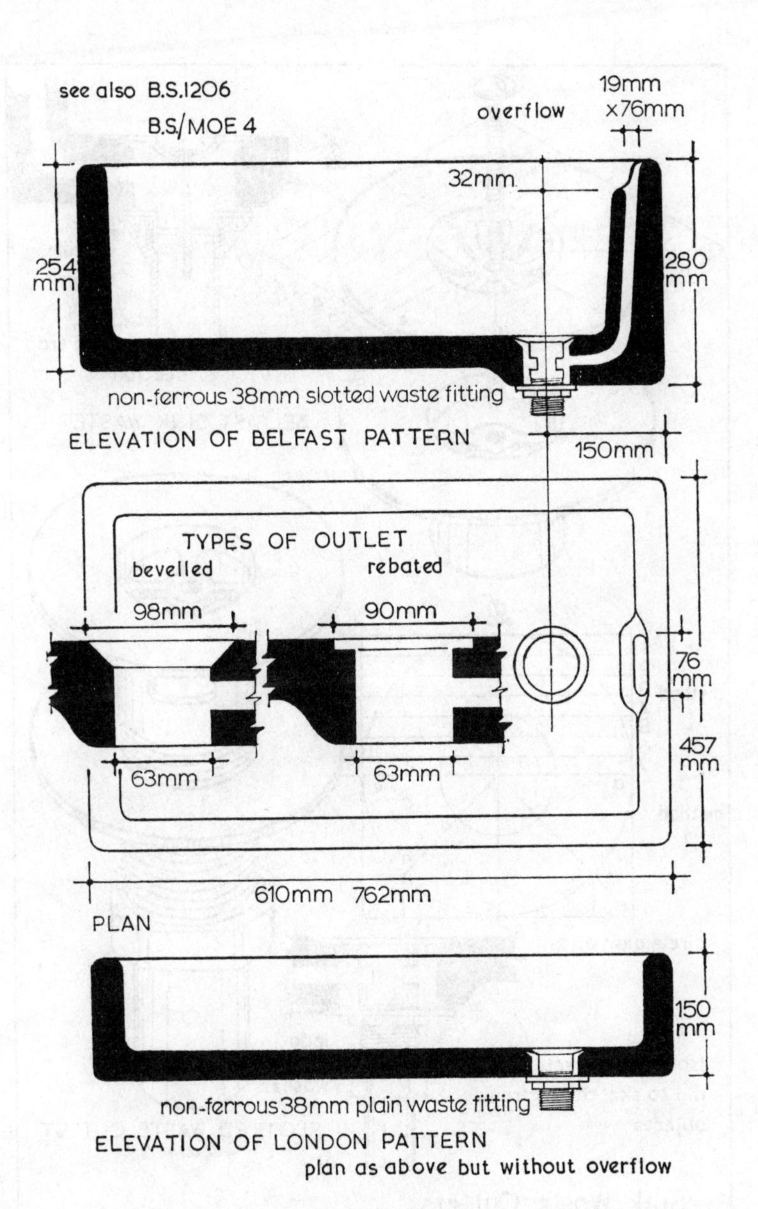

Fireclay Sinks

FIG. 66

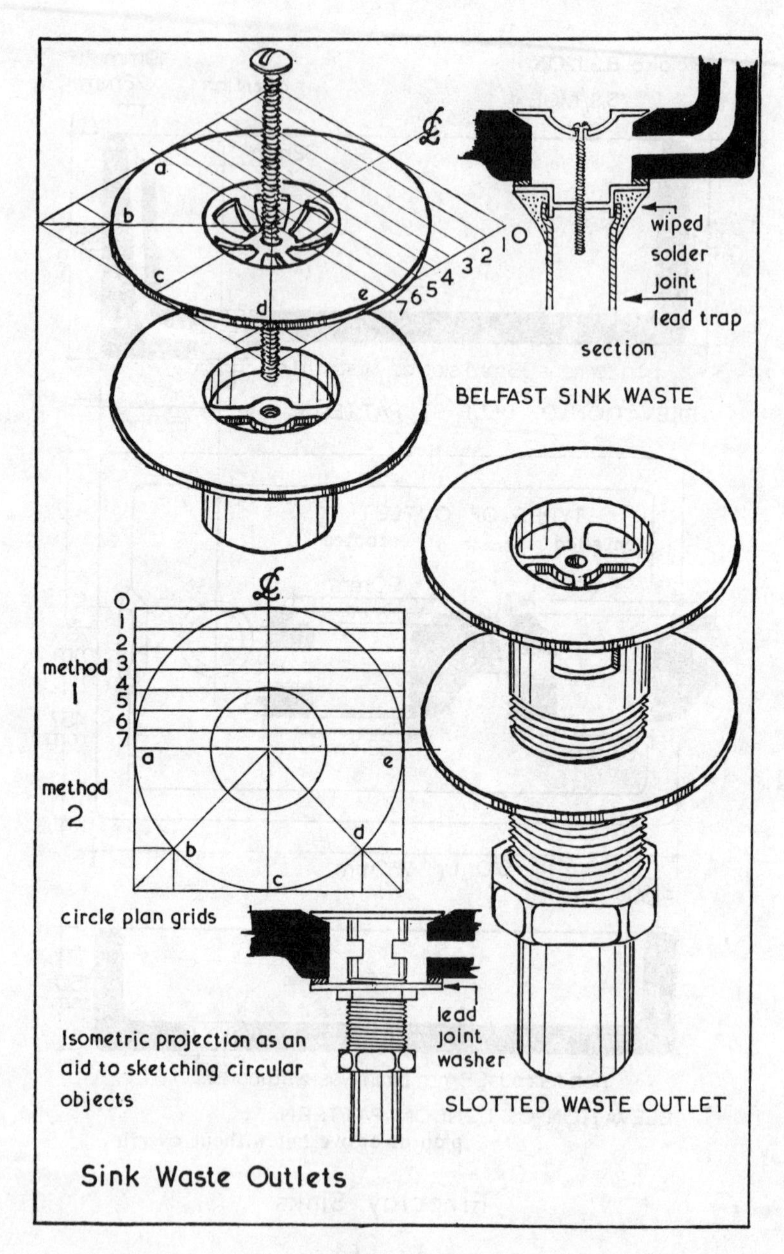
℄
a
b
c
d
e
7 6 5 4 3 2 1 0
wiped
solder
joint
lead trap
section
BELFAST SINK WASTE
℄
0
1
2
3
4
5
6
7
method
1
method
2
a
e
b
d
c
circle plan grids
lead
joint
washer
Isometric projection as an aid to sketching circular objects
SLOTTED WASTE OUTLET
Sink Waste Outlets

Fig. 67

laundry. The ample inside dimensions of the sink, especially its depth, are convenient for the washing and rinsing of clothes. Now that so many homes have washing machines, the depth of the Belfast sink is no longer such an attractive feature and for this reason the shallow *London* pattern which was the only choice some fifty years ago is now making its comeback.

Fixing height

The height at which the kitchen sink should properly be fixed has been the subject of much argument. A common rule in the trade has been to set the sink support brackets 610 mm above floor level, so that the top edge of the sink is at a height of 864 mm (2 ft 10 in). This has proved reasonably convenient for most people, but it must be remembered that there is no standard height for human beings, and therefore, in order to satisfy the various needs of the 'long, and the short, and the tall', a sink of adjustable height seems to be called for. This is by no means impossible, but unfortunately, since much tiredness results from working at the wrong level, no one has bothered about it.

Working-top heights in kitchens have been standardised at 914 mm (3 ft), and sink tops have to conform with this height in order to match up with other items of equipment, such as cookers and refrigerators.

Clause 313 of C.P. 305 'Sanitary Appliances' recommends that the fixing height to the top of single draining-boards should be 914 mm (3 ft). Presumably, short people will have to continue to make do by standing on a box.

Ceramic lavatory basins: BS 1188

'Ceramics' or 'pottery' includes all objects made from clays and then hardened by heat in a specially constructed kiln—a process called firing. The plasticity of the clay, together with the skill of present-day pottery designers and craftsmen, has brought remarkable changes in the design of sanitary ware. Hygiene, durability and beauty are combined, and lavatory basins in particular now appear in a wide variety of new designs although the traditional pattern, as shown in Fig. 68, is still available and will continue to provide good service for some time yet.

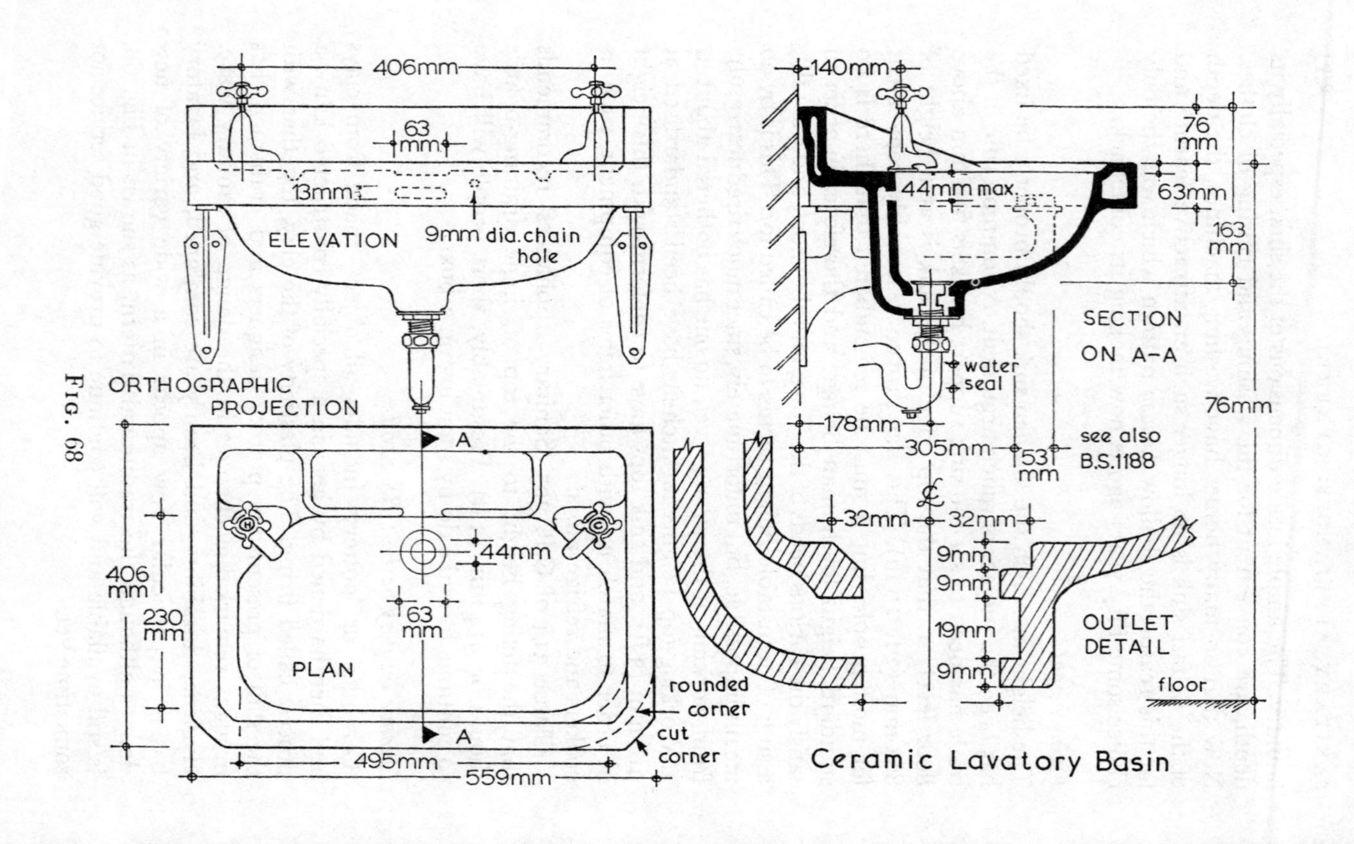
406mm
63 mm
13mm
9mm dia. chain hole
ELEVATION
ORTHOGRAPHIC PROJECTION
A
A
44mm
63 mm
406 mm
230 mm
PLAN
rounded corner
cut corner
495mm
559mm
140mm
76 mm
44mm max
63mm
163 mm
SECTION ON A–A
water seal
76mm
178mm
305mm
53 mm
see also B.S.1188
32mm
32mm
9mm
9mm
19mm
9mm
OUTLET DETAIL
floor
Ceramic Lavatory Basin

FIG. 68

Lavatory basin design

BS 1188 prescribes the quality of material and workmanship to be used in the manufacture of wall or pedestal-mounted lavatory basins. It also lays down important dimensions, such as the width, depth, height, and size and position of the tap and waste holes. Apart from these, each manufacturer is free to develop his own designs. The basic patterns of lavatory basin are available in two sizes—635 mm by 457 mm (25 in × 18 in) back to front, and 559 mm by 406 mm (22 in × 18 in). The second is more suitable for the small bathroom.

The appliance will have a combined overflow and waste fitting of the slotted type shown in Fig. 67, but with a standard diameter of 32 mm instead of the sink's diameter of 38 mm.

The *waste outlet fitting* may be bevelled or rebated according to the type of basin outlet provided (see enlarged diagram of outlets, Fig. 66). The advantage of the bevelled waste fitting and outlet is that the feather edge of the fitting merges with the inside surface of the basin, giving a smooth, easy, clean line without place for dirty water to lodge.

The better overflow outlet for the basin, as for the sink, is the weir type, since it is easy to keep clean. Many basins, however, have slotted overflow outlets as shown in Fig. 68, and these are clearly more difficult to get at for cleaning.

Fixings

Clause 308 of C.P. 305 'Sanitary Appliances' deals with lavatory basin design, and Clause 321 of the same Code of Practice deals with fixings. These may consist of brackets to a wall, a ceramic pedestal, or a combination of both. Where pedestals are used, it is a good idea to use brackets to secure the basin in place against the wall.

If a pedestal basin is to be used, then the height of the pedestal will determine the fixed height of the basin—usually about 762 mm to 812 mm. If bracket-supported basins are used, the fixing height may be varied to suit the kind of building in which they are to be installed. The following table suggests the kind of scale that might be employed:

* mm ÷ 25·4 = inches

Use	*Fixing height to top front edge*
Infant Schools (5- to 8-year-olds)	610 mm
Junior Schools (8- to 11-year-olds)	686 mm
Secondary Schools (11 to 15 plus)	787 mm
Adults	812 mm

BS 1188 recommends that basins for domestic use should be fixed at 787 mm (31 in).

A lavatory basin will require hot and cold water services, a waste plug and chain, and a waste pipe and trap. These are dealt with on page 253.

Fixing taps to basins

Basin taps are known as 'pillar taps' (see Fig. 69). The water service is connected to the threaded vertical shank or pillar of the tap, which passes through the basin top. A squared shoulder beneath the flange of the tap locates with the square tap hole in the basin, fixing the direction of the tap nozzle, and preventing the tap from turning while the water is running. For a key to the numbers in Fig. 69, see pages 305–306.

There is some difference in size between the shoulder square of the tap and the tap-hole square in the basin. This difference has to be made good in order to secure the tap and provide a waterproof joint between tap and basin. This may be done by cutting and fitting narrow strips of lead, and using them to pack the spaces, thus ensuring a snug fit for the tap. This packing should not be too tight or the basin might crack as the tightening backnut pulls the tap into the basin hole; it should not, on the other hand, be too loose or the packing will be useless.

When the packing seems just right, some bed-jointing material is applied around the flange, the packing, and the upper part of the tap pillar. The tap is then offered into the basin hole, and the backnut screwed home against the underside of the basin top. Red and white lead putty, white lead alone, glazier's putty, or even plaster and cement, can be used

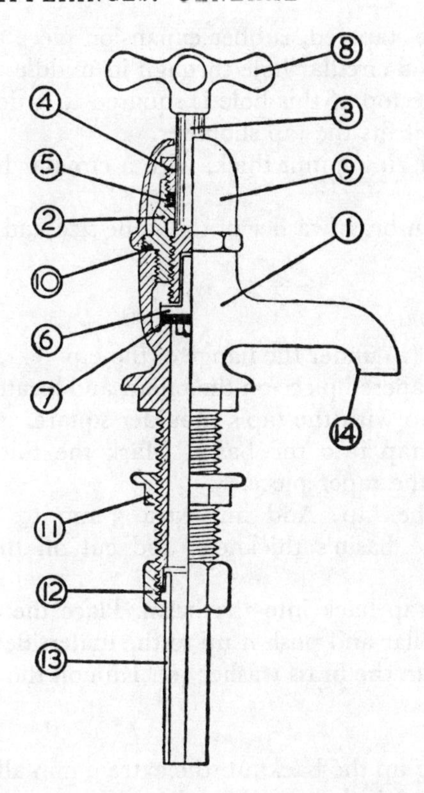

Pillar Tap to B.S.1010

FIG. 69

for bedding material, but all have disadvantages. Plaster and glazier's putty are not particularly good waterproofers. Cement sets rock-hard, and so do red and white or white lead putties after a time, so that it is usually impossible to remove a defective tap from a basin without breaking the basin in the process.

A better method of fixing taps is to use tap-fixing sets made for the purpose. These are obtainable from ironmongers and builders' merchants, and contain:

(a) A 3 mm-thick rubber ring with a square hole to fit the tap shoulder.

(b) A square, tapered, rubber expansion piece about 32 mm long. This has a circular hole through its middle to receive the tap pillar. The top of this hole is squared to a depth of about 3 mm so that it fits the tap shoulder.
(c) A rubber ring 3 mm thick, with a circular hole to fit the tap pillar.
(e) A 1·5 mm brass washer of the same size and shape as the ring (c).

Fixing instructions

1 Place ring (a) under the flange of the tap.
2 Place the tapered piece on the pillar, and locate the squared hole on the top with the tap's shoulder square.
3 Offer the tap into the basin. Mark the thickness of the basin top on the taper piece.
4 Remove the tap. Add an extra 3 mm to the distance marked as the basin's thickness, and cut off the remaining piece.
5 Offer the tap back into the basin. Place the circular ring (c) over the pillar and push it up to the underside of the basin. Follow this with the brass washer (e). Run on the backnut and screw it home.

On screwing up the backnut, the extra 3 mm allowed on the taper piece is pushed upwards and made to expand, filling up the gaps between the tap square and the larger square hole in the basin.

The result—a quickly made, clean, waterproof, positive fixing that can be dismantled at any time without danger of damage to the expensive basin.

Ceramic W.C. pans: BS 1213

A W.C. pan is of little use on its own. To function properly it must have a seat, a flushing cistern and a flush pipe. This complete assembly is known as a W.C. suite, and may be one of three kinds:

High level suite. The height from the floor to the top of the

* 3 mm ≃ ⅛ in

flushing cistern can vary between 1828 mm (6 ft) and 1970 mm (6 ft 6 in). A flush pipe 32 mm in diameter will be used.

Low level suite. The height from the floor to the top of the flushing cistern will be about 914 mm (3 ft) and a flush pipe 38 mm in diameter will be used.

Combination or close-coupled suite. The flushing cistern bolts directly on to a specially designed W.C. pan, and the flush pipe as a separate piece of equipment is no longer necessary.

Sketches of the three types of W.C. suite are shown in Fig. 70. For sketching methods, see Chapter 22.

W.C. pan design

With the 'washdown' W.C. pan the soiled contents are removed by the momentum of the flushing water. In this respect it differs from the 'siphonic' W.C. pan, where the action uses atmospheric pressure to empty the pan. The construction, operation, and advantages of siphonic pans are dealt with later. In the meantime the washdown pan, which is extremely common and quite satisfactory for most domestic purposes, is worthy of closer attention.

The pan should be made in one piece to ensure a hygienic continuity of line and to avoid unnecessary jointing work. Generally pans are in one piece, but it is interesting to notice that two-piece pans are available. With such pans the basin and trap portions are separate, and the trap portion can be turned to discharge in one direction while the basin portion is fixed to face in any other direction that may suit the room. These two-piece or rotatable trap pans can be useful in overcoming a difficult situation, especially in the conversion of old houses where even a one-piece pan with an outgo turning to right or left may not quite meet the need.

W.C. pans are more commonly of the pedestal type shown in Fig. 70, but they may be of the 'corbel' type.

Corbel W.C. pans have stout ceramic projections for building into the wall behind the pan. Alternatively, where the wall is

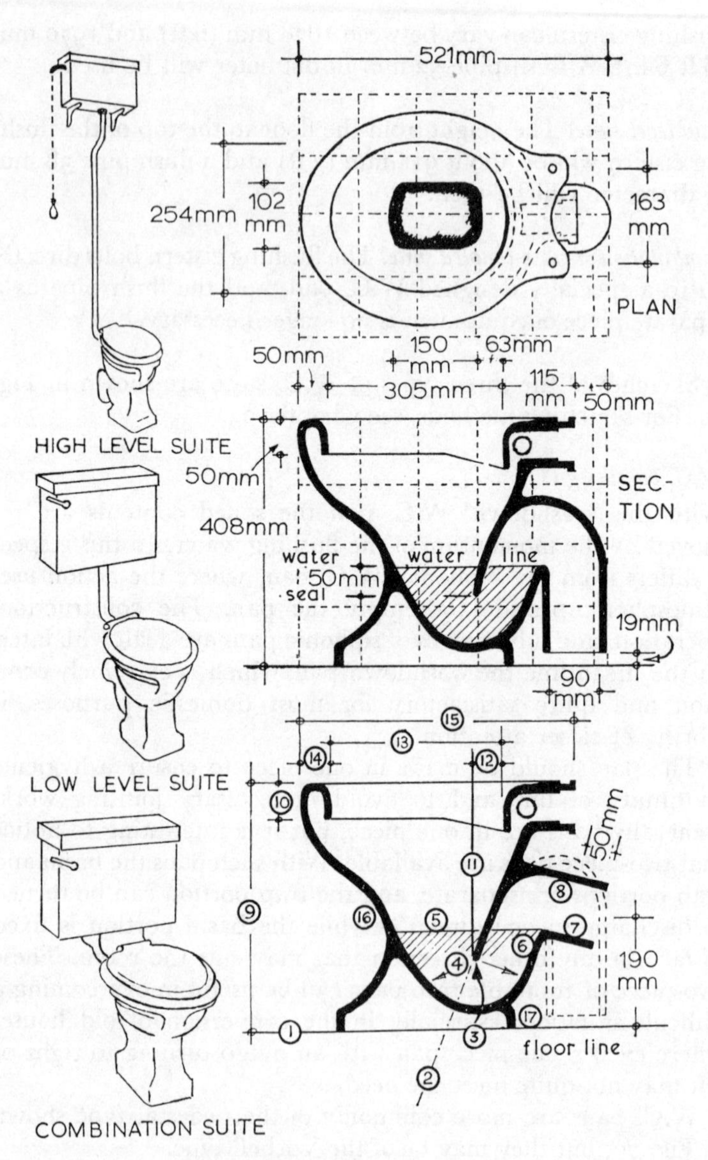
521mm
254mm
102 mm
163 mm
PLAN
50mm
150 mm
63mm
305mm
115 mm
50mm
HIGH LEVEL SUITE
50mm
SECTION
408mm
water seal
50mm
water line
19mm
90 mm
LOW LEVEL SUITE
50mm
190 mm
floor line
COMBINATION SUITE
Washdown WC Pan

FIG. 70

not sufficiently strong, for example if it is a plasterboard partition or a similar lightweight construction, special metal 'chair' brackets may be used to support suitable corbel pans.

The advantage of the corbel pan is that it has no pedestal support and therefore the floor beneath the pan is quite clear of any obstruction. It is obvious that this is a more hygienic arrangement, and for this reason such pans are much used in hospitals.

Whether pedestal or corbel, the bowl or basin of the pan should be large enough and so shaped as to prevent as far as possible any soiling of its inside surfaces or the floor around it.

Pedestal pans may be obtained with traps shaped either like an 'S' or 'P'. The dotted line in the diagram at the bottom of Fig. 70 shows the 'S' trap outgo. It will be seen that the outgo finishes 19 mm above the floor level. If a joint were made at or below floor level, the consequences of an unnoticed leak might be unpleasant.

Siphonic W.C. Pans

The wash-down W.C. pan as described (pages 245 to 247) has long been a first choice for low cost housing. It has the advantages of simplicity in design which makes for a comparatively low cost of production. It also has a simple full bore trap and outgo which practically eliminates all chance of blockage even if the pan is abused.

However, rising standards of living result in prospective house owners wanting comfort and convenience of a higher order than those provided by equipment which, although effective, can be improved upon at some extra cost.

The siphonic W.C. pan is such an appliance. Basically, it differs from the washdown pan which relies upon the momentum of a falling body of water to discharge the soiled contents of the pan, in that the soiled contents are discharged with the assistance of atmospheric pressure—an application of the simple principle of siphonic action.

The extent to which the siphonic pan utilises atmospheric pressure to discharge the soiled content of the pan varies according to the type of siphonic pan used.

Siphonic W.C. pan types may be classified as of the single trap or the two-trap kind.

Single trap or one trap siphonic W.C. pans are illustrated in Fig. 71 (diagrams 1 and 2).

Two trap or double trap siphonic W.C. pans are basically all as shown in Fig. 71 (diagram 3). The only difference likely to be met with is a variation of the means whereby the siphonic action is started by the flush water as it falls down the flush pipes of this particular type of siphonic pan.

The *working* of each kind of siphonic W.C. pan must be understood before critical comparison can be made and an informed choice of pan pattern finally decided upon.

The *bulbous outgo* pattern (Fig. 71, diagram 1) is now virtually obsolete. This type of siphonic pan is still manufactured but is not popular. The reason for this may be that, in order to promote the siphonic evacuation of the pan content, a small outgo diameter has to be incorporated in the design. This diameter is 63 mm (2½ in) which is considerably less than the 90 mm (3½ in) outgo of the washdown (non-siphonic) pan.

The *siphonic action* in the bulbous outgo pan is secured by the restriction of the specially designed outgo and is simply described as follows:

When the pan is flushed the restriction and special shape of the bulbous outgo 'D' causes that part of the branch soil pipe beneath the bulb to become solidly charged with water and the 'pull' exerted by this column of water causes a partial vacuum at 'E'. The atmospheric pressure bearing on water area 'F' then expels the contents of the pan.

Note that siphonic action does *not* occur until most of the flush water has entered the pan and found its way into the branch soil pipe. This will be referred to in the description of the working of the single trap pan (diagram 2) and again in comparison with the working of both of the single trap siphonic pans with the siphonic action generated in the two-trap kind of pan.

In the type 2 pattern, the siphonic action is created, again by some restriction occasioned by the special shape of the outgo portion 'G' as follows:

When the pan is flushed by the passage way 'G' becomes solidly filled with water, the resulting 'pull' creating a partial vacuum at 'H' and causing the atmospheric pressure at 'J' to expel the contents of the pan.

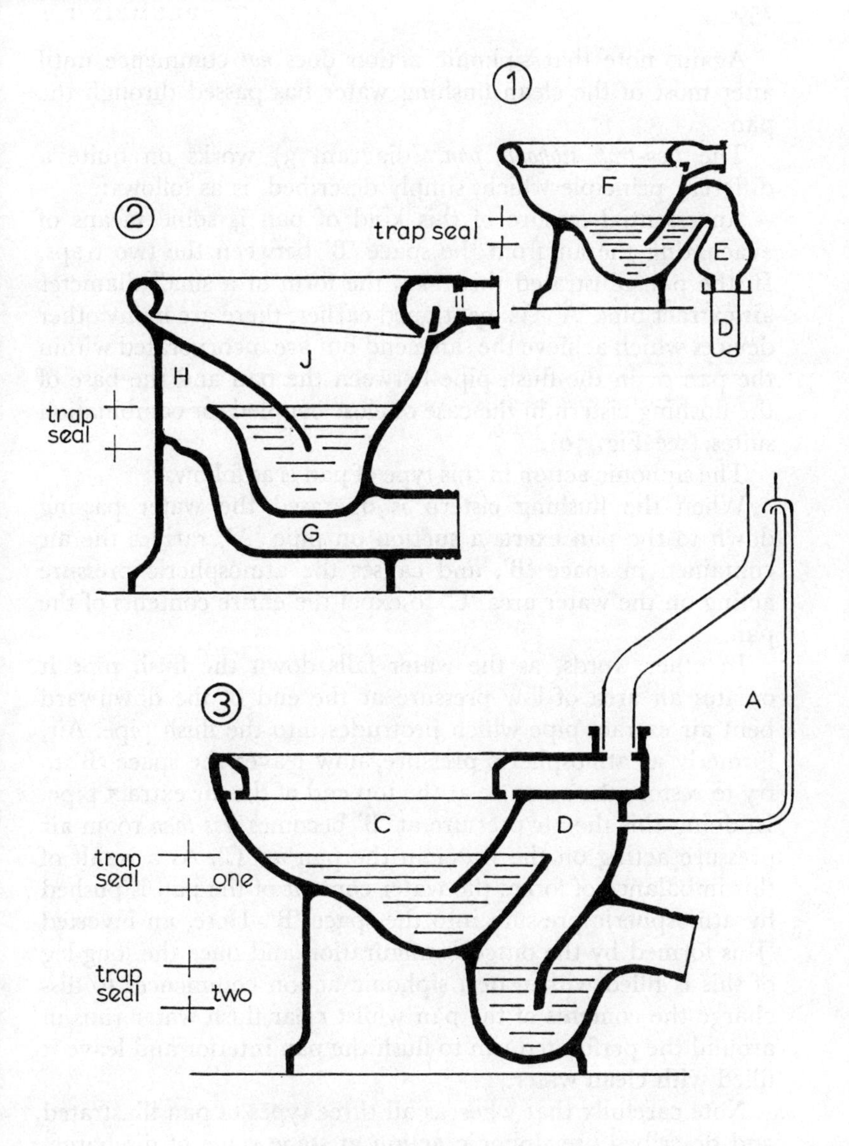

Siphonic W.C. Pans

FIG. 71

Again, note that siphonic action does *not* commence until after most of the clean flushing water has passed through the pan.

The *two-trap siphonic pan*, (diagram 3) works on quite a different principle which, simply described, is as follows:

An essential feature of this kind of pan is some means of evacuating the air from the space 'B' between the two traps. In the pan illustrated this takes the form of a small diameter air extract pipe 'A'. As mentioned earlier, there are many other devices which achieve the same end but are incorporated within the pan or in the flush pipe between the pan and the base of the flushing cistern in the case of close coupled, or combination suites (see Fig. 70).

The siphonic action in this type of pan is as follows:

When the flushing cistern is operated the water passing down to the pan exerts a suction on pipe 'A', rarifies the air contained in space 'B', and causes the atmospheric pressure acting on the water area 'C' to expel the entire contents of the pan.

In other words, as the water falls down the flush pipe it creates an area of low pressure at the end of the downward bent air extract pipe which protrudes into the flush pipe. Air, formerly at atmospheric pressure, now leaves the space 'B' to try to restore the pressure at the top end of the air extract pipe. In doing this the air pressure at 'B' becomes *less than* room air pressure acting on the water in the pan at 'C'. As a result of this imbalance of forces the water content of the pan is pushed by atmospheric pressure into the space 'B'. Here, an inverted 'J' is formed by the outgo configuration and once the long leg of this is filled with water, siphonic action commences to discharge the contents of the pan whilst clear flush water runs in around the perforated rim to flush the pan interior and leave it filled with clean water.

Note carefully that whereas all three types of pan illustrated and described use siphonic action at some stage of discharge, only the two-trap kind starts to evacuate the soiled content of the pan *as soon as* the flush cistern is operated. The other two rely upon the entry of considerable portion of the flush water into the pan before siphonic action occurs. It will be seen, therefore, that the two-trap pan in which there is least mingling of clean flushing water with soiled pan content, gives the most positive and most effective cleansing action with certainty of clean water content left in the replenished first tap of the pan on completion of flush. This results from the different method of starting the siphonic action, as previously described. This kind of pan with its 76 mm diameter outgo is more expensive than other kinds but it may be regarded as the 'Rolls Royce' of W.C. pans and, where comfort, hygiene, convenience and quiet operation take precedence over cost, then this is the kind of pan to recommend.

Features common to all siphonic W.C. pans are:

1 *High flushing efficiency* resulting from atmospheric pressure assistance in discharge. Compare with the noisy momentum discharge method as essential to the washdown W.C. pan design.

2 *Silent in action* because the siphonic action, once started, is motivated by air pressure, not the fall of a large body of water.

3 *Large water area* presented to the user. This effectively reduces the risk of soiling the pan sides. Compare with the restricted water area (150 mm × 100 mm) in the washdown type of pan. The water areas in siphonic pans will vary according to type. For example, Type 1 shown in Fig. 76 would be a little less than Type 2 whereas Type 3 with an oval on plan water area of about 250 mm × 200 mm, offers the biggest area.

You will remember that the comparatively small water area in the wash-down pan is a circumstance of its simple design which uses the force of a falling body of water to cascade on the soiled pan content and physically 'push' it out. Hence, the force of the push must be well directed and the area acted upon be relatively small to take full advantage of it. On the other hand, siphonic pans have their soiled content 'pushed' out by air pressure acting on the surface of the trap water. Hence, the bigger the water area the greater the total pressure of air which can be brought to bear. However, the siphonic pans would work just as well with smaller water areas; it is just that their principle of operation enables a more hygienic retention of a greater water area to reduce risk of fouling the sides of the pan.

It may be argued that the lesser diameter outgoes of siphonic pans could become potential sources of blockage nuisances.

In fact this is not the case. The very powerful 'suction' effect of the siphonic discharge ensures effective scouring of the outgo and positive removal of all the pan content.

Furthermore, the higher cost of the siphonic pan prohibits its common use in lower cost housing and it is not likely to be abused in higher-class dwellings.

The wash-down pan will continue to serve for most general purposes including schools, factories, offices, public conveniences, and so on, but for higher class work where quiet operation is an essential design requirement, the siphonic pan, particularly the two trap kind, will always be a wise specification.

W.C. seats

These may be of polished, non-absorbent wood or of plastic. The latter is more popular nowadays. If solid in construction and not hollow in cross section they are strong and certainly hygienic.

Seats are commonly of the 'ring' type, completely covering the rim of the pan when in the down position. Since W.C.'s in domestic dwellings serve also as urinals there is a risk of fouling the front portion of the seat unless it is lifted when used for this purpose. The 'horseshoe' or 'open ring' seat has its front portion cut away, since it serves no useful purpose, and the seat cannot then be fouled by urine however inconsiderate or careless the user.

W.C. pan fixings

Pedestal pans are secured to the floor by brass wood screws, which pass through holes provided in the pedestal base.

Newly laid timber floors are liable to shrink as they dry, and therefore move. A pan secured to such a floor, and jointed at its outgo by a rigid cement and sand joint, will suffer severe strain if floor movement occurs. Many pans have been cracked and made useless through this bad practice.

Outgo joints for pedestal W.C. pans fixed to timber floors should be 'mastic'; that is, slow- or non-setting. Red and white lead putty serves very well. It offers a waterproof joint material with sufficient flexibility to 'give' until the wood floor has settled. Alternatively, proprietry mastic jointing materials may be used. 'Plumbers Mait', for example. Pans fixed to concrete floors may have similar outgo joints, or they may be jointed by a cement and sand composition in the proportion of one volume of sand to one volume of cement. The use of new cement should be avoided since the heat generated in the chemical reaction which takes place on mixture and setting can give rise to sufficient expansion to damage the vitrified china-ware pan.

In either case, one ring of tarred gaskin should be inserted firmly and deeply into the outgo socket of the soil pipe. This will prevent jointing compound from entering the pipe and possibly causing a blockage.

Modern practice tends towards the application of preformed pan connectors in either plastic or neoprene rubber. These are positive in fit, quick to use, and easily dismountable should need arise. They are, of course, flexible too.

Flush pipe joints are best made with rubber cones specially designed for this purpose. Normally, no additional material is required to make a satisfactory joint.

Traps for sanitary appliances: BS 504 and BS 1184

The chief aims in the drainage of waste from sanitary appliances are the rapid removal of soiled matter and the thorough cleansing of the waste pipe by each discharge of water through it. With proper design, suitable materials and good workmanship this can be achieved, so that the waste drainage is free from persistent bad smells. But, however carefully installed, there will be short periods of time when all waste-drainage systems carry some unpleasant smell because of the very nature of the waste content—for example, of a W.C. pan. Even waste pipes from basins, baths and sinks, which do not carry offensive solid matters, may become coated inside with soap, scum fats, and other deposits which can give rise to unpleasant smells.

To prevent the possibility of any smell entering the building by way of a waste pipe, a trap is fitted as close to the sanitary appliance as possible. A trap is a device holding a quantity of water, which forms a barrier against the passage of air from the waste pipe into the building.

Trap design

A good sanitary trap must fulfil the following conditions:

1 It must be made of non-corrosive material. (A corroded and leaking trap would lose it water barrier and cease to be effective.)
2 It must be strong enough to serve its purpose under all normal conditions.
3 Inside it must be of smooth bore, since this will help to make it self-cleansing.
4 Its size must be suited to the appliance it serves. Too small a trap diameter would slow up the waste discharge. Too large a trap would become coated with deposits likely to smell and thus defeat the object of the trap. Neither would be self-cleansing, since the smaller one would not pass enough water to clean the waste pipe and the larger would not be sufficiently scoured.
5 It must retain a sufficient quantity and depth of water to form a good 'water seal' or air barrier.
6 There must be an easy means of access to the trap so that any blockages that arise can be cleared. This may be done either by providing a cleaning eye or by making the base removable.

Trap types and styles

Fig. 72 on page 255 shows the basic idea in trap design—a pipe bent in the form of a 'U'. This tube-type trap clearly illustrates the principle of the device. The dip tube trap is an adaptation of the U-tube trap, but it is not as efficient as far as self-cleansing is concerned.

Dip-type traps are of two kinds. The 'dip tube' is shown in Fig. C, and the 'dip partition' type in Fig. B. It can be seen that whereas the tube-type trap (A) allows an unrestricted and self-cleansing flow of waste water, the dip partition trap (B)

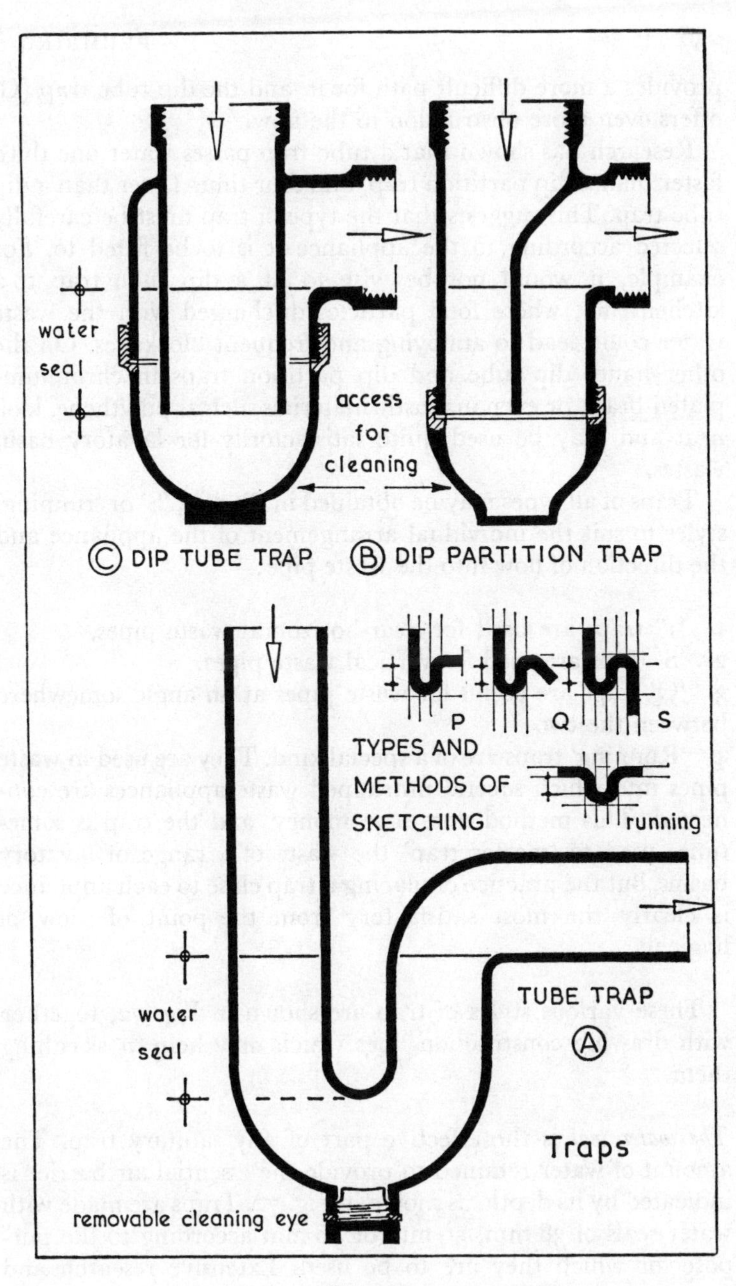
water
seal
access
for
cleaning
C DIP TUBE TRAP
B DIP PARTITION TRAP
P
Q
S
TYPES AND
METHODS OF
SKETCHING
running
water
seal
TUBE TRAP
A
Traps
removable cleaning eye

FIG. 72

provides a more difficult path for it, and the dip tube trap (C) offers even more obstruction to the flow.

Research has shown that a tube trap passes water one third faster than a dip partition trap, and four times faster than a dip tube trap. This suggests that the type of trap must be carefully selected according to the appliance it is to be fitted to. For example, it would not be wise to fit a dip tube trap to a kitchen sink, where food particles discharged with the waste water could lead to annoying and frequent blockages. On the other hand, dip tube and dip partition traps in chromium-plated brass, or even in plastic materials such as polythene, look neat and may be used quite satisfactorily for lavatory basin wastes.

Traps of all types may be obtained in 'P', 'Q', 'S' or 'running' styles to suit the individual arrangement of the appliance and the direction of flow into the waste pipe:

1 'P' traps are used for near-horizontal waste pipes.
2 'S' traps are used for vertical waste pipes.
3 'Q' traps are useful for waste pipes at an angle somewhere between the two.
4 'Running' traps are of a special kind. They are used in waste pipes into which several untrapped waste appliances are connected. This method does save money, and the trap is sometimes used to 'master trap' the waste of a range of lavatory basins. But the practice of placing a trap close to each appliance is clearly the most satisfactory from the point of view of hygiene.

These various styles of trap are shown in Fig. 72, together with drawing construction lines which may help in skeching them.

The water seal is the effective part of any sanitary trap. The amount of water retained to provide the essential air barrier is indicated by its depth, as shown in Fig. 72. Traps are made with water seals of 38 mm, 50 mm or 76 mm according to the purpose for which they are to be used. Extensive research and experiment backed by practical experience has shown that a minimum water seal of 25 mm is an adequate safeguard against the passage of air from the waste to the room in which it is fixed. The depth of water seal for traps to waste sanitary appliances which discharge into a gulley trap outside the building will be 38 mm. On the other hand, those connected direct to a soil pipe will have a 'deep seal' trap, with a water seal of 76 mm.

W.C. pans have traps 'built' into them as part of the pan (see Fig. 70). Such traps are called 'integral', meaning 'part of'. In all cases W.C. pan traps have a water seal depth of 50 mm.

These up-to-date developments in plumbing research and their effect on sanitary design will be fully dealt with in *Plumbing 2 and 3* of this series. For the time being it is sufficient to know and remember the purpose of a trap; how it works, and why; the various types and styles of trap available; and the fact that each sanitary appliance should be fitted with this barrier between the appliance and waste of drainpipes.

The flushing cistern

Most Water Undertaking's bye-laws now prescribe that flushing cisterns should have a capacity of not more than 2 gal (9 l). If the best use is to be made of this water, and the soiled contents of the washdown pan are to be completely removed in one flush of the cistern, care must be taken that flush pipes do not wander too far in the near horizontal direction before reaching the pan. Bends in the flush pipe will also retard the flow of the flush water. Flush pipes should, in other words, be as near vertical and as free from bends as possible. (See Chapter 5, *Plumbing 2*.)

W.C. pan manufacturers are aware of this problem, and take care to design the flushing rim of the pan so that it gives an effective action. As a general rule the pan and cistern which make up a suite are coupled up and tested for proper discharge before they are sent off to the stockist or building site. It may be difficult to get an effective, cleansing flush with unmatched pans and cisterns.

22

Sanitary appliances: sketching methods

It is helpful to have a system or method for sketching. In Chapter 2 a simple method of sketching tools was described and illustrated. Sanitary appliances and many other plumbing fittings are more difficult to sketch than the simple tools, but, with perseverance, one or other of the following methods should prove helpful and produce pleasing results.

Gridding

This method can be used when a picture of the object to be sketched is ready to hand. The picture is covered with a 'grid' of squares of any convenient size. If the grid cannot be drawn directly on to the picture, then squares drawn on tracing paper placed over the picture will do just as well.

Suppose that a small picture of a W.C. pan is available (Fig. 70), and that another, larger, sketch of it is required. First, the small picture is 'gridded' with squares. Then a square or shape large enough to contain the sketch is drawn on the sketching paper and gridded with the same number of squares as was the original, smaller, picture. Clearly, the squares in the larger grid will be bigger than the squares on the small picture, but they will be bigger *in proportion*. Now, if lines are drawn in the bigger squares just as they appear in the ones gridding the original picture, a larger, *proportionate*, sketch will result.

In a like manner a larger picture can be reduced to a smaller sketch simply by gridding the large picture and transferring the picture freehand to the smaller grids on the sketch paper, which has already been prepared to the required size.

Proportion and outstanding dimensions

Obviously the 'grid' method can only be used when a picture of the object is available for gridding and copying from. This is not always convenient or even possible. In examinations, for instance, one is not allowed to have pictures or books of reference. In this and many other cases; for example, when sketching an object as it stands—some method other than 'gridding' must be used. In such cases, the first step is to gain a sense of proportion—to get the height, width, and so on correctly related one to the other. The next step is to obtain by measurement, or from memory, the most notable dimensions of the object to be sketched.

When there is an actual object to examine, these salient dimensions can be measured directly with a rule. If the object is not in a position to be measured, its proportions can be gauged by eye. It might be clear that the length is just about twice the height, for example. This easily seen fact can just as easily be transferred to the paper, and a good *proportionate* sketch will result.

But, once again, if objects are not available—for example in an examination room—what then? In such cases one has to create a mental 'picture' of the object, and from this picture make a proportionate sketch. Sometimes a sketch can be built up from dimensions remembered from past experience or from past practice in sketching similar objects.

Examples of both methods are shown in Fig. 70. The gridded example has already been dealt with, but the proportionate and outstanding dimensions method, shown at the bottom of the figure, needs some explanation. The full steps suggested for the first trial run are set out below. With practice, and as memory and skill improve, many of these steps can be left out until eventually a W.C. pan can be drawn without any effort at all. Both methods are applicable to all forms of sketching likely to be encountered in plumbing work.

Note: The numbered sequence of steps below are also to be found on the diagram, Fig. 70.

1 Draw the floor-line.

2 Draw the slope-line (about 104°) to represent back of pan.
3 Mark distance from floor about 127 mm up slope-line.
4 From this mark (3) draw a semi-circle 90 mm in diameter to form the bottom of the pan trap.
5 Draw a horizontal line 50 mm above mark (3) to show the water level in the pan, and to show the trap seal, which has a depth of 50 mm.
6 Continue the trap outgo parallel to the slope-line (2).
7 Turn the outgo where line (6) cuts water-line (5).
8 Draw the top of the outgo 90 mm away, and parallel to line (7).
9 Mark the height of pan from the floor to the top of the rim and draw a horizontal line to represent the rim.
10 Mark out the depth of the flushing rim—50 mm—and draw a horizontal line to show this.
11 Set up a vertical line from mark (3).
12 Mark a point 63 mm to the right of line (11) at rim-level.
13 From this point (12) measure 305 mm to the left to represent the length of the bowl opening at the top of the pan.
14 Add 50 mm for the flushing rim at the front.
15 Mark in overall front to back length of the pan top (front edge to end of flush pipe inlet).
16 Draw freehand curves to complete the front of bowl and pedestal.
17 Add the back of pedestal—and the sketch is complete.

These dimensions are actual pan sizes. In your sketch they can be reduced to any scale.

Should this sequence of 'steps' seem too laborious just try it once and then judge by the result. If faithfully followed it will certainly bring success and the pleasure that comes of success. Remember that with practice the number of steps can be reduced until they are no longer necessary as aids to sketching.

Remember, also, that it has been said that sketching is the language of the workshop. This is very true, and all the diagrams in this book of sinks, lavatory basins and traps offer useful material for practice.

23

Domestic water supply: pipes

The installation of a domestic water supply involves a great deal more than just running a water-pipe into the building and on to the various taps. The right kind of pipe must be chosen and fixed in such a way that it cannot be damaged by ordinary wear and tear. Pipes must be so placed that they are not likely to freeze or burst in cold weather, and they must be securely fixed so that they will not vibrate and thus cause an irritating noise. Above all, there must be no chance of the water being wasted, misused or contaminated once it has entered the pipework system of the building.

Regulations

The model water bye-laws 1966 of the Ministry of Housing and Local Government give guidance about the materials which can be used for water services, and the way they should be installed for satisfactory service.

The Local Water Undertaking's bye-laws may be a copy of the model bye-laws, if these have been adopted by the Local Authority and published under its own seal and title.

Local authorities do, however, have the right to make their own bye-laws or amend the model bye-laws to suit the particular needs of their area. If they have done so, and the bye-laws have been approved by the Ministry of Housing and Local Government, then it is these bye-laws alone that are legally enforceable in the district, and they must be worked to regardless of the model bye-laws.

It is important, therefore, before going on with any design or work connected with water services, to find out just what are the bye-laws of the Water Undertaking in the area. Copies of these are available at the offices of the Water Undertaking.

Materials for pipework

Pipes are needed to convey water from the main in the roadway to within the building; for distributing water throughout the building; for the discharge of water from overflowing cisterns by way of 'warning pipes'; and for flushing pipes which convey flush water to sanitary fitments such as W.C.s, urinals and slop sinks. Pipes are also needed for the conveyance of gas and the disposal of soiled waste waters, and these will be dealt with under separate headings.

The basic requirements of a pipe are that it should be durable, capable of withstanding the pressures likely to be imposed on it, easily jointed, and neatly and simply fixed. For potable waters; that is, water for drinking or cooking, the pipe must be non-toxic. That is, it must not harmfully affect the purity of the water passing through it.

Materials commonly used for this work include lead, copper and mild steel. Plastics materials such as polythene and rigid polyvinyl chloride tubes have only recently been used in the cold-water services. Because of its high resistance to corrosion, polythene tube is being extensively used above and below ground in districts where the soil conditions or the water conveyed are known to be corrosive. In such cases, polythene's property of resisting corrosion means that it is chosen although its mechanical properties may be less good than those of some metal pipes.

Bye-laws governing water services in areas not so affected by corrosive conditions generally include provision for materials other than lead, copper or mild steel. Such provisions usually insist that the material shall be 'suitable and of sufficient strength to withstand not less than double the pressure to which the pipe would be subjected under working conditions'. It should be noted that the Water Undertaking has the right to decide what is 'suitable,' and this fact further emphasises the wisdom of seeking the Undertaking's advice before proceeding with any material other than those known to satisfy the requirements of its bye-laws.

Lead pipes are generally accepted by Water Undertakings for use on water services provided that their quality and strength comply with the requirements of BS 602.

Lead pipe is made by what is known as an extrusion process. Molten lead is placed in a hydraulic press, pressure is applied, and the lead is extruded or squeezed out in pipe form between a die which shapes its outer diameter and a mandrel which determines its inner diameter. The low mechanical strength of lead demands that pipes of lead be made with thicker walls than either copper or mild steel pipes designed for the same working pressures. The resulting larger outside diameters of lead pipes are sometimes thought to be ugly in comparison with the others. Lead is, however, durable; easily manipulated, especially in awkward places; easily jointed; and resistant to corrosion in normal soils and waters. These properties will often outweigh superficial criticism of its bulk, particularly when it is to be used underground or above ground in ducted work or awkward situations—for example, behind panelled baths.

*Lead pipe is manufactured to BS requirements which lay down the purity of composition, the truth of cylindrical bore, the uniformity of wall thickness and freedom from defects or surface blemishes. All BS lead pipe up to 2 in (50 mm) bore diameter is marked on a raised ribbon throughout its length with the BS number, the internal diameter of the pipe, and the weight/yd.

Specifications, or descriptions of lead pipe sizes for water services, quote the internal diameter of the pipe and its weight/yd. For example, BS 602 $\frac{3}{4}$ in × 11 lb indicates ordinary lead pipe for underground water services, which has a diameter of $\frac{3}{4}$ in and weights 11 lb/yd. As specified, this pipe would be suitable for pressures not exceeding 108 lb/sq. in (744·5 kN/m^2) = (7·445 say 7·5 bar).

Warning pipes used to indicate waste of water through a defect or misuses of fittings do not have to withstand pressures. Much lighter pipes are therefore installed for the purpose. A $\frac{3}{4}$ in warning pipe might be specified as BS 602 $\frac{3}{4}$ in × 4 lb, and, similarly, flush pipes in lead might be specified as BS 602 1$\frac{1}{4}$ in × 6 lb or BS 602 1$\frac{1}{2}$ in × 7 lb (38 mm × 3·5 kg/m).

* See chapter entitled 'The International Metric System' and worked example in chapter entitled 'Plumbing Calculations'.

Silver-copper-lead alloy pipes to BS 1085 are generally acceptable and have all the properties of the ordinary lead pipe. They also possess a greatly improved mechanical strength which permits them to be made in lighter wall thicknesses so that they look neater and more economical in material. The working and jointing of this alloy, which is a mixture of lead with very small amounts of silver and copper, are the same as for ordinary BS 602 lead pipe.

A service pipe is one which is subject to the pressure of the Water Undertaking's supply main, or would be if the service pipe stop taps were open.

A distributing pipe conveys water to taps from a storage cistern within the building and is subject only to the head pressure imposed by the water in the cistern. On page 54 'head' pressure was defined as the weight of water column measured vertically from the water level in the storage cistern to any point below, in this case the lowest part of the distributing pipe.

Head pressure may be given in m head or kN/m^2.

Alternatively, head pressure might be expressed in *bars* (1 bar = 100 kN/m^2).

1 m head equals 9·8 kN/m^2.

$$\frac{\text{Intensity of pressure } (kN/m^2)}{9{\cdot}8\ kN/m^2} = \text{head in metres}$$

Light gauge copper tubes are generally accepted for water services fixed above ground if their quality, gauge, and dimensions meet the requirements of BS 659. Their dimensions are made to very fine engineering tolerances to ensure not only that the tubes have a uniform gauge, but also that they will accept a wide variety of fittings by different manufacturers, all of which have tube housings with exactly the same internal diameter. A badly made tube could not be used with these standard-sized fittings. Light gauge copper tubes are durable, neat in appearance, and of adequate strength for all normal purposes.

TABLE 12 RECOMMENDED MINIMUM WEIGHTS OF LEAD PIPE FOR VARIOUS USES

Position, pressure and type of pipe		*Nominal internal diameter*				
		Minimum weights in lb/yd run				
		½ in	¾ in	1 in	1¼ in	1½ in
Service pipes buried underground						
Up to 150 ft head: 65 lb/sq. in	BS 602	6	9	13	16	20
	BS 1085	6	9	13	14	16
Up to 250 ft head: 108 lb/sq. in	BS 602	7	11	16	28	—
	BS 1085	6	9	13	21	—
Up to 350 ft head: 152 lb/sq. in	BS 602	9	15	21	—	—
Service pipes fixed above ground						
Up to 150 ft. head	BS 602	4	5	8	14	20
	BS 1085	4	5	7	10	15
Up to 250 ft head	BS 602	5	11	16	28	—
	BS 1085	4	7	13	21	—
Up to 350 ft head	BS 602	9	15	21	—	—
	BS 1085	6	12	21	—	—
Distributing pipes fixed above ground						
Cold water services						
Up to 75 ft head	BS 602	4	5	7	9	—
	BS 1085	4	5	7	9	—
Up to 150 ft head	BS 602	4	5	8	12	—
	BS 1085	4	5	7	10	—
Hot water services						
Up to 60 ft head	BS 602	4	5	8	12	—
	BS 1085	4	5	8	12	—
Up to 120 ft head	BS 602	5	10	18	28	—
	BS 1085	4	9	16	23	—
Flushing and warning pipes	BS 602	—	4	5	6	7

See also British Standards 602 and 1085.

Consult Local Water Undertaking's requirements. They may differ in some cases from these recommendations.

Metrication awaiting the revision of BS 602.

Conversions: in × 25·4 = mm.
lbs/yd run × 0·45 kg/lb × 1·094 yd/m run = kg/m run.

Their qualities of lightness and rigidity, and the fact that they are easy to bend, joint and fix, all combine to make them quick, efficient and economic to install in above ground water services, both hot and cold.

The tensile strength of L.G. copper tube is very high, and though they may have quite thin walls, tubes of up to 50 mm nominal bore will withstand working pressures of up to

1400 kN/m^2(= 14 bar), a pressure unlikely to be exceeded in normal building practice.

The tubes are manufactured by a cold-drawing process. A cylindrical billet of copper is heated and formed into a tube shell much larger in diameter and wall thickness than the finished tube is required to be. The shell is allowed to cool, and one end is then reduced so that it can enter a cold-drawing machine. The shell is drawn through a die. This drawing process stretches the tube and thus reduces its diameter. The process is repeated, using a smaller die each time, until a tube of the required diameter and wall thickness, or gauge, is obtained. The finished tubes are carefully examined,, straightened if necessary, and tested for bore truth by passing steel balls through the tubes. They are then cut to random lengths of 5·5 to 6 m, bundled, and stacked for delivery.

It is interesting to notice how this cold-working affects the physical properties of the copper and causes it to become work-hardened. You will remember that work-hardening increases the tensile strength of a material but reduces its ductility or 'stretchability'. During the cold-drawing process the part-worked tubes are annealed or restored to a soft state by the application of heat. L.G. copper tubes are finished in ½ hard temper.

L.G. copper tube's well-known property of rigidity is developed during the carefully controlled process of manufacture. The final cold draws are allowed to impart the right degree of hardness to give the tube rigidity and improve its strength.

Lighter gauge tubes having thinner walls than light gauge tubes to Table I BS 659 offer economy in cost due to the reduced weight per metre run of tube. These copper tubes to Table III BS 659 are adequately strong and may be bent with springs of suitable o.d. or they may be bent in an adjustable bending machine. The reduced wall thickness results in a slightly increased bore diameter, hence the need for a special spring for this kind of tube.

Lighter gauge tubes may be jointed all as described for light gauge tubes on page 281, but note, if 'Kingley' swaged compression fittings are used do make sure that you use a swaging tool designed for Table III tube.

Hard drawn thin wall copper tube to BS 3932 is another economy form of copper tube. Unlike Table III tube which is ½ hard temper and may be bent, hard drawn thin wall tube cannot be bent. Fittings for this tube must be of the solder capillary or of the non-manipulative compression kind (see Fig. 76).

Copper tubes to be buried underground are generally accepted for water services if they comply with the requirements of BS 1386. These tubes are made by the same drawing process as L.G. copper tubes, but longer drawbenches are used so that tubes up to 36 m in length can be drawn. The tube is supplied in 18 m or 36 m coils which are easily manipulated and can be quickly installed, with a minimum of joints below ground.

For underground water services BS 1386 copper tube is used in 'soft' temper, and one of the final processes of manufacture is the annealing process which makes it so.

The outside diameters of BS 1386 and BS 659 copper tubes are the same, so that standard copper tube fittings can be used with either. You may have noticed some slight variation in the inside diameters of these tubes. The BS 1386 has a nominal bore of 13 mm, whereas the BS 659 tube bore is slightly larger. The wall thickness, or gauge of the soft BS 1386, is accordingly greater than that of the ½ hard BS 659 tube, and this extra thickness makes the strength of the two in resisting bursts comparable.

Mild steel tube is generally included in the list of materials suitable for water services, but it does not find universal favour with water-supply engineers for use in cold-water services. It is sometimes used, where the Water Undertaking permits, because it is said to be cheaper than copper or lead, but careful comparison of the cost of labour and material will show that in any sizeable installation mild steel tube is no cheaper, and is indeed often more expensive, than other materials.

The durability of mild steel is not equal to that of copper or lead, and the bye-laws insist that every mild steel tube, and every fitting used in its jointing, shall be protected against corrosion. Generally, galvanised tubes and fittings are adopted to comply with this requirement.

Galvanised mild steel tubes are extensively used in the domestic hot-water services where rigidity and mechanical strength are an advantage. Where used for water services, whether hot or cold, galvanised mild steel tubes must comply with BS 1387 and be of what is known as medium-quality grade. Where used for underground cold-water services, galvanised tubes to the same BS must be of heavy quality.

Mild steel tubes are colour coded by 50 mm wide painted bands. Brown indicates *light* tube as might be used for gas installations in its black (ungalvanised) form. Blue indicates *medium* tube as might be used for central heating work in its ungalvanised form or for above ground work where galvanised. Red indicates *heavy* tube as would be used for steam in ungalvanised form or for underground service pipes in its galvanised form.

M.S. tubes are manufactured from strips or ribbons of mild steel bar heated and shaped on the tube-making machine. The strip edges are prepared and butted, and sometimes lapped, for the automatic seam-welding operation. The process is continuous and all operations, from heating the strip to cutting the final tube to length, are done automatically by the tube-making machine.

Stainless steel tube used for domestic hot and cold water installations, is of the austenitic type. This is commonly referred to as 18/8 stainless steel and indicates a composition of 18% chromium and 8% nickel. Stainless steel sinks and hot store cylinders are made from the same material.

This new form of water pipe is covered by BS 4127: 1967 'Light gauge stainless steel tubes' with outside diameters and tolerances of the tubes being all as for light gauge copper tubes to BS 659.

The minimum tensile strength of S.S. tube is 508 Mega Newtons/m^2 which, compared with copper at 254 MN/m^2, itself very strong, is a measure of the strength of S.S. tube and its ability to withstand very high pressures before bursting. Pressures in 15 mm o.d. tube of over 55 MN/m^2 have been recorded without damage to the S.S. tube.

The tube is sufficiently ductile to withstand 'belling' although present tube jointing techniques employ either solder capillary fittings or non-manipulative compression fittings as made for, and used with, BS 659 copper tubes (see Fig. 76).

However, the ductility of S.S. tube can be used to advantage in conjunction with 'Kingley' fittings which cannot be pulled off the tube. This form of joint incorporates no loose rings but has a swage formed in the pipe end by a simple tool provided by the makers of the joint (see Fig. 76, Type B Manipulative fitting).

When solder capillary fittings are used with S.S. tube, it is important to have the correct flux. This is obtainable from stockists of the tube and is specially prepared to suit both S.S. and copper or brass solder capillary fittings. One other important point to bear in mind when using solder capillary joints on S.S. tube is that the flame should be applied to the fitting and *not* to the tube. Surplus flux should be carefully and completely removed on the completion of the joint.

Stainless steel tube may be bent with a spring in the 15 mm o.d. and 22 mm o.d. sizes but care in spring selection is advised. A spring suitable for thin walled copper tube to BS 3932 is necessary. This is on account of the larger bore resulting from the thin wall thickness of S.S. tube.

Machine bending, using conventional light gauge copper tube benders is the best and is recommended for 22 mm o.d. tube unless only easy radius bends are required. Again, because the tube is made of thinner gauge material than BS 659 Table I copper, it is necessary on conventional bending machines to bring the roller nearer to the sliding shoe back former as one has to when using BS 659 Table III copper tube which also has thinner walls than Table I copper. Fixings for stainless steel tube are similar to those as described for copper tube on page 292.

The Stainless Steel Development Association, of 7 Old Park Lane, London W.1, publishes a booklet, *Stainless Steel for Domestic Water Services*, and this is available free.

Polythene tubes manufactured to BS 1972 are not at present generally accepted by all Water Undertakings for use above or below ground. Moreover, different Water Undertakings have different regulations concerning their use. It is advisable to enquire whether the particular district demands heavy gauge tubing or normal gauge; if indeed either is permitted. As has been said, in some districts corrosive soil conditions or the corrosive nature of the water have meant that polythene tube is used because of its resistance to corrosion and its non-toxic properties. Polythene cannot be used for sustained flows of hot water, and so it cannot be used for domestic hot-water pipes.

Polythene is manufactured from ethylene gas, which is obtained from coal or crude petroleum oils. Heat and pressure are applied to the gas, bringing about a physical change, and raw polythene is produced. This is then granulated into small chips, and melted in a hydraulic press. The molten material is extruded in the form of pipe. Polythene tube is available in two wall thicknesses or gauges for water services. Its light weight means that it can be made in easily transportable coils, up to 150 m in length.

The material has low mechanical strength. It has a high coefficient of expansion (see Table 7, page 92). It tends to 'creep', and in consequence its fixing requires extreme care. It is a non-conductor of electricity and cannot be used as an earth for electrical appliances as the metal water service pipes which rise from the earth usually are. Not that the Water Undertakings would object to this, but if a short polythene 'insert' were used to repair an existing metal rising main, the continuity of the conduction of electricity along that pipe would be broken. An electrical fault in an appliance 'earthed' to the metal pipe above the polythene insert could mean that it remained electrically 'alive', with possibly dangerous consequences.

However, polythene has many advantages apart from its resistance to corrosion which make it very suitable for the purposes already mentioned. It is also used for waste pipes from laboratory sinks, and it is becoming more and more popular for this sort of purpose.

Polythene is elastic and is a poor conductor of heat. These two properties make it resistant to frost bursts, because the tube does not lose heat as quickly as does a metal pipe, and therefore freezing is delayed; and because, if freezing does occur, the pipe will stretch, without bursting, to accommodate the expansion of the frozen water. When the ice thaws, the elasticity of the tube makes it possible for it to resume its normal bore and shape.

The following table indicates some of the important details of tube to BS 1972, as amended 1961.

Nominal bore	*Nominal outside diameter (inches)*	*Approx weight per 100 ft*	*lb/per sq. in working pressure at 20°C*
		Normal gauge tube	
½ in	0·686	7 lb	130
¾ in	1·00	14 lb	115
1 in	1·25	175·5 lb	90
		Heavy gauge tube	
½ in	0·840	15 lb	210
¾ in	1·096	20 lb	150
1 in	1·346	25·5 lb	120

It is light (S.G. 0·9) and can be obtained in coiled lengths of up to 150 m. This, together with its resistance to frost burst, makes it very suitable for installation in isolated farm buildings or cattle troughs by mole plough.

BS 1972 awaits metrication at time of going to press. Nominal bore sizes will be given in inches. o.d. sizes will be given in metric and working pressures will given in *bars* (1 bar = 100 kN/m^2). See examples 11, 12, and 14 in chapter entitled 'Plumbing Calculations'.

Mole ploughs

A mole plough is a device which bores a tunnel under the ground, through which polythene and BS 1386 copper tubes can be drawn, thus saving the time and cost of digging pipe trenches. The plough has a vertical blade which slices through the ground, and at the bottom of this blade is fixed a pointed cylindrical 'mole' which bores the tunnel.

A starting hole is dug, and the plough blade, adjusted to bore at the required depth, is lowered into the hole. The plough, which travels on its own wheels, is then coupled to a farm tractor, the tube to be laid is attached to the mole, and as the plough is pulled along the mole bores the tunnel, drawing the tube after it. Intermediate holes have to be dug according to the length of the tube, to allow another tube to be connected to the mole. They also make it possible for one to inspect the couplings when the water is turned on, to make sure they are watertight.

Warning. Unless you wish to see your newly laid polythene tube disappearing back into its hole like some frightened worm, you will make allowance for the fact that it is stretched by the 'drag' imposed by long pulls. 230 mm at the end of each length is not too much to allow for the tube's elastic contraction.

24

Domestic water supply: joints for pipework

Joints for lead pipes

The bye-laws require that joints in lead pipe shall be of the kind known as the wiped soldered joint, or some other joint equally good. The wiped joint is the most common, but the soldered spigot joint, which is very economical on solder, is accepted by many Water Undertakings for use above ground. It is very useful for bath and basin tap connections to lead pipe.

The wiped soldered joint is made by preparing the ends of the pipe, fitting them carefully together, fluxing the prepared joint area with tallow, and heating it with a blowlamp until it will accept solder, and 'tin'. Sufficient solder is then added to make the joint. This solder mass is heated until it melts and runs off into a 'catch cloth' of moleskin, held underneath the joint. The molten solder is then quickly transferred to the pipe-joint area, and a moleskin 'wiping cloth' is used to shape and wipe the joint to the proper form while it is still a pasty mass. This is a skilled operation, and can only be learnt with practice.

Fig. 73 shows three applications of a plumber's wiped soldered joint.

Tools required:

1 Mandrel. A cylindrical boxwood tool used to restore the cylindrical bore of the larger thin-walled lead waste pipes. The mandrel is given a light smear of tallow to lubricate its passage through the pipe. It is inserted into the pipe and driven through with a drive stick. The drive stick might be of wood or it might be a piece of gas barrel one end of which is threaded. On this end is screwed a socket filled and headed with lead; this will protect the wooden mandrel from damage. Alternatively, a

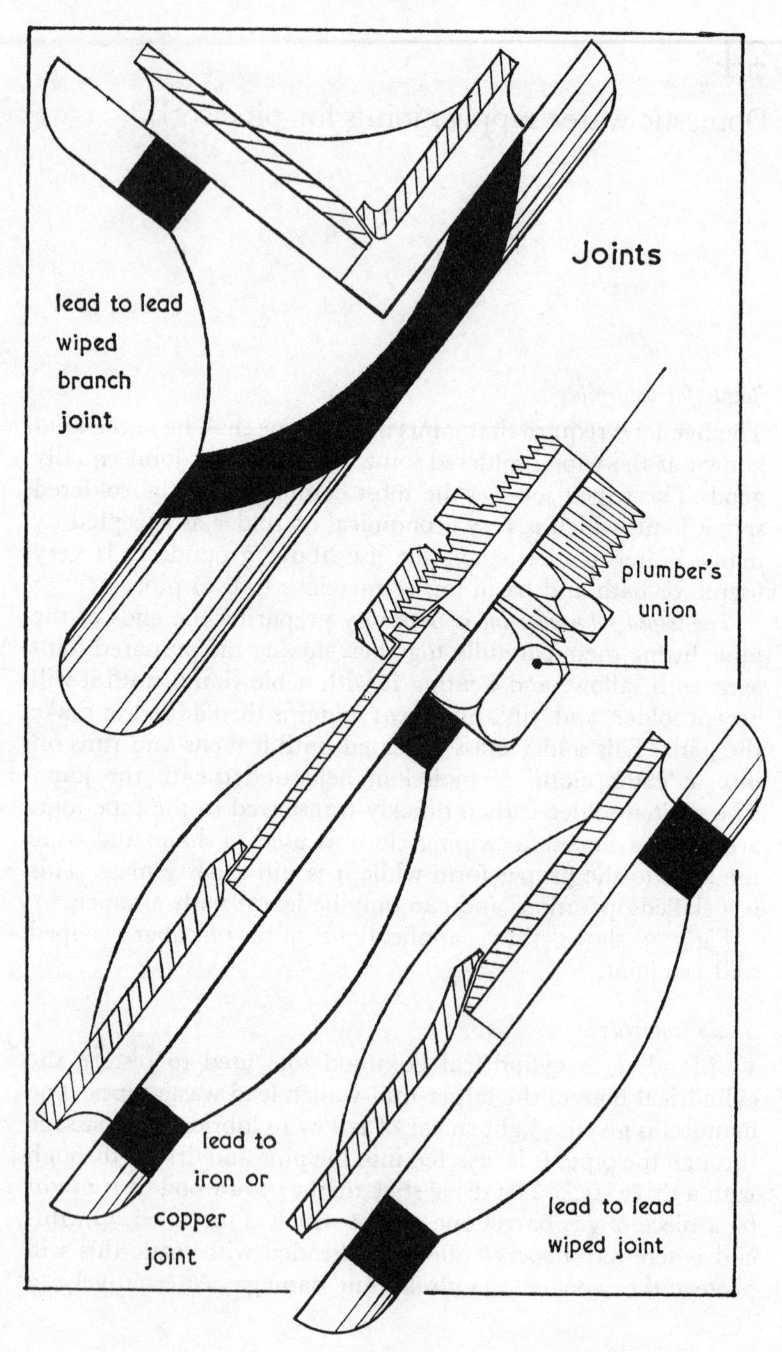
Joints
lead to lead
wiped
branch
joint
plumber's
union
lead to
iron or
copper
joint
lead to lead
wiped joint

FIG. 73

cap could be used in place of the lead-filled socket, and a piece of rag or cotton waste placed just behind the mandrel so as to protect it from the blows of the metal cap. A mandrel is not necessary for lead pipes used for water services since the wall thickness of these is such that the bore seldom becomes flattened or distorted.

2 Dresser. This would be used in conjunction with the mandrel to dress the external wall surface of the pipes.

3 Saw. To cut the pipe to length.

4 Rasp. To square the pipe end, and to taper the spigot end so that it fits snugly into the 'bell' made by the tanpin in the end of the socket pipe (see Fig. 74).

5 Tanpin, or turnpin. This is a boxwood tool used to make a 'bell' on the socket end of the pipe. It is a simple cone-shaped tool, and is available in a wide variety of sizes to suit the many different sizes of lead pipe used in plumbing.

6 Cardwire. This is used to scratch clean the surface of the lead pipe before plumber's black is applied. Lead is naturally a greasy material, and the plumber's black, which is a water mixture, will not readily take to greasy surfaces.

7 Chalk or powdered chalk. This is applied to the cardwired surfaces in order to absorb any traces of grease that might be left.

8 Plumber's black, soil, or tarnish. This is painted on to the pipe over and well beyond the joint area. Solder will not stick to these blacked or 'soiled' areas, so when the actual joint area has been cleaned of plumber's black, the edge of that which is left defines the joint area.

9 Shave-hook. This is used to shave clean the areas to be soldered. The prepared 'bell' end of the pipe and the rasped tapered end are carefully fitted together and the length of the wiped joint is marked. The ends are then dismantled and a pair of dividers or a pair of compasses are used to mark the length to be shaved on each pipe end. This is done by rotating the pipe and allowing the point of the adjusted compasses to mark round the pipe wall.

10 Dividers or compasses are needed for the process just described.

11 Small bossing mallet or hammer. When the joint has been

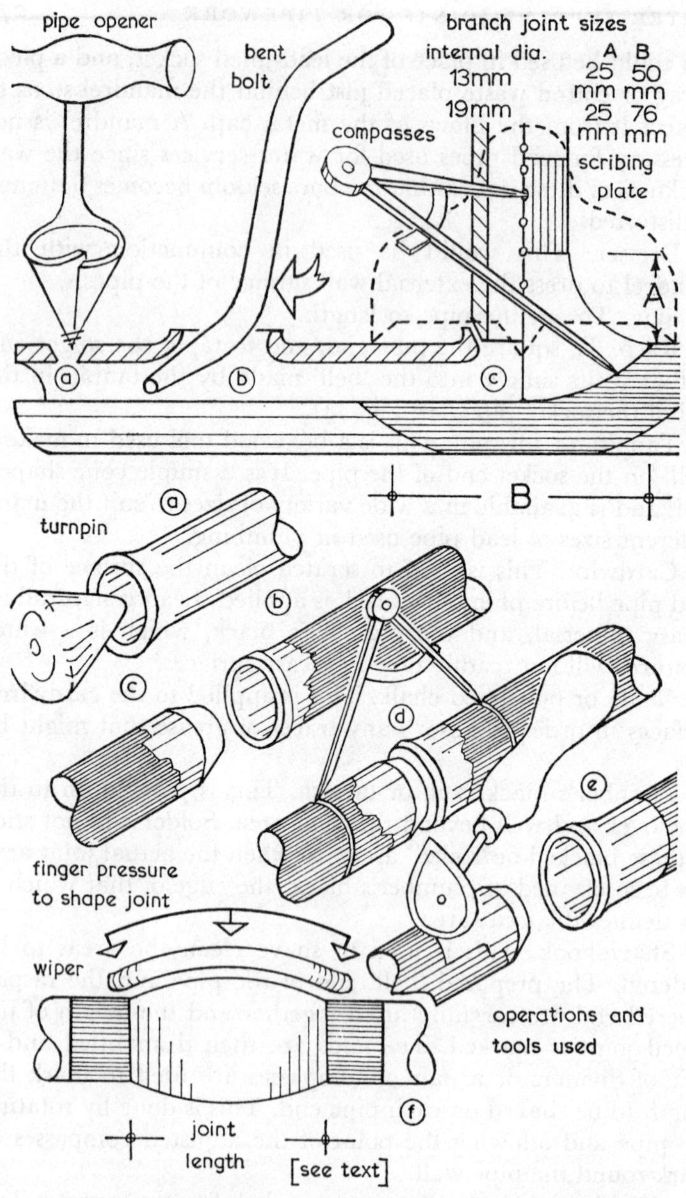

The Preparation of Wiped Solder Joints

FIG. 74

prepared and shaved ready for wiping, the two ends can be finally mated together. The 'bell' of the socket pipe is dressed neatly and tightly to the wall of the spigot pipe so as to prevent solder from entering the pipe bore during the joint-wiping operation.

12 Pipe fixings. These may be special devices purchased from the tool stockists for the purpose, or they may be an improvised fixing. It is important that adequate fixing be provided before, during, and after the joint has been wiped, until it finally sets solid.

13 Moleskin wiping cloths. These cloths are made of a tough material with 'ribs' not unlike the ribs in corduroy, and they are used to manipulate the molten solder. They can be made from moleskin cloth bought by the yard or they can be bought ready-made from tool stockists. Moleskin cloths should be treated with great care, and are best kept in a tin or bag, not thrown carelessly into the tool-bag along with all the other pieces of equipment. When a wiper is getting nicely 'broken in' it is also getting worn out. It is a good idea to have several wipers in different stages of breaking in. It is best to have a new one which you can start breaking in on less important work, and one well broken in ready to replace the wiper which is best at the moment but will soon become hopelessly worn out.

14 Blowlamp. Generally these are of the type which use paraffin fuel, but butane gas blowlamps are now becoming popular since they are easier to light, are clean to use, and more economical in time.

TABLE 13 JOINT LENGTHS—WIPED SOLDERED JOINTS

Internal pipe diameter	*Joint length*	*Wiping cloth width*
13 mm and 19 mm	70 mm	76 mm
25 mm and 32 mm	76 mm	90 mm
38 mm and 50 mm	82 mm	90 mm or 100 mm

* mm ÷ 25·4 = inches

Branch joints in lead pipe require much the same technique as the wiped solder joints which joint lead pipes end to end. Again, practice in preparation and wiping is the only way to gain the necessary skill, but Fig. 74 shows the important features of a branch joint. They are as follows:

1 The bossed branch opening in the main pipe must be carefully worked so as to avoid obstruction in the main pipeline.
2 The branch entry must not enter or obstruct the main pipeline.
3 The 'cup' to receive the branch should be worked well up to ensure a good entry for the branch pipe.

Tools required:

All the tools listed for the preparation and wiping of the underhand or the upright wiped plumber's joint will also be used for branch joints. The following additional tools will be needed:

1 Pipe-opening tool. This is sometimes called a 'scallope'. It resembles a gimlet used for boring holes in wood except that its end flares out to form a wide cutting edge. The leading 'gimlet' screw enters the wall of the lead pipe followed by the flared part of the tool, which scoops out a hole large enough for the next operation. If a pipe-opening tool is not quickly available, then a hole drilled in the pipe wall with a 8 mm bit in an ordinary engineer's drill will serve the purpose quite well. Take care not to let the bit ride right through the pipe wall and across so that it cuts the inside of the pipe wall opposite.
2 Bent bolt. This is a short, round steel bar about 13 mm in diameter. One end of the bent bolt is tapered down to a diameter of about 6 mm. Both ends are bent a little—hence the name 'bent bolt'. It is used with a hammer to work up the 'cup' in the main pipe which is to receive the tapered spigot end of the branch pipe. Begin by inserting the small end of the bolt into the hole made in the main pipe and levering up the lead pipe wall. This will increase the size of the hole and begin the upward flow of lead in the formation of the cup. As soon as possible the larger end of the bolt is used, and well-directed

blows on to it from the hammer continue the working of the cup until it is the required size and shape. A tanpin lightly driven into the cup will finally true up any slight irregularities and ensure a good fit for the circular branch pipe entry.

3 Scribing plate. This is a metal plate with three points on its bottom edge. The outer two dig just enough into the lead main to hold the plate in position while it is in use. The centre point locates the centre of the plate with the hole in the main. Vertically up from this centre point is a series of small holes. Using the compasses, adjusted for the required size of joint, select one of these holes so that when the compasses are swung the point traces the line of joint on the blacked area of the main pipe. The shave-hook is then used to shave clean all lead areas within the marked joint-line. If a scribing plate is not handy, the handle of a hammer can be placed over the branch opening, and will receive one point of the compass.

The soldered spigot joint may be used to join 'cap and lining' connections to bath or basin taps; for joining brass or copper trap unions to lead waste pipes; or for joining brass overflow connections to lead warning pipes.

The joint is easily prepared and needs a special tool kit which will suit 13 mm to 32 mm lead pipes and costs about £3. Tinman's solder, Grade A to BS 219, is used. About 25 mm of this solder will be enough to make a good joint on 25 mm diameter pipe.

It is the fact that this joint needs so little solder that makes it so economical; about 40 one-inch joints can be made with $\frac{1}{2}$ kg of tinman's solder, whereas one 25 mm wiped soldered joint requires $\frac{1}{2}$ kg ($\simeq$ 1 lb) of plumber's solder.

The preparation of this joint and the tools used in the process are shown in Fig. 75. It is important that the whole space between the brass or copper lining and the inside of the prepared lead pipe should be filled with solder which 'tins' and unites the two. A ring of solder will appear at the top of the joint when it is correctly made, but one must be careful, for sometimes the solder ring will appear before the air has escaped from the space below. This will prevent the solder from completely filling the space and the result will be a poor joint.

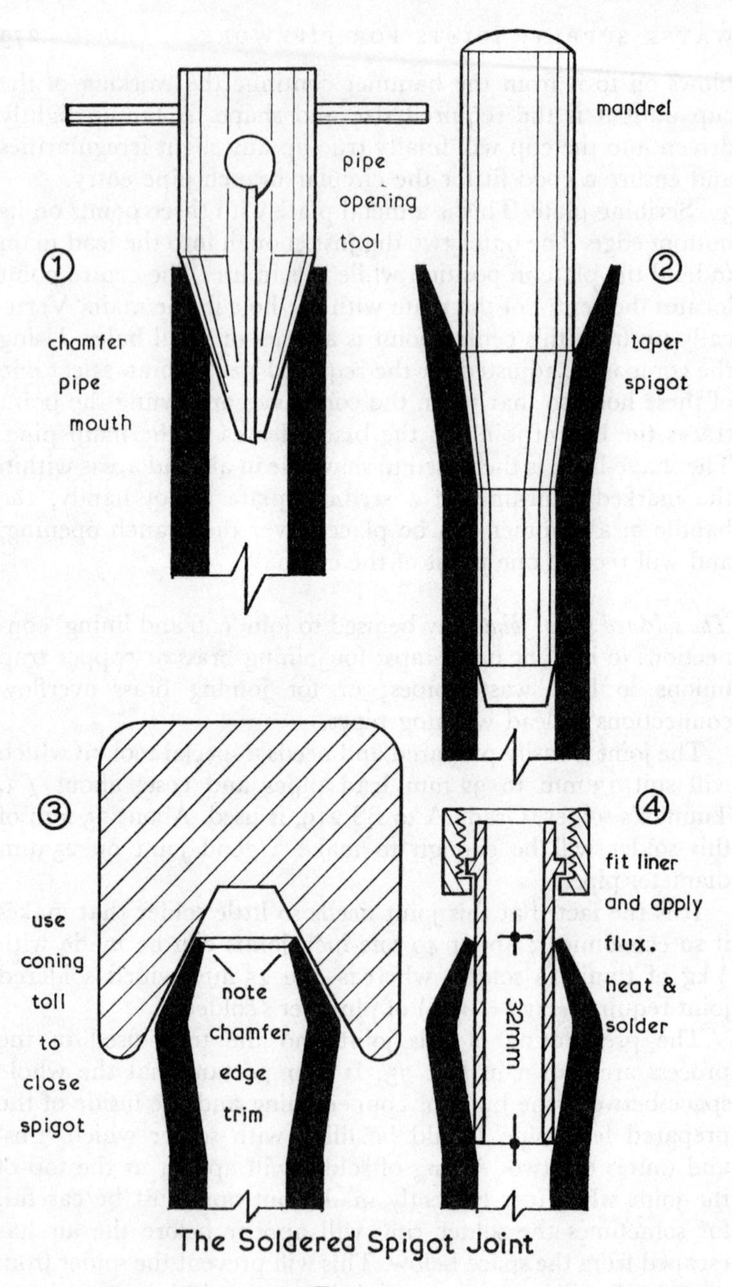

pipe
opening
tool
mandrel
①
②
chamfer
pipe
mouth
taper
spigot
③
④
fit liner
and apply
flux.
heat &
solder
32mm
use
coning
toll
to
close
spigot
note
chamfer
edge
trim
The Soldered Spigot Joint

FIG. 75

The blowlamp should be kept moving up and down the joint area to keep the solder molten and to expand the air, forcing it out so that the solder can run down and take its place.

Joints for copper pipes

Copper tubes may be jointed by compression fittings or by solder capillary fittings.

Compression fittings fall into two distinct groups—Type A and Type B. Type A fittings are known as non-manipulative fittings and require only that the tube end should be cut square and to length (see Fig. 76). They require only a soft copper ring or some other device, which is compressed between the inside of the fitting and the outside wall of the tube, so that there is little or no deformation of the tube wall.

Type B fittings are known as manipulative fittings. They are designed in such a way that during the making of the joint the tube end is shaped or worked so that in the final operation it can be 'squeezed' and securely held between the various components of the fitting (see Fig. 76). This manipulation of the tube end ensures that the fitting cannot pull off the tube, and for this reason most Water Undertakings insist upon Type B fittings being used for copper water services which are to be buried underground.

Copper services underground are generally in BS 1386 tube because, as has already been said, this soft tube is made in longer lengths, needs fewer joints, and is therefore economical. It is clear that the relatively soft tube would 'give' to Type A compression fitting, but equally, since it is soft and easy to work, it is very suitable for the Type B joint of the 'Olive' kind (not shown in Fig 76) which also gives a sound joint—an important matter for pipework laid under the ground where an undetected leak could cause a considerable loss of water.

Gunmetal is an alloy of copper and tin. You will remember (Table 11, page 231) that it has a very high copper content—much greater than that of brass, of which compression fittings are normally made for work above ground. Gunmetal is more resistant to corrosion than brass, and for this reason fittings of

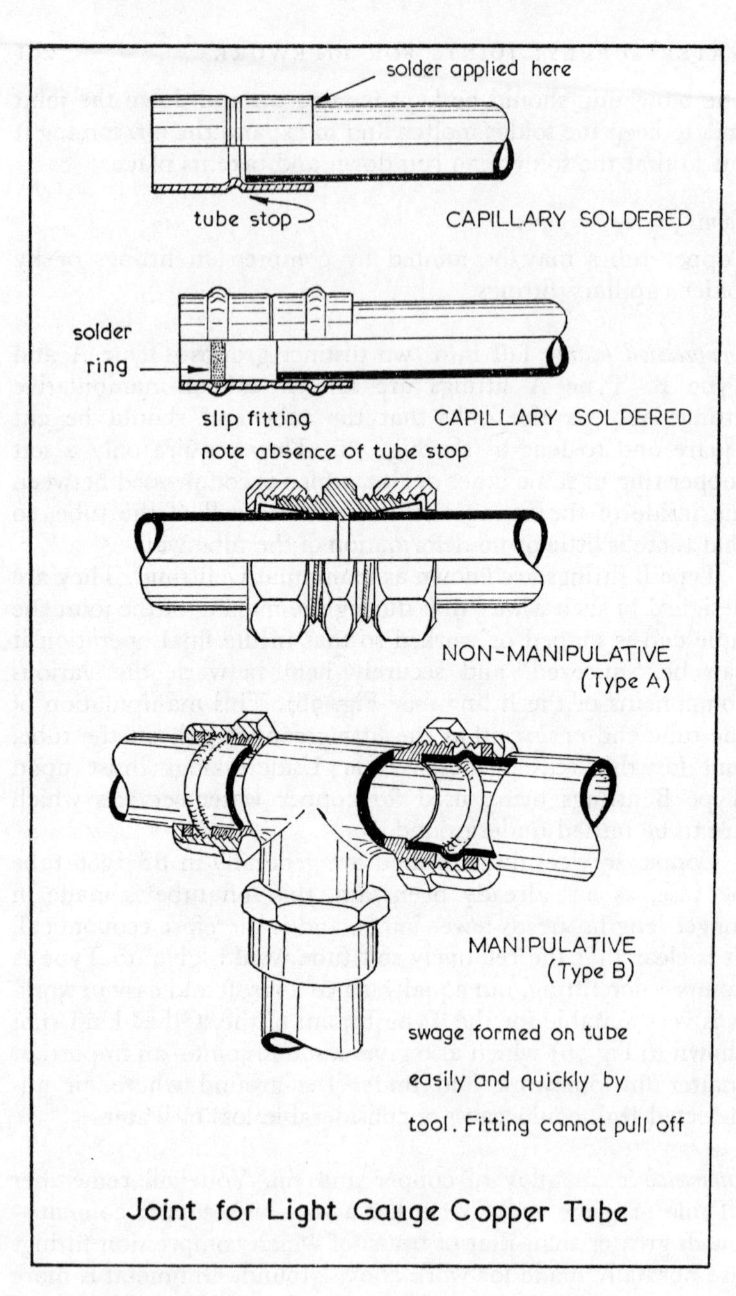
solder applied here
tube stop
CAPILLARY SOLDERED
solder
ring
slip fitting
CAPILLARY SOLDERED
note absence of tube stop
NON-MANIPULATIVE
(Type A)
MANIPULATIVE
(Type B)
swage formed on tube
easily and quickly by
tool. Fitting cannot pull off
Joint for Light Gauge Copper Tube

FIG. 76

gunmetal might well be used for copper services underground where the soil is known to have corrosive tendencies.

Underground copper tubes to BS 1386 may be jointed by solder capillary fittings where the Local Water Undertaking permits.

Solder capillary fittings are made of good quality brass, or of copper. The choice of material depends upon the importance of appearance against cost, and whether the fitting is for use underground or above ground (see Fig. 76).

When two dissimilar metals are in contact in a substance which can conduct an electric current, they will work like the two plates of a voltaic electric cell, producing very small amounts of electric current. The flow is created at the expense of one of the plates, which is electro-chemically corroded away. A moist, acidic soil could form such an electrolyte or conductor of electricity; and the copper tube with brass fittings could provide the dissimilar metals. Hence, while brass fittings are quite satisfactory for all work above ground, and for underground work where the soil is known not to be electrolytically corrosive, where there is evidence of acidic tendencies it is best to use copper fittings in order not to introduce a second metal into the installation. Solder capillary fittings admittedly introduce tin-lead solder alloy, which is a dissimilar metal, but this alloy has so nearly the same electro-chemical properties as the copper tube that serious electrolytic action between the two is very unlikely.

Solder capillary fittings (Fig. 76) may have a ring of solder already inserted in their sockets, in which case the copper tube end is cleaned with fine steel wool, fluxed, and inserted into the socket of the fitting until it hits the tube stop in the fitting. Heat is then applied, usually from a blowlamp, until the solder melts and flows by capillary attraction all around the closely fitting outer wall of the tube and the inner wall of the fitting. This is just one example of the way in which capillarity may be useful to the plumber.

Some of these fittings will have no solder ring insert, and solder will have to be fed from outside to make the joint. Most of these solderless capillary fittings are 'pre-tinned' inside.

They need to be carefully examined and some will perhaps have to be cleaned before the joint is assembled for soldering.

Fine sandpaper wrapped round a pencil provides an easy way of cleaning inside a fitting, but do not overdo this or you may so increase the capillary gap between tube and fitting that it will not work as it should. In most cases, one or two brisk twists with the sandpaper or fine wire wool will be sufficient. In some cases 'active' fluxes are used, and mechanical cleaning is reduced to a minimum; but do take care to wash off all surplus flux after the joint is made or it will continue to act on the tube face, leaving unsightly corrosion marks.

Capillary action draws the solder into the joint, and it is necessary that there should be a fine and equal gap all round between the tube and the fitting. Soft BS 1386 tube might accidentally lose its true cylindrical bore and this could prevent, or make difficult, its entry into the socket of the fitting. A special steel tool, resembling an inside and outside drift, is obtainable to restore the inside and outside cylindrical shape of copper tube to be used underground, and this should be employed where the tube is to be jointed by solder capillary fittings.

Bronze welded joints, which are made with oxy-acetylene welding equipment, special bronze welding fluxes, and copper-zinc alloy filler rods, are generally accepted by Water Undertakings for use in copper water services.

Weldable fittings are specially made for this purpose and may be of brass or gunmetal. They have sockets to receive the tube end just like other forms of copper tube fittings, but there is also an annular groove at the top end of the socket which forms a gap between the fitting and the tube wall into which the bronze welding rod can be introduced. Alternatively, the tube end may be 'belled' out and the weld deposited in the cup formed in this way. This method is cheaper and quite satisfactorily if expertly done. Bronze welded pipe-joints are dealt with in *Plumbing 2*.

Blowlamps used for jointing copper commonly burn paraffin. Some, however, use petrol, and although all blowlamps must be used with great care, these should be treated with special caution. Be very careful not to become so interested in the actual job of jointing that you put the lamp down where its flames can play on the woodwork of doors, skirtings or furniture.

Butane blowlamps burn a liquid fuel which gasifies and burns at atmospheric pressure. The fuel is stored in special replaceable 'bottles' on to which the burner of the lamp is screwed so that it looks very like an ordinary paraffin blowlamp.

The advantages of this newer type of blowlamp are, first, that it is economical. It needs no pre-heating; a turn of the gas valve allows gas to flow to the burner where it ignites the moment a flame is applied. The second advantage is that it is safe. As soon as a joint is complete, a turn of the gas valve shuts off the lamp, saving fuel and avoiding any risk of starting a fire.

Joints for mild steel pipes

Mild steel tubes are jointed by fittings which are screwed tightly on to screw threads provided on the pipe ends. The threads are standard in pattern and number of threads/in, and the fittings are threaded to the same British Standard Pipe Thread (B.S.P.T. as BS 21) to ensure that fittings of different kinds, and made by different firms, are interchangeable.

Pipe threads are cut on to the pipe ends with special thread-cutting tools called stocks and dies. There are usually four dies, housed in the adjustable die stocks, and these are the actual cutting tools. The stocks also house the adjusting mechanism, and have long handles which provide the leverage necessary to work the dies round the pipe, so that they cut the thread.

The joint is formed as follows. A fitting is tried on the screwed pipe end to check that the threads are correctly cut, and that they are engaging properly with the fitting. Jointing compound is then applied to the thread on the pipe, and strands of hemp are carefully wound into the threads. As the fitting is finally screwed home, the hemp is ground and compressed between the screw threads of the pipe and fitting so that a watertight joint is made.

Alternatively, a specially prepared plastic tape may be used in place of jointing compound and hemp. This tape, made of one of the polythene group of plastics, is 13 mm wide and supplied in 12 m reels. It is simple, quick and clean to use. Sufficient tape to pass round the circumference of the pipe thread, with about 13 mm for overlap, is torn off the reel. One end of the tape is held on the thread while the free end is wound tightly round so that it takes up the shape of the threads. The 13 mm overlap secures the tape in place by sticking to the tape below. When the fitting is screwed home, the tape, locked between the pipe and fitting threads, automatically and positively seals the joint.

The plastic tape jointing material is suitable for virtually all screw-threaded joints in any material. It is resistant to corrosion, self-lubricating, and it never hardens. Joints made with it can be tighter because of its lubricating property, and the ease with which joints can be undone even after long periods of time is a great advantage in maintenance and repair work.

Normally, the dies cut a tapered thread on the pipe end, and the fittings, elbows, bends, sockets and so on are parallel threaded.

Over-tightened, parallel-threaded, malleable cast-iron fittings may stretch beyond their limit and split. This might be noticed because it would suddenly become easier to turn the fitting with the pipe wrench. On the other hand, it might go unnoticed until the pipe were filled with water and a leak showed the split.

Joints for polythene tube

Copper tube compression fittings can be adapted for polythene tubes. This has been a very common practice and will no doubt continue, but considerable thought has been given to the development of joints designed to suit the particular qualities of the plastic tubes which are now becoming available.

Joints for plastic tubes will be more fully dealt with later on, but in the meantime Fig. 77 shows one type of polythene joint where a copper tube compression fitting is adapted for the purpose. Notice the copper internal liner which supports

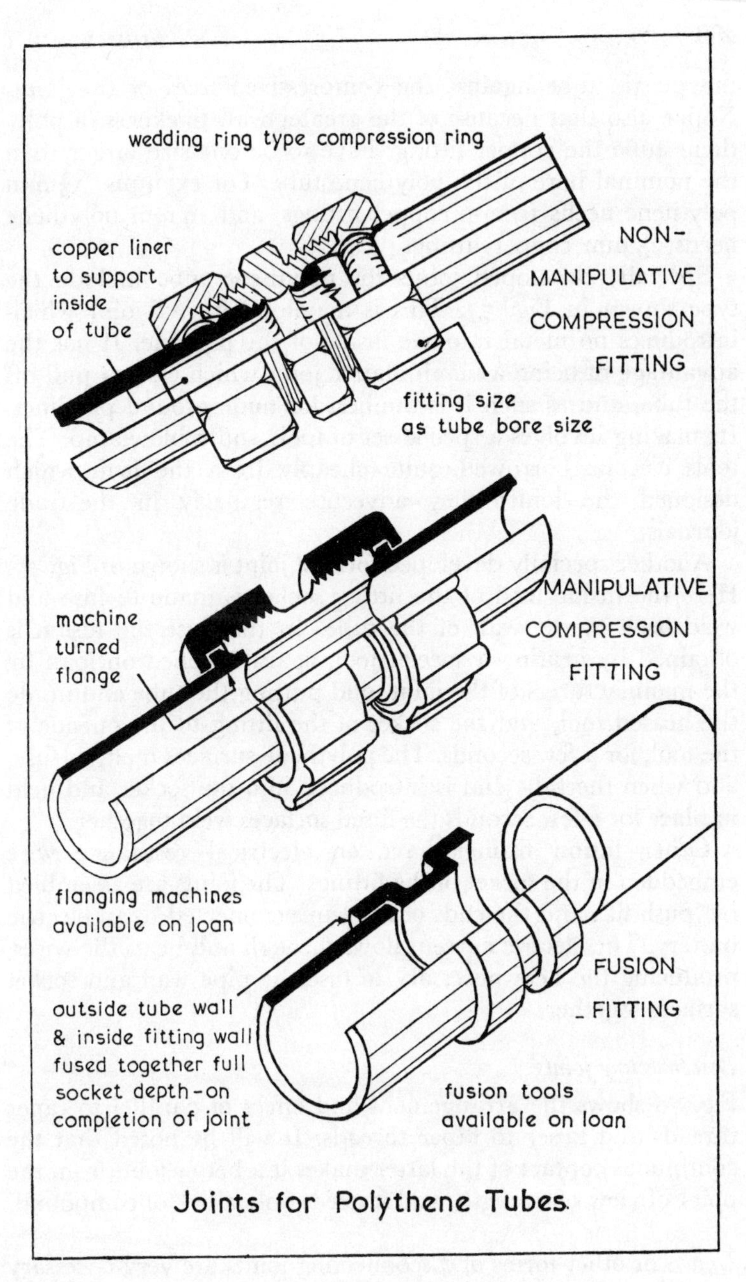
wedding ring type compression ring
copper liner
to support
inside
of tube
NON-
MANIPULATIVE
COMPRESSION
FITTING
fitting size
as tube bore size
MANIPULATIVE
COMPRESSION
FITTING
machine
turned
flange
flanging machines
available on loan
FUSION
FITTING
outside tube wall
& inside fitting wall
fused together full
socket depth-on
completion of joint
fusion tools
available on loan
Joints for Polythene Tubes.

FIG. 77

the plastic tube against the compressive forces of the joint. Notice also that because of the greater wall thickness of polythene tube the copper fittings have to be one size larger than the nominal bore of the polythene tube. For example, 13 mm polythene needs 19 mm copper fittings, and 19 mm polythene needs 25 mm copper fittings.

Specially developed joints for polythene tube include the type shown in Fig. 77. This is a union form of joint which introduces no metal into the inside of the pipeline. It has the advantage of being a manipulative joint which cannot pull off the tube, and as such it is the best for underground pipelines. Its making involves a special set of tools and a blowlamp. The tools can be borrowed quite cheaply from the firm which designed the joint; they advertise regularly in the trade journals.

Another specially developed form of joint is shown in Fig. 77. Here the inside wall of the fitting socket is made to fuse and weld itself to the wall of the tube. In this case the fusion is obtained by heating a special tool, again supplied on loan by the manufacturers of the joint, and putting the tube end inside the heated tool, and the socket of the fitting to the outside of the tool, for a few seconds. The polythene surfaces melt, or fuse, and when the tube end is introduced into the socket and held in place for a few seconds the fused surfaces weld together.

Other fusion fittings have an electrical resistance wire embedded in the socket of the fittings. The joints are assembled by 'push fit', and the ends of the wires connected to an electric battery. The electric current flows through and heats the wires, producing the heat necessary to fuse the pipe wall and socket surfaces together.

Disconnecting joints

Fig. 78 shows the arrangement and effect of parallel to taper threads and taper to taper threads. It will be noted that the continuous contact of the latter makes it a better joint from the point of view of strength, and it needs only a lick of compound.

Unions or other forms of disconnecting joints are very necessary in all pipework. If they are properly placed and used they are

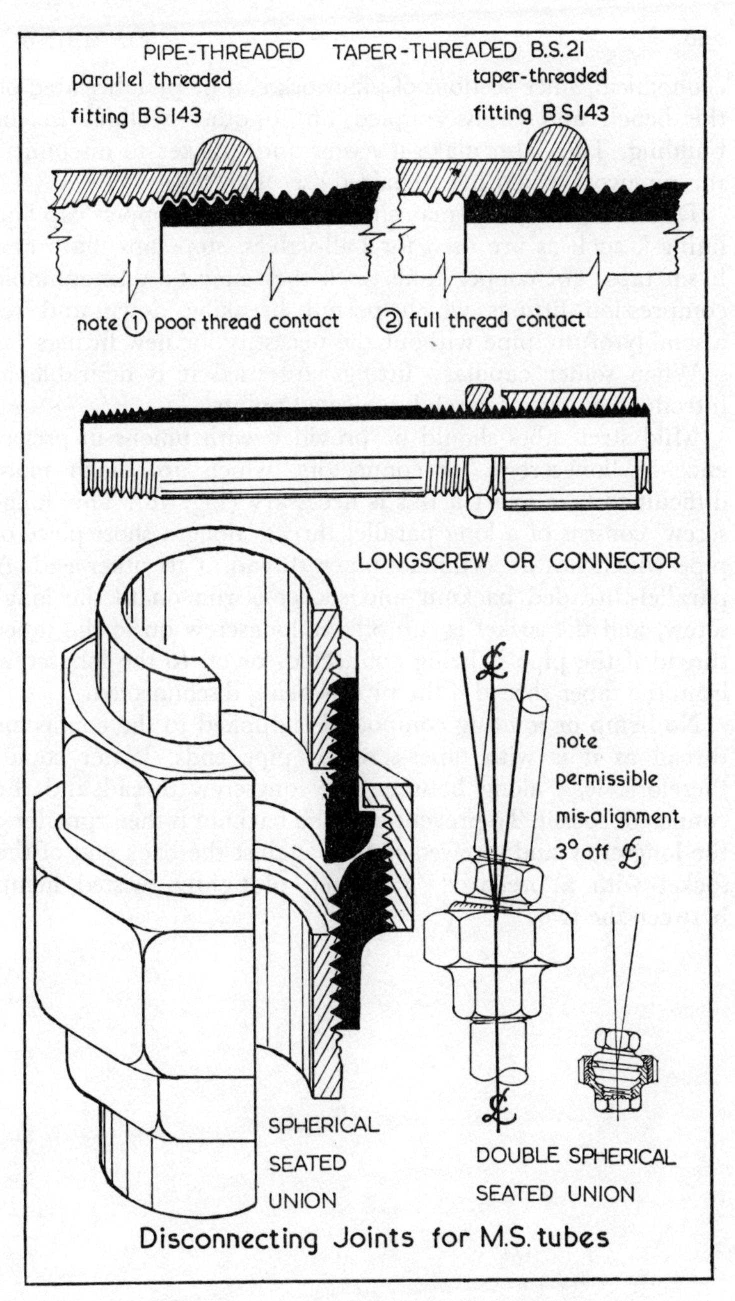
PIPE-THREADED TAPER-THREADED B.S.21
parallel threaded
fitting BS 143
taper-threaded
fitting BS 143
note ① poor thread contact
② full thread contact
LONGSCREW OR CONNECTOR
note
permissible
mis-alignment
3° off ℄
SPHERICAL
SEATED
UNION
DOUBLE SPHERICAL
SEATED UNION
Disconnecting Joints for M.S. tubes

FIG. 78

economical, since sections of pipework can be prefabricated on the bench and easily coupled up to other sections in the building. They also make it easier and quicker to disconnect the pipework when it is repaired (see Fig. 78).

Disconnecting joints may simply be brass or copper 'cap and linings', such as are used for ball-valves, stop taps, bath and basin taps. For copper tube work they may be dismountable compression fittings which permit breaking down and re-assembly of the pipe without the necessity for new fittings.

When solder capillary fittings are used it is desirable to introduce unions at carefully selected points.

Mild steel tubes should be provided with unions in preference to 'longscrews' or connectors, which are much more difficult to take apart if this is necessary (Fig. 78). The 'longscrew' consists of a long parallel thread along a short piece of pipe which has an ordinary taper thread at its other end. A parallel-threaded backnut and socket is run on to the longscrew, and the socket is run off the longscrew on to the taper thread if the pipe is being connected, or on to the longscrew from the taper thread if the pipe is being disconnected.

No hemp or jointing compound is applied to the connector thread as it is with taper-screwed pipe ends. Water could, therefore, leak along between the longscrew thread and the connector socket. To prevent this, the backnut is then run along the longscrew and screwed tightly against the back end of the socket with a prepared 'grommet' of tightly twisted hemp between the two.

25
Domestic water supply: installation

Arrangement of pipes
The model water bye-laws require that pipes shall be fixed so that they have sufficient support to prevent sagging, vibration, or mechanical damage, and in such a way that air-locks, which could diminish or stop the water flow, cannot occur.

Air-locks, as you will remember (page 67), happen because air is matter and occupies space. Since no two amounts of matter can be in the same space at the same time, it folllows that if a pipe is run so that air accumulates at a high point without means of escape, then water will not be able to flow through this 'air-lock'.

Air-locks do not give much trouble in pipes which carry water at supply mains pressure, for this pressure is generally so great that it is able to force the air out of the nearest tap. But if the water is supplied to the taps from a storage cistern inside the building, then the pressure might not be great enough to move the air.

Domestic hot-water systems with a boiler and hot-store tank or cylinder are always fed from a cold-feed cistern inside the building in order that there should be a reserve of water in case the mains supply should be temporarily shut down.

Some Water Undertakings insist that all *cold taps* must be fed from a cold-storage cistern within the building; all taps, that is, except one which is taken off the rising service main to provide a pure supply of drinking water. In every case, it is vital that the plumber should fix the pipes so that air can escape easily and quickly. In the case of hot-water systems, all pipes must rise to the air-vent pipe, which leaves the top of the cylinder or hot-store tank.

Where cold taps are fed with water through distribution pipes from a storage cistern, pipes must rise from the taps to the connection at the cold-storage cistern. Air can then rise up these pipes and escape through the water in the cistern. Any accumulation of air in the short risers to bath and basin taps will quickly vent itself as soon as the taps are opened.

Pipe fixings

If pipes are not sufficiently supported or fixed, mechanical damage may occur.

Lead is relatively soft, heavy, and of low tensile strength. If fixing clips are too far apart the lead between them will sag. This will 'drag' on the clips, making them cut into and damage the pipe walls. Tinned fixing clips for lead pipes should not be farther apart than 610 mm (2 ft) in the horizontal direction nor more than 760 mm (2 ft 6 in) in the vertical direction. Wherever possible, continuous support for lead pipes is the ideal. Fig. 79 shows some typical fixings for lead service pipes.

Light gauge copper tubes and *mild steel tubes* are rigid and, to a great extent, self-supporting. Fewer fixings are needed for these, and they may be spaced as follows:

TABLE 14

Tube	*Nominal bore*	*Outside diameter (o.d.)*	*Horizontal spacings*	*Vertical spacings*
Light gauge copper	½ in	15 mm	1220 mm	1830 mm
	¾ in	22 mm	1830 mm	2440 mm
	1 in	28 mm	1830 mm	2440 mm
Mild steel tube	½ in 15 mm	—	1830 mm	2440 mm
	¾ in 20 mm	—	2440 mm	3050 mm
	1 in 25 mm	—	2440 mm	3050 mm

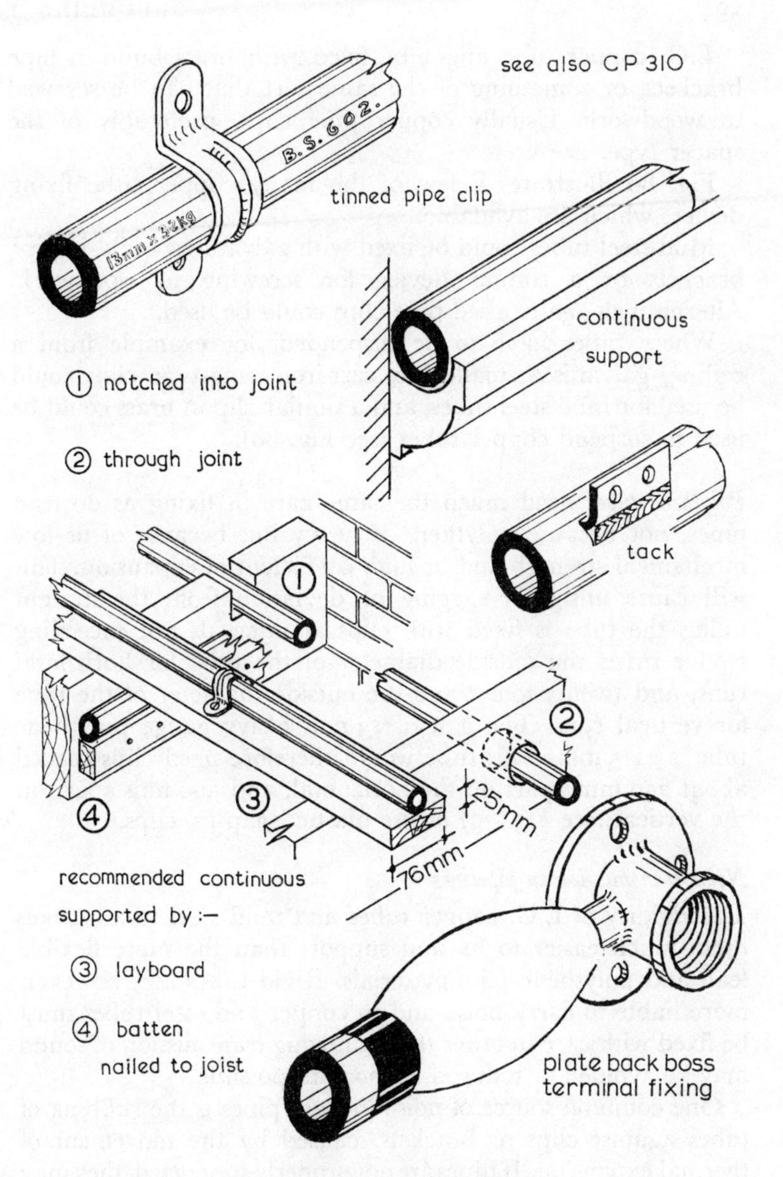
see also CP 310
B.S. 602.
13mm x 3.2kg
tinned pipe clip
continuous support
① notched into joint
② through joint
tack
25mm
76mm
recommended continuous supported by :—
③ layboard
④ batten nailed to joist
plate back boss terminal fixing
Fixing for Lead Pipe

FIG. 79

L.G. copper tube might be fixed with brass build-in pipe brackets, or something of the same sort that can be screwed to woodwork. Usually copper pipe clips, preferably of the spacer type, are used.

Fig. 80 illustrates a few of the many copper tube fixing devices which are available.

Mild steel tubes could be fixed with galvanised build-in pipe brackets or a similar device for screwing to woodwork. Alternatively, galvanised pipe clips could be used.

Where pipes have to be suspended, for example from a ceiling, galvanised, malleable, cast-iron, ring-type clips could be used for mild steel tubes, and a similar clip in brass could be used to suspend copper tubes (see Fig. 80).

Polythene tubes need much the same care in fixing as do lead pipes, not because polythene is heavy but because of its low mechanical strength and its high coefficient of expansion. This will cause unsightly sagging or deviation from the straight unless the tube is fixed with clips at intervals not exceeding twelve times the outside diameter of the tube for horizontal runs, and twenty-four times the outside diameter of the tube for vertical runs. The o.d. of 13 mm heavy gauge polythene tube is 21·5 mm. This tube would therefore need clips spaced about 250 mm apart on the horizontal, and 500 mm apart on the vertical (see Fig. 79) or use plastic 'snap-in' clips.

Noise transmission in pipelines

The rigidity of L.G. copper tubes and mild steel tubes makes them much easier to fix and support than the more flexible lead and polythene pipe materials. Rigid tubes are, however, more liable to carry noise and so copper and steel tubes must be fixed with care in order that annoying transmission of sound may be avoided or reduced as much as possible.

One common source of noise in rigid pipes is the rubbing of tubes against clips or brackets, caused by the movement of thermal expansion. If pipes are not properly supported, they may vibrate at the clips or brackets, and this can also cause a noise.

'*Water hammer*' is a common source of annoying noise in water

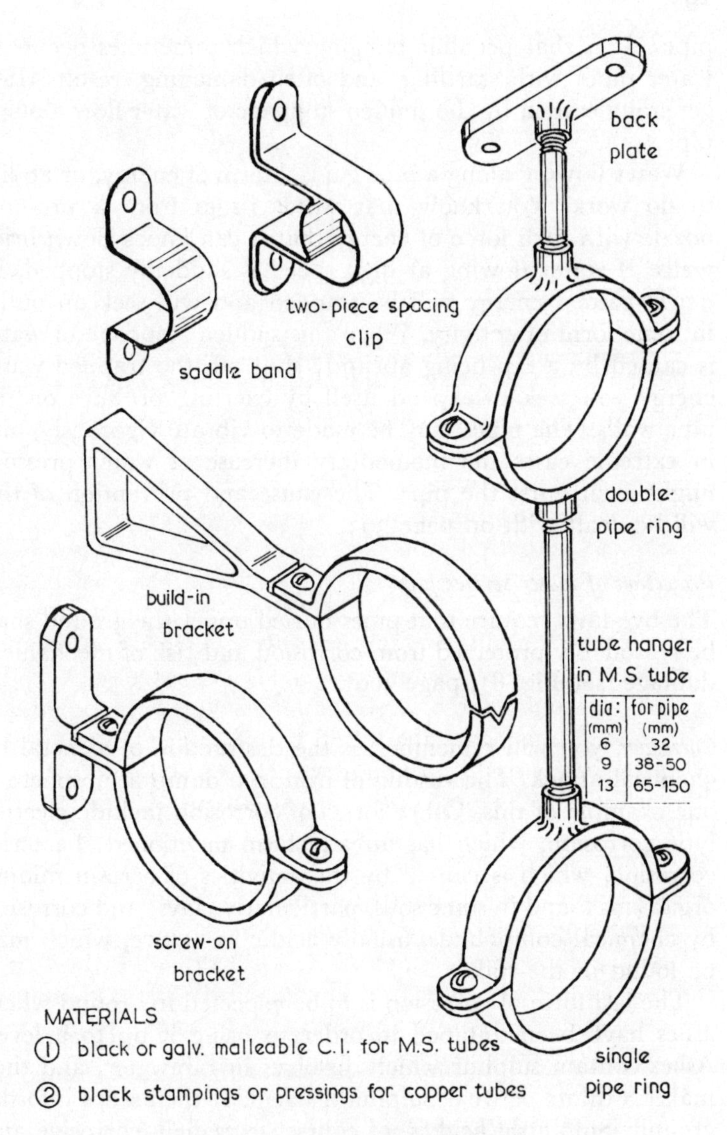

Hard Metal Pipe Fixings

FIG. 80

pipes. It is that peculiar banging which sometimes occurs in water pipes with startling, and often damaging, results. It is generally caused by the sudden stoppage of water flow along a pipe.

Water flowing along a pipe gains a form of energy, or ability to do work. You know that water issues from a fire-hose nozzle with such force of energy that it can knock down brick walls. If water flowing at high speed is suddenly stopped, its newly gained energy will be trapped and will seek an outlet in some form of activity. When this sudden stoppage of water is caused by a tap being abruptly shut off, the trapped water energy will seek to expend itself by exerting pressure on the pipe walls. The pipes may be made to vibrate vigorously, and in extreme cases the momentary increase of water pressure might even burst the pipe. The cause and prevention of this will be dealt with on page 303.

Protection of water service pipes

The bye-laws require that pipes buried under the ground shall be reasonably protected from corrosion and risk of mechanical damage (see Fig. 81, page 300).

Corrosion, you will remember, is the destruction of a metal by chemical attack. The rusting of iron in a damp atmosphere is one example of this. Other forms of corrosion include electrolytic corrosion, which has already been mentioned; bacterial corrosion, which is caused by the activities of certain minute organisms found in some soils, particularly clays; and corrosion by chemical compounds, usually acidic in nature, which may be found in the soil.

The last form of corrosion is to be expected in ground where ashes have been dumped in order to bring it up to a level. Ashes contain sulphur which dissolves in rainwater, and thus makes a dilute or weak sulphurous acid. It then seeps into the ground. Sulphuric acid is, of course, extremely corrosive and dilute sulphurous acids can be just as harmful even though they need longer to take effect.

Lead and copper pipes are more resistant to corrosion than mild steel pipes, and polythene is known to be very resistant to corrosive attack, but all pipes must at all times be protected against corrosion and physical damage.

It is sometimes suggested that pipes to be buried in corrosive soils should be laid on and surrounded by 150 mm of sand. This is all very well, but the acidic ground water could still seep through the sand and attack the pipes. If it is possible in such circumstances, it is a wise precaution to prepare the pipe trench to a greater depth than is necessary, and to fill the bottom 150 mm with large stones or similar material which will collect ground water and drain it away from the pipe. On the top of this underdrainage of stones, finer stones could be laid, and on the top of these the 150 mm sand bed for the pipe. When the pipe has been laid and covered with sand, the trench could be filled with limestone chippings. These chippings, being alkaline, would neutralise or cancel out the acidity of any ground water which did stray towards the pipes.

In some severe cases, pipes have been embedded in bitumen. A simple trough of boards nailed together to form a 'V' is placed beneath the pipe and then filled with hot-run bitumen so that the pipe is completely surrounded. This method is rather messy, and its success depends upon the continuity of the bitumen and the absence of cracks through which corrosive water could attack the pipeline.

Pipes may be wrapped in specially prepared anti-corrosive bandages of protective material. This is very often done since it is a relatively cheap, quick and effective way of permanently protecting pipes from corrosive conditions. One kind of protective wrapping is made from hessian or cotton bandage soaked in petroleum jelly. In certain soils, however, where corrosive bacteria are known to exist, these materials are useless since the bacteria will actually feed on them. In such cases the protective wrapping should be specified as being made of glass-fibre bandage, since the bacteria cannot thrive on this (see also Table 1, page 65).

Precautions against mechanical damage must be taken when pipes are laid. *Careless backfilling* of pipe trenches might result in sharp stones or flints falling directly on to the pipe. These could cut holes in the pipe if driven into it by the weight of the soil above. All pipes buried in the ground must be protected against this form of damage with sand, or at least with carefully selected fine soil, which is laid under the pipes and around them to a depth of 150 mm before the ordinary soil is backfilled in.

Sinking ground, or ground subsidence, could be caused by the natural movement of the earth, or it could be caused by earth settling in a deeper trench running at an angle across the shallower pipe trench. In either case, the subsidence could impose a severe strain on the buried pipe and its joints.

In the first case, where a general settling might be expected, the pipe should be laid on and surrounded by sand. This will allow the pipe to move slightly and adjust itself to any ground movement. If the pipe is laid not dead straight, but in a series of smooth curves, any strain on the pipe can be taken up by the spare pipe in the curves, and thus accommodated or smoothed out.

In the case of local settling of a lower trench, the easiest answer is to lay a concrete slab to 'bridge the gap'. The service pipe can then lie over this 'bridge' and be fully supported even though the soil below might continue to sink a little.

Shrinkage in clay soil is a common cause of underground pipe damage. Damp clay is, of course, sticky stuff that can be moulded easily. A pipe laid in clay will become moulded into the solidifying clay, which will hold it in a vice-like grip. When clay soil dries out in hot weather, it shrinks. You will have noticed the cracks caused by such shrinkage in summertime. The cracks, often 25 mm or more across at the surface, extend downward into the ground for 1·5 to 2 m. A pipe firmly embedded in the clay will be severely stretched if a clay shrinkage crack runs directly across it.

Many cases of burst underground pipes have been found to have been caused by lack of understanding of the behaviour and extent of clay soil shrinkage. All pipes laid in clay should first be encased in sand, whether they are wrapped against corrosion or not.

Vibration on roads caused by traffic can damage pipes. A form of work-hardening occurs, brought about by the constant 'hammering' of the ground under the wheels of heavy traffic. In severe cases, the service pipes should be threaded through specially laid conduits of stoneware pipes, which will protect the service pipe from the shocks from vibration. Fig. 81 illustrates these various precautions.

Precautions against frost damage must also be taken by the plumber. A pipe blocked with ice cannot deliver water. If only for this reason, frost damage is a great nuisance to the householder, and in some circumstances, indeed, the stoppage of the water supply to domestic hot-water appliances could be quite dangerous.

When water pipes freeze they are likely to burst because of the expansion which occurs when water becomes ice (see page 44). This expansion is quite irresistible—something has to give. Sometimes the ice formation can expand along the length of the pipe, but this is seldom possible and the whole expansive force is usually directed outwards against the pipe walls. Polythene pipes can stretch to accommodate this pressure, but metal pipes tend to split. When the pipe does burst, no water will pour out at the time because the pipe is full of ice. It is not until the thaw comes and the ice melts that the water will flood out, damaging both building structure and furnishings.

The bye-laws require that so far as is reasonably practical all water fittings and pipes inside or outside a building shall be so placed or protected that the possibility of frost damage is reduced.

The British Standard Code of Practice 99 'Frost precautions for water services', price 2*s*. 6*d*. recommends various ways in which this might be achieved. Most bye-laws now state that these recommendations must be adopted; but they do not say what the recommendations are. Therefore to comply with this particular bye-law one must obtain and study the contents of C.P. 99, which will be fully dealt with in *Plumbing 2*.

For the time being it is enough to realise the wasteful and damaging effects of frozen pipes; to remember that water pipes can freeze only when exposed to air temperatures of 0°C or

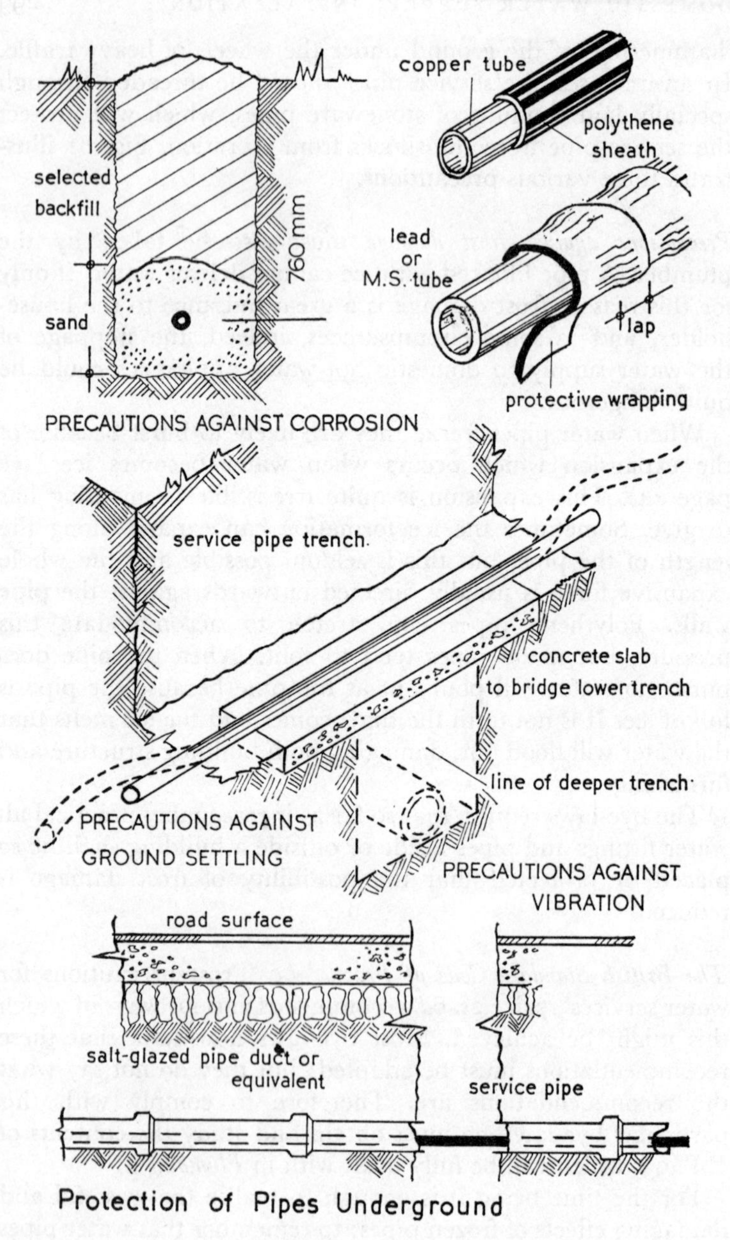

Protection of Pipes Underground

FIG. 81

less; and to adopt the following simple, commonsense, precautions:

Basic precautions against frost damage:

1 Fix pipes away from places where the air temperature is liable to fall to freezing point or below.

2 Insulate pipes from contact with cold walls by fixing them about 19 mm off the wall surface.

3 Insulate all unduly exposed pipes from contact with cold air by wrapping them with suitable 'lagging' or thermal insulating material.

4 Provide means of completely draining the entire water system in cold weather if the building has to be left unheated.

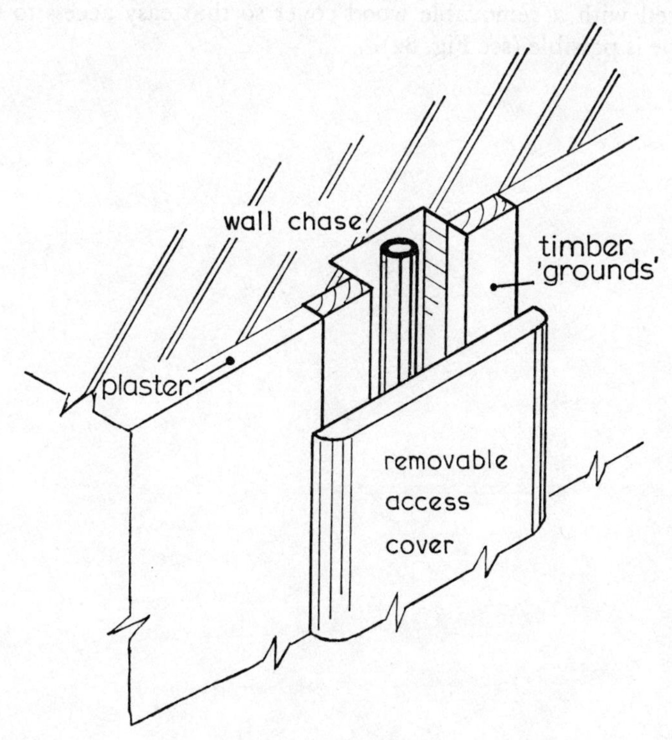

Provision for access to pipes

FIG. 82

Access for maintenance and repair

Remember that pipes do sometimes need maintenance or repair. A pipe buried in a wall or under a concrete floor is difficult to get at, to say the least. Should a pipe burst in such a position it could do serious damage to the building before the leakage of water was detected and put right; and, clearly, the work of getting to the pipe would be difficult and costly.

Some form of pipe duct should be used to house pipes which, for some reason or other, cannot be mounted on the surface. Under concrete floors, an earthenware pipeline makes a good and fairly inexpensive duct. Where a pipe has to be fixed below a wall surface, it should be fixed in a properly made chase fitted with a removable wood cover so that easy access to the pipe is possible (see Fig. 82).

26

Domestic water supply: cocks, taps and valves

In the south tea is brewed; farther north it is mashed. The result in both cases is the same, but the different names or terms used to describe its making can be confusing.

Dictionaries define the word 'cock' as 'a tap'; the word 'tap' as 'a fitting from which a liquid may be drawn'; and the word 'valve' as 'a fitting for controlling the flow of liquids'. As you can see, there is plenty of room for confusion between the various names for fittings controlling the flow of water. Until a standard method of identification has been worked out, this is bound to continue. The important thing is to know the differences in construction of plug cocks and screwdown valves, and to understand how these differences affect their use.

Design and function

Plug cocks have been used for centuries, and are a very simple form of tap. A tapered plug, with a hole through it, turns in a tapered hole in the body. When the plug is turned the hole lines up with the inlet and outlet holes in the plug cock body. Through this passage the water flows.

As you will see from this description, and from Fig. 83, water can be made to flow or stop by only a 90°, or quarter turn of the plug. This gives a rapid on-off action.

On page 296 it was said that water hammer occurs when the flow of water is abruptly stopped. Any form of tap which opens and closes quickly is liable to cause this, especially plug cocks, since they open and close particularly quickly. For this reason they must not be used on pipes carrying water at town mains pressure, where fast water flows are to be expected. They must only be used for low-pressure work—on oil-drums, tea-urns,

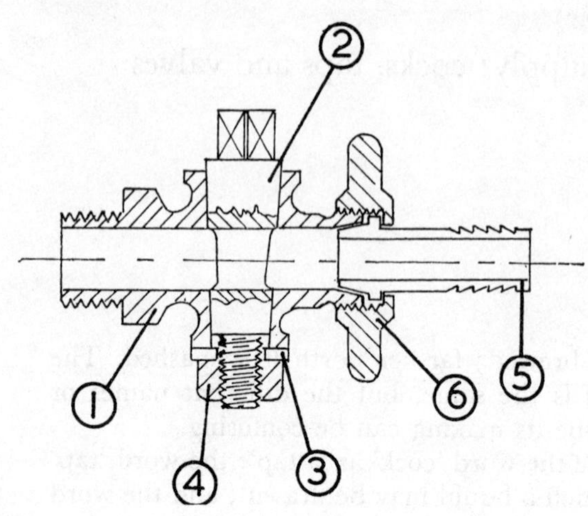

FIG. 83

and so on—or where once they are opened they will not be closed until all water is drawn off and the flow ceases; for example, in drain-down cocks for emptying hot-water systems.

Screwdown valves, as their name suggests, are taps with valves, commonly called 'jumpers', to which a seating washer is fixed. The valve is operated by a screw-threaded spindle which must be turned several times before the tap is fully open or closed. The screwdown action therefore gradually slows up the water flow and allows the energy built up in motion to die down before the flow is finally stopped.

Warning. Screwdown taps have a packing gland (Fig. 84) to prevent water from leaking along the operating spindle. If this gland is allowed to wear, or is badly packed so that it becomes loose, the spindle could be 'spun' down with a flick of the finger, thus creating a sudden stoppage of flow, and possibly water hammer.

Special screwdown taps used in hospitals and surgeries have a rapid-action screw thread which needs only a quarter turn to shut it off. This enables the doctor to wash his hands and turn off the tap with his elbow so that his hands can remain sterile. Again, these rapid-action taps could cause water hammer on the higher pressure systems and so this special type of tap should be fed by low-pressure distribution pipes from storage cisterns, and not directly off the town main.

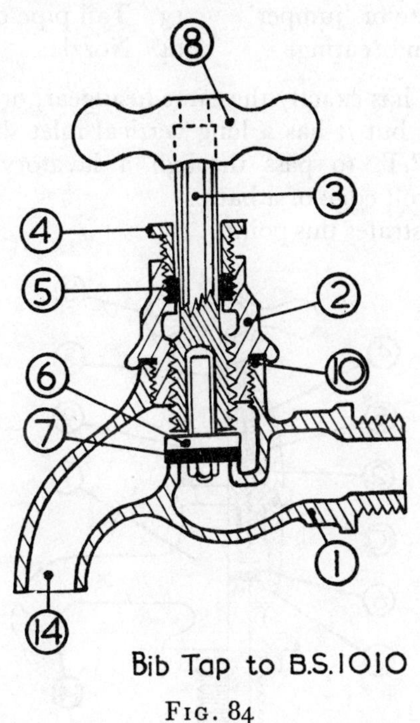

FIG. 84

A bib tap, bib valve or bib-cock is a type of screwdown tap used where a tapped boss, or other connection for the tap, is at right angles to the wall face. This is the sort of tap that is usually fixed above fireclay kitchen sinks. It is fitted with a 15 mm or 20 mm* B.S.P.T. connection of the male or external kind. Fig. 84 shows the important features of BS 1010 bib tap.

All tap diagrams have numbers indicating their various parts. To make it easy to refer to them, the key to these numbers is as follows:

* *Note:* 15 and 20 mm represent metric size nominal bore mild steel tube formerly ½ and ¾ in UK dimensions.

1 Body
2 Headgear
3 Spindle
4 Packing gland nut
5 Packing
6 Valve plate or 'jumper'
7 Washer and seating
8 Capstan (or crutch) turn
9 Easy clean cover
10 Headgear—body washer
11 Backnut
12 Union nut or 'cap'
13 Tail pipe or liner
14 Nozzle

A *pillar tap* has exactly the same headgear, or working parts, as a bib tap, but it has a long vertical inlet shank 15 mm or 20 mm B.S.P.T. to pass through a lavatory basin top or through the roll edge of a bath.

Fig. 85 illustrates this point.

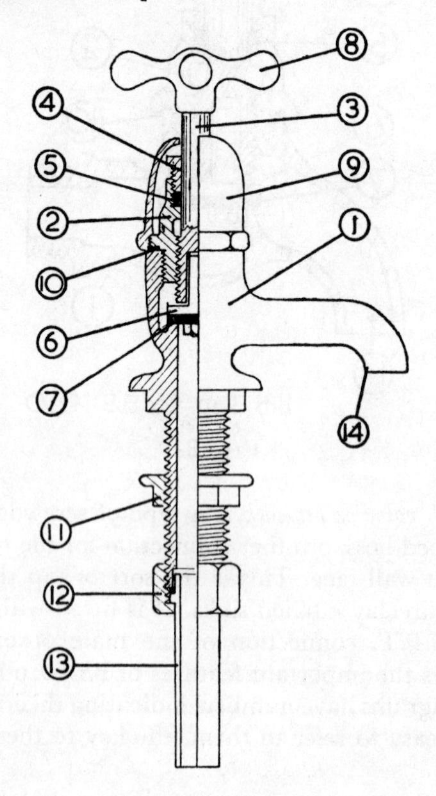

Pillar Tap to B.S.1010

FIG. 85

Globe taps also have the same headgear as bib taps, but are fitted with a horizontal female-threaded connection, usually 20 mm B.S.P.T., so that they will fix through the side or end wall of a bath. Fig. 86 shows a typical globe tap.

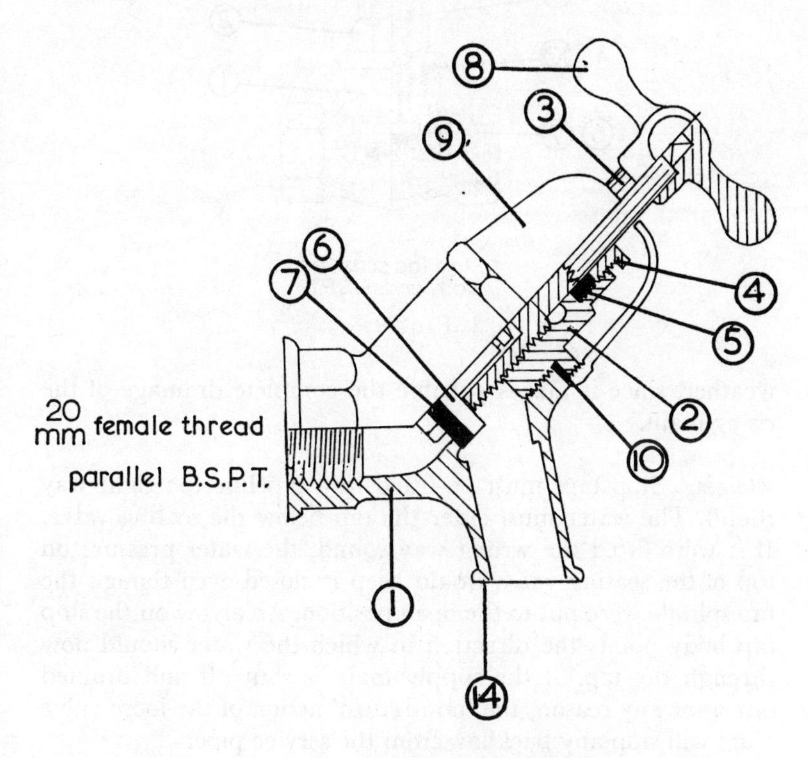

Globe Tap to BS1010

FIG. 86

Stop taps or stop-valves again have headgear just like that of a bib tap, but a different body to suit the particular purpose of the valve. They are used to form a part of a water pipeline and to control the flow of water along it (see Fig. 87).

Some stop taps have an emptying-down tap or washer-seated screwdown plug as part of the complete combined stop tap-drain tap. This fitting is neat, and can be very useful in frosty

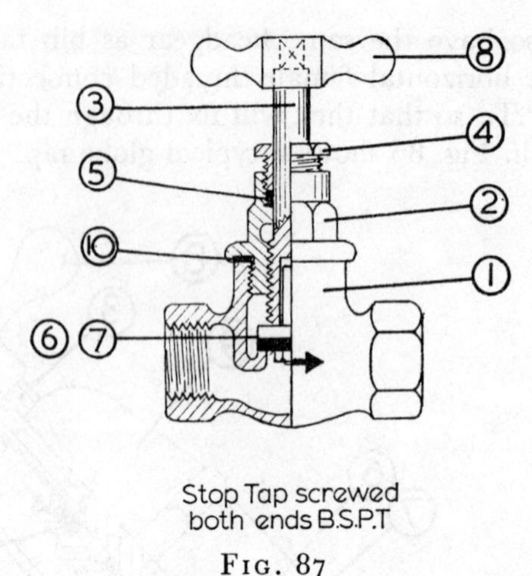

Fig. 87

weather, since it makes possible the complete drainage of the rising main.

Warning. Stop taps must be fitted in a pipeline the right way round. The water must enter the tap below the seating valve. If it were fitted the wrong way round, the water pressure on top of the seating valve would keep it closed even though the tap spindle were put to the open position. An arrow on the stop tap body points the direction in which the water should flow through the tap. If the supply main is shut off and drained down for any reason, the 'non-return' action of the loose valve plate will stop any backflow from the service pipe.

Draw-off taps for use on service pipes; that is, pipes subject to mains pressure, have loose valve plates for the same reason. The pressure in the pipe will be great enough to lift the jumper as the tap spindle is put to the open position.

Draw-off taps for domestic hot-water services, and cold-water taps fed by distribution pipes, which are subject only to the 'head' pressure of the cold-storage cistern, are fitted with pegged valves. The valves are free to turn in the spindle so that the seating washer is not worn away in use, but the pegging ensures that as the tap is opened the valve plate is automatically lifted off the seating. Otherwise the jumper might stick down in the closed position, and the low water pressure would not be able to lift it.

Fullway gate valves are used where it is important that there should be the least possible restriction to water flow, as with low-pressure distributions from storage or feed cisterns. A gate

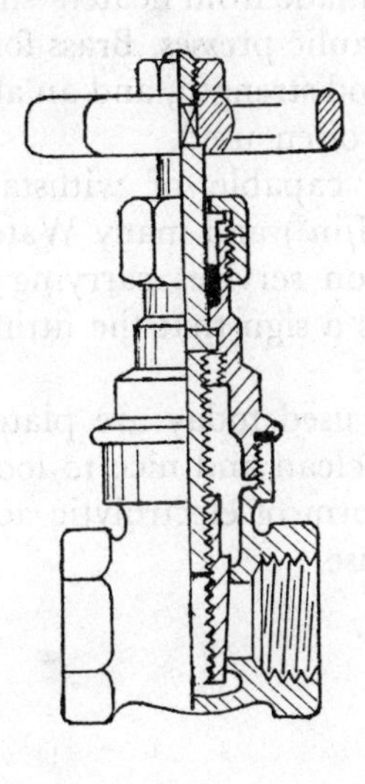

Fullway Gate Valve

FIG. 88

valve is operated by a screw spindle, generally of the non-rising type. That is, instead of rising out of the fitting when the valve is opened, as the screwdown tap spindles do, it rotates without rising and the wedge-shaped valve 'climbs' up the spindle screw thread. Gate valves are supplied with female-screwed end connections in B.S.P.T. and in a wide variety of sizes.

Fig. 88 shows a sectional view of a gate valve. Compare the headgear design, and the water-flow path, with those of the screwdown valve.

Materials for cocks, taps and valves

The dimensions and manufacturing tolerances of draw-off taps; that is, of bib taps, globe, pillar and stop taps, must by bye-law comply with the design requirements of BS 1010. All taps below 50 mm nominal size, and that means all domestic draw-off taps, must be made of a corrosion-resistant alloy.

Taps are usually made from heated 'slugs' of brass which are hot-pressed in hydraulic presses. Brass for this process needs to be ductile and of good strength, and an alloy of 60 parts copper to 40 parts zinc is often used.

Taps should be capable of withstanding a pressure of 2070 kN/m^2 (300 lbf/in^2) and many Water Undertakings insist that all taps used on services carrying their water must be stamped by them as a sign that the fitting is approved for use in the area.

Most of the taps used today are plated with chromium to make them easy to clean and nice to look at. This is done by electro-plating—a form of electrolytic action, which is in this way put to a good use.

27
Domestic water supply: ball-valves

Ball-valves are designed to control the entry of water into a cistern, closing off the supply when the water level in the cistern has reached a predetermined level.

Ball-valves are classified as:

(a) High pressure valves capable of closing against pressures of 1380 kN/m^2 (200 lbf/in^2).
(b) Medium-pressure valves for pressures of up to 690 kN/m^2 (100 lbf/in^2).
(c) Low-pressure valves capable of closing against pressures of up to 276 kN/m^2 (40 lbf/in^2).

All classes of valve should be capable, while held in the closed position, of resisting a pressure of 2070 kN/m^2 without leaking or showing signs of 'sweating'.

There are many variations of the *Portsmouth*, or *horizontal plunger* ball-valve, but the best is undoubtedly the Portsmouth BS 1212.

The BS 1212 has the advantage of being easily detachable so that it can be got out from the cistern, replaced, adjusted and tested to see if it is watertight and if the shut-off water level is correct (Fig. 89). It also has devices for adjusting the water level, which cut out the need for strong-man tactics in bending the ball-valve float arm.

It has renewable seatings. If a seating orifice becomes defective, it can be replaced easily and at little cost so that the valve remains serviceable. These renewable seatings are also interchangeable. This means that the small-holed orifice used for

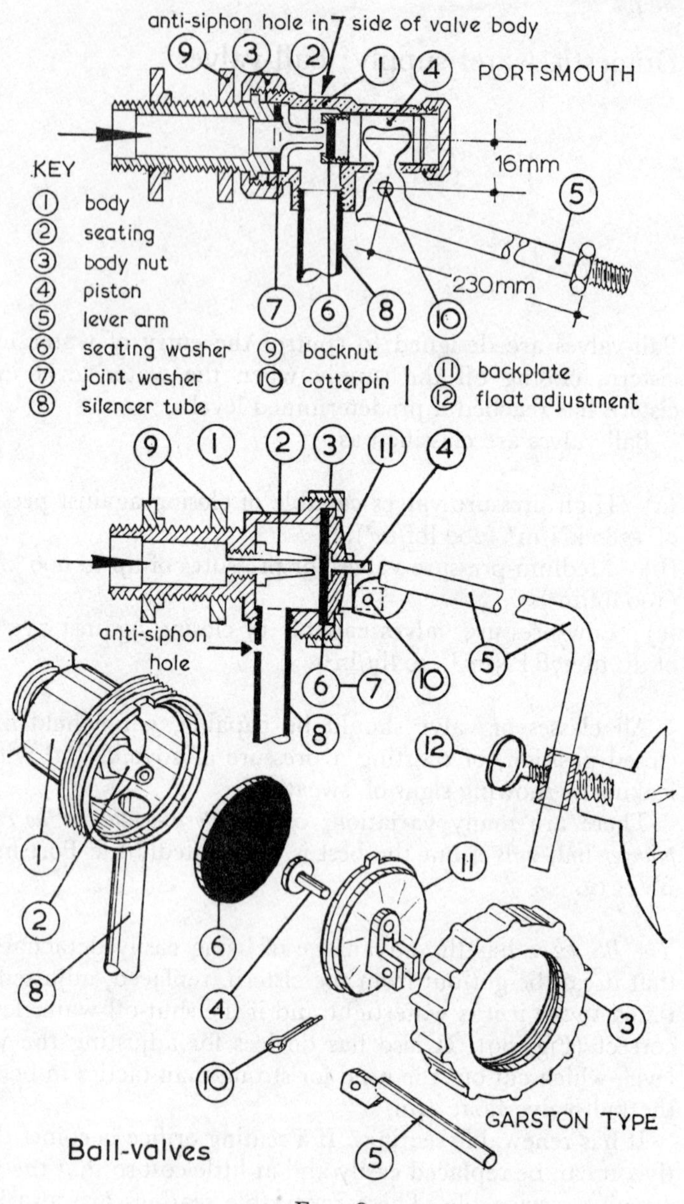
anti-siphon hole in side of valve body
PORTSMOUTH
16mm
230mm
KEY
1 body
2 seating
3 body nut
4 piston
5 lever arm
6 seating washer
7 joint washer
8 silencer tube
9 backnut
10 cotterpin
11 backplate
12 float adjustment
anti-siphon hole
GARSTON TYPE
Ball-valves

FIG. 89

high-pressure supplies can be changed for an orifice with a larger hole if the valve is needed for a medium- or low-pressure pipeline. Thus, with a few ball-valve bodies and a supply of seatings with various sized holes, a ball-valve can be quickly adapted to suit high-pressure, medium-pressure or low-pressure services.

Nylon—a chemically inert plastics material with a high resistance to mechanical wear—is now used to make interchangeable seatings for the BS 1212 valve in districts where the water or its flow tend to corrode, or wear, metal seatings.

The 'Garston' ball-valve was developed at the Government's Building Research Station at Garston. Its designers set themselves the task of producing a ball-valve which would be free of the corrosive troubles already mentioned, and free also from the nuisance caused by lime deposits in temporary 'hard' water districts. These deposits of 'fur' tend to interfere with the proper working of the valve and sometimes they make it stick in the open or closed position.

Fig. 89 shows clearly how this has been simply yet successfully done. Notice: (a) that no water touches any working or moving part, (b) the simple and convenient water-level adjustment device, and (c) the specially large, shaped inlet chamber which breaks the speed with which the water enters, and reduces mechanical wear on the seating orifice, which, incidentally, is again renewable.

Re-washering is easily and simply done by unscrewing the large nut which secures the ball-valve body to the inlet, which is in turn fixed to the cistern. You will notice that in this respect the Garston ball-valve resembles the BS 1212 ball-valve.

All ball-valve parts must be of non-corrosive material. Generally, the BS 1212 valve is a hot-pressing which is afterwards drilled, threaded and dressed to the dimensions laid down in the Standard. Its lever arm and cotter pin are of drawn or extruded brass rod. The seating washer should be of good quality rubber 3 mm thick.

The Garston ball-valve is obtainable as a brass pressing or in plastic. In either case the lever arm and cotter pin will be of similar material to that used in the Portsmouth ball-valve.

Silencer tubes are commonly fitted to ball-valves to reduce the noise of inrushing water. A thread is provided inside the ball-valve outlet to receive the short length of tube which forms the silencer tube and conveys the inflow water to the bottom of the cistern. In this way the noise of splashing water is avoided, as during most of the filling action the outlet of the tube is under water.

Under certain circumstances—when, for example, a street main is temporarily closed for draining down—a 'backpull' on the service could occur and polluted water be drawn into the service pipe and main. Precautions against this form of 'back siphonage' must be provided by air holes in the outlet of a ball-valve or silencer tube, and these air-vent holes must be above the working water level of the cistern.

In the BS 1212 ball-valve the air hole is in the side of the ball-valve body. In the Garston ball-valve no provision for an air vent is made, and if silencer tubes are used with this type of valve they must have an anti-siphon hole drilled in them above the level of the water in the cistern when it is full.

Strange whistling noises have been known to come from the air hole of silencer tubes, but if the hole is arranged so that it faces the nearest wall of the cistern (as shown in Fig. 89) this noise is avoided.

Floats, which provide the initial closing effort on all ball-valves, are spherical in shape and are quite commonly made of copper. Other materials for floats include polythene spherical floats, expanded ebonite, and expanded vinyl-chloride.

See also Chapter 15, *Plumbing 2*.

28

Domestic cold-water supply

Even on metered supplies, which are hardly ever used in houses, water costs only about 20p per 1,000 gal and is thus man's cheapest commodity. This does not mean that anyone can afford to waste it, but rather the reverse, since the demand for piped water tends to exceed the national sources of supply.

Water Undertakings' bye-laws prescribe the materials and methods that should be used in the installation of water supplies. They aim to conserve water by preventing the waste, undue consumption, misuse, or contamination of supplies.

Having satisfied yourself that your materials and proposed methods of installation meet the requirements of the bye-laws, you can proceed with the work of laying the underground water service pipe, and the fixing of the internal pipework and fittings.

External arrangements

Suppose that the Undertaking's supply main has been tapped, and a short length of service pipe already laid up to a stop tap situated just outside the boundary of a detached house standing well back from the road. Incidentally, the length of service pipe between the main and the boundary stop tap is sometimes called a 'communicating pipe', but this term is not universally adopted by all Water Supply Undertakings.

There is no hard and fast rule as to how the work should be organised. The order of work is dictated by conditions on the site, and many other things.

You might proceed as outlined below, checking all the good installation practice points as you go. (The numbers appearing in the description relate to relevant details as shown in Fig. 90, page 316.)

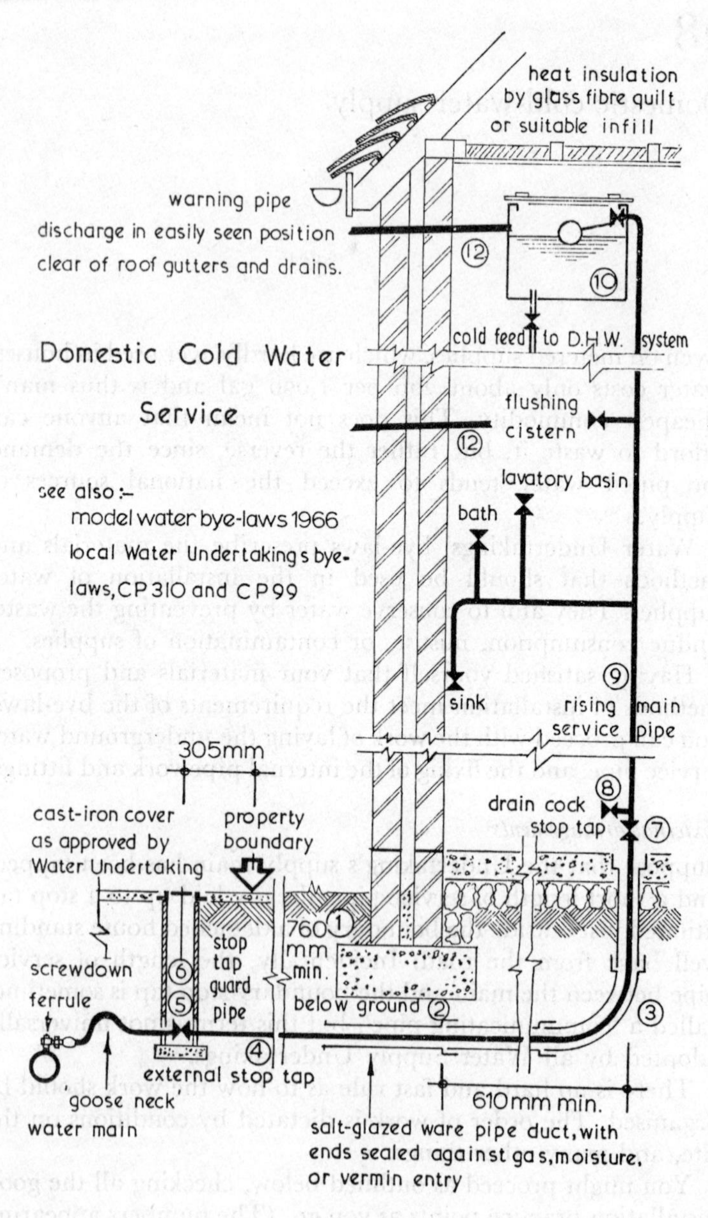
heat insulation
by glass fibre quilt
or suitable infill
warning pipe
discharge in easily seen position
clear of roof gutters and drains.
12
10
cold feed to D.H.W. system
Domestic Cold Water
Service
flushing
cistern
12
lavatory basin
see also:-
model water bye-laws 1966
local Water Undertakings bye-
laws, CP 310 and CP 99
bath
9
sink
rising main
service pipe
305mm
8
drain cock
stop tap
7
cast-iron cover
as approved by
Water Undertaking
property
boundary
1
760
mm
min.
below ground
2
3
screwdown
ferrule
6
5
stop
tap
guard
pipe
4
external stop tap
goose neck
water main
610mm min.
salt-glazed ware pipe duct, with
ends sealed against gas, moisture,
or vermin entry

Fig. 90

The pipe trench (1) will have a level bottom, free from sharp stones, and be prepared in the way described on pages 296, 297 and 298, according to the soil conditions. It will be deep enough at all places to ensure that the new underground service will have at least a 760 mm (2 ft 6 in) cover of soil to protect it from frost damage (2). Some Water Undertakings demand a 915 mm (3 ft) depth of cover.

The ducted service entry (3) of salt-glazed ware pipes, or the equivalent, will be already in place. The service pipe, with the union for the inside stop tap already wiped on, will be threaded through the duct until about 150 mm of it stands above the inside floor level.

The service pipe (4) will then be uncoiled along the prepared pipe trench bottom, and finally connected to the boundary stop tap (5).

The infilling of the pipe trench can now go on as soon as the householder's stop tap has been fitted inside the house, and the service filled with water and examined to make sure that there are no leaks.

A stop tap guard pipe (6) to BS 1185 (1944) will be placed on a suitable concrete base beneath the boundary stop tap. The guard pipe will make access to the stop tap easy at all times, and will be fitted with a cover to a pattern approved by the Water Undertaking.

Internal arrangements

Internal work can now be done. Having fitted the internal stop tap (7) which the householder normally uses for turning off the water and draining down in frosty weather if he intends to leave the house unoccupied and unheated, a drain down tap (8) must be fitted immediately above the stop tap. Alternatively, the stop may have a drain tap incorporated in it—a neater, cheaper, and altogether better arrangement.

The rising service main (9) will then be prepared and fixed as described on pages 292–94. Notice that it is run up an inside wall, not the inside of an outside wall. In this position on an inside wall, the pipe is less likely to lose heat, and is therefore less likely to freeze.

The rising main service can be arranged so that it feeds all cold supplies to sink, basin, bath, W.C. flushing cistern, and then supplies the ball-valve of the cold-feed cistern (10) for the domestic hot-water supply system. Alternatively the rising main could supply only one tap, usually over the kitchen sink, from which drinking water supplies could be drawn, and then go on to feed the ball-valve of a combined cold storage and feed cistern which would downfeed through distribution pipes to all other cold taps and flushing cisterns in the house. Possibly the cold-store cistern and the domestic hot-water system's cold-feed cistern could be separate. If not, then a cold feed from the common cistern will supply the hot-water system with cold water. In this case the inlet of this pipe and the main cold-water distribution pipe must be at the same height—at least 25 mm—above the bottom of the cistern.

The choice of system will be largely decided by the bye-law requirements of the Undertaking supplying the water. For the present, assume that the Undertaking permits all taps and cisterns to be fed from the rising service main.

The cold-feed cistern (10) for the hot-water system must of course be watertight, and since 1 litre of water weights 1 kg*, and the cistern will hold several litres, it will need to be adequately supported. It will need to have a snugly fitting but not airtight lid to prevent dust and dirt from getting into the cistern and making the water impure. It will also need thermal insulation if it is exposed to cold air, for example in a roof space, to protect it from damage by frost.

Usually cisterns are made of hot-dip galvanised mild steel to BS 417, but they could be of timber lined with lead or copper sheet. Alternatively, in the smaller sizes they could be made of asbestos, cement or copper; or of one of the newer plastic materials such as glass fibre. The choice of material for the feed cistern and pipework will depend on the chemical properties of the water and particularly on whether it is corrosive to

* 1 kg = 2·2 lb

metals. Cost may have to be considered, and, again, the wishes of the Water Undertaking may have to be taken into account (see page 261, and the section dealing with the metal solvent action of water on metals, page 111, and also *Plumbing 2*).

The capacity of the cistern must be sufficient to meet the needs of the household. Later on the various factors which help to decide what capacity is right in given circumstances will be dealt with. For the time being, remember that the smallest cold-storage capacity permitted by the Model Water Bye-laws (1966) is 25 gal (114 l), and that where the cistern is used both as a cold store and as a feed for a hot-water system its actual capacity, that is, to working water level, must not be less than 50 gal (228 l). Some Undertakings require a minimum capacity of 228 l and 364 l respectively (see *Plumbing 2*).

In order to store a fair reserve of water against possible interruption of the main supply, the cistern will normally have a bigger capacity than these minimum sizes.

Warning pipes (11) must be fitted to W.C. flushing cisterns and to cold-storage and feed cisterns. The warning pipe must have an inside diameter of not less than 19 mm; it must be fitted so that its overflowing level is not less than twice its diameter below the top edge of the cistern, and at least 25 mm above the working water level of the cistern.

The warning pipe must discharge so that any overflow due to a defective ball-valve can be easily seen and quickly put right, and any waste of water prevented.

29

Domestic hot-water supply

An adequate supply of hot water is to be regarded as an essential of modern living. There are a great many types of system, and when one is to be chosen the following points must be carefully considered:

i The type of fuel to be used (oil, gas, electricity, or solid fuel)
ii The quantities of hot water needed; the temperature that will be required; the time available in which to heat the water; and the frequency with which the water will be needed
iii The extent to which automatic control is desirable, as opposed to the manual work involved in stoking and so on
iv The extent to which atmospheric pollution by smoke is to be avoided.
v The cost of the system components, accessories and pipework, and the cost of installing these
Finally, but by no means least,
vi The cost of running the system when it is installed.

When these points have been discussed with the customer and decided, the plumber must plan the system that has been chosen. He must consider the size of the component parts, since these must be large enough to meet all the demands likely to be imposed on the system. He must work out pipe sizes and the most efficient and economical way to arrange them.

Eventually, when all the factors have been carefully weighed and all the plumber's technical know-how applied, a suitable design emerges and he can order his materials. Now begins the interesting work of installing the system—a process that will need as much practical and technical skill as the actual design.

As an introduction to this very interesting and challenging aspect of plumber's work consider a simple 'boiler-cylinder' hot-water system, which is very popular and quite satisfactory for small domestic dwellings. At the moment one is concerned chiefly with its layout, the names of its component parts, and their functions. Discussion of the actual system design must be left to a later book in this series, together with many other important points which will arise as this simple system is examined.

A diagram of the system is shown in Fig. 91, page 322. It has three major components: the boiler to heat the water; the hot-store cylinder to store the heated water until it is needed; and, at a higher level, a cold-feed cistern to provide the head pressure necessary to push the water out of the taps (see pages 43 and 67). This also provides a reserve of cold feed water in case the supply mains are shut off for a time.

Because the heated water is stored until it is drawn off for use, this arrangement is called a 'storage' type of domestic hot-water supply. Compare it with an instantaneous gas water-heater, where the water is heated as it passes through the appliance and not stored, and you will see the difference between the 'storage' and 'non-storage' systems.

The boiler

When a back boiler is put in the grate or in inset room heaters the fire heats both it and the room. Boilers in solid-fuel cookers share the heat with the cooking service. Boilers which do nothing but heat water, apart from perhaps providing a small amount of heat in the kitchen, are independent of all other heat services, and are called independent boilers.

Independent boilers are usually made of cast iron or mild steel, and the more expensive ones have an enamelled jacket. This not only improves their appearance but also their efficiency, for the jacket is designed to prevent unnecessary loss of heat from the boiler to the surrounding air.

Suppose the system has an independent boiler. It must be

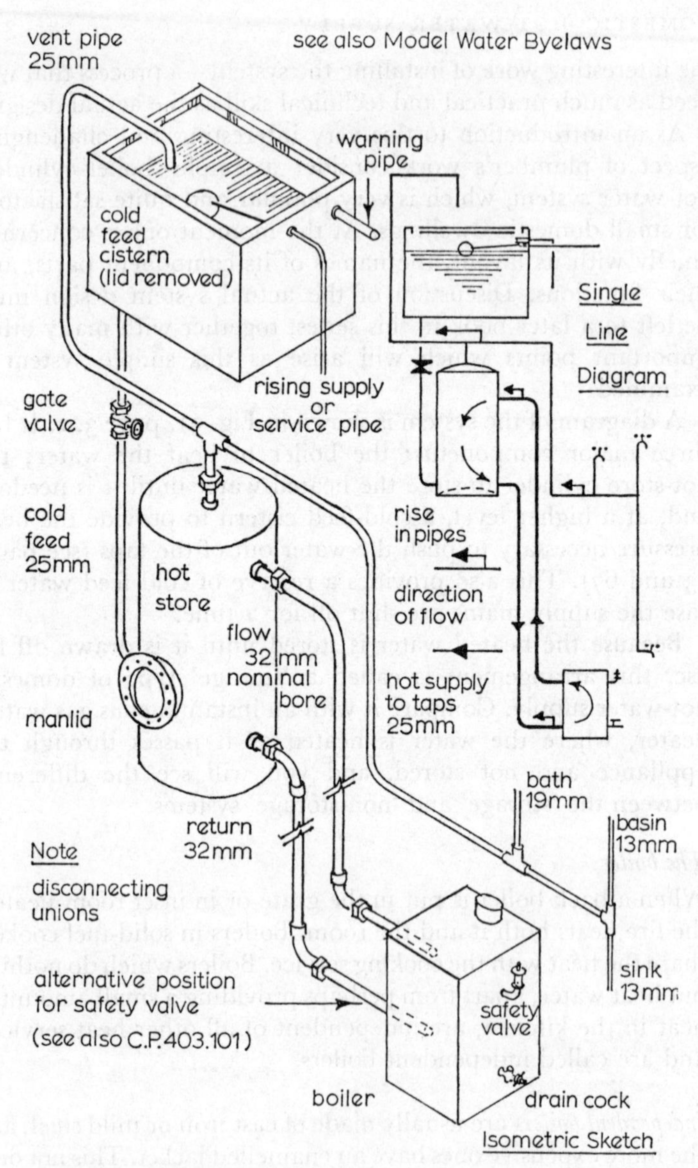

Boiler-Cylinder Direct D.H.W. System

FIG. 91

carefully put together, according to the maker's instructions, and erected on a level fireproof base. When all the necessary work in connection with the smoke pipe and flue has been attended to, one can proceed with the placing and fixing of the other components.

The hot-store cylinder

The bye-laws of the local Water Undertaking prescribe minimum sizes for hot-store vessels and cold-feed cisterns. A 136 l hot-store cylinder will satisfy their requirements and will provide adequate storage of hot water for a normal household.

It may be constructed from galvanised mild steel to BS 417, or from sheet copper to BS 699. The choice of material will largely depend upon the character of the water to be stored. Copper, being more resistant to corrosion, is commonly used in soft water districts where the water may be aggressively corrosive (see page 111). Copper cylinders can be used in 'hard' water districts, but care should be taken to see that there is easy access for the periodic removal of the fur deposits which are bound to form inside the cylinder when used with such water.

The cylinder should be mounted upright so that the lighter, heated water can 'layer out' or stratify over the heavier colder layers to a greater depth than they could if the cylinder were fixed horizontally. (This is fully dealt with in *Plumbing 2*).

The cylinder will most probably be fitted to an airing cupboard in the bathroom above the kitchen. It will be conveniently supported by the bathroom floor but it should not stand directly on the floor. It should be supported off the floor by two strips of wood. This will allow air currents to pass under the cylinder bottom and so prevent moisture in the atmosphere from condensing into water, which could corrode the metal cylinder.

The cold-feed cistern

This must have a minimum 'actual' capacity (i.e. the water capacity as measured to its working water level, and not the amount it would hold if filled to its top edges, which is called its 'nominal' capacity) of 114 l. If it is to supply cold-water taps as well as feed cold water to the hot-water system, then its 'actual' capacity must be at least 228 l. (The siting and sizing of cisterns will be dealt with in a later book.)

Suppose the cistern has an 'actual' capacity of 250 l. When filled to its working water level, its water content alone will weigh

$$250\ \text{l} \times 1000\ \text{g/l}$$
$$= \textit{250 kg} \text{ or nearly } \tfrac{1}{4} \text{ tonne}$$

The need for proper support of this weight is clear.

The cistern should be fitted with a close-fitting but not air-tight lid to prevent dust or dirt from entering and contaminating the water. It must be fitted with a suitable ball-valve (see page 311), and, of course, a warning pipe to convey any overflow water to a spot where it can be clearly seen, so that if any defect occurs to prevent the ball-valve from stopping the water flow when it has reached the proper working water level in the cistern, it is obvious at once.

Pipework

This consists of the cold-feed pipe; the primary circulating pipes; the vent pipe; and the hot-water supply pipes.

The cold-feed pipe, which feeds cold water to the system by way of a connection near the bottom of the cylinder, should be 25 mm in diameter. Its inlet end to the cold-feed cistern should stand at least 25 mm above the cistern bottom. The pipe should be fitted with a Fullway gate valve (see page 309) which makes it easier to washer taps, de-scale the boiler (in hard-water districts) or do any other repairs without having to wastefully drain down the entire content of the cold-feed cistern. No feeds to other services are permitted off the cold-feed pipe (see Water Undertakings bye-laws or the model water bye-laws, 1966). The cold feed must feed the hot-water system only.

The primary circulating pipes which connect the boiler and hot store form a circuit round which heated water moves, or circulates. The reason for this circulatory movement of heated water is fully described on pages 102 and 104, and in Fig. 16.

The cooler, denser water in the system falls by gravity down the pipe from the bottom of the cylinder to the boiler. It 'returns' down this pipe to the boiler where it is reheated; hence this pipe is called the *return pipe*.

The dropping return water pushes the heated water out of the boiler and makes it 'flow' up the pipe, which rises off the top of the boiler and joins the cylinder near the top; hence this pipe is called the *flow pipe*.

Flow and return pipes should not be less than 25 mm bore, and for hard-water districts where the bore is likely to be reduced by deposits of fur, 32 mm bore pipes are to be recommended. They must be carefully arranged with a continual rise from the boiler to the cylinder so that air-locks are avoided (see page 67).

The vent pipe is fitted to rise continually and without dips from the crown at the topmost point of the cylinder. It finishes over the cold-feed cistern. It must rise by 40 mm for each metre of head on the system, plus about 150 mm just to be on the safe side, above the working water level of the cistern. This is clearly shown in the diagram of the system. The additional height of vent pipe is intended to accommodate the increased height of lighter hot-water column that can be supported by the heavier cold feed and return water columns.

The end of the vent pipe should not dip below the top edge of the feed cistern. There have been cases where this rule has been ignored, causing considerable trouble and expense, and even an element of danger. If the end of the vent pipe dips into the water, or the water level rises to cover it, and the water surface in the cistern freezes over, as sometimes happens when cisterns are fixed with no proper regard for frost damage, the vent becomes 'sealed' and cannot serve its prime function—the release of air from the system. Neither can it allow air to enter the system, and, as will be shown in a later volume, this can contribute to the collapse of a cylinder in certain circumstances.

Discharge of water from this vent pipe is most unlikely in a properly designed, installed and operated system. A small hole in the cistern lid is quite satisfactory, especially if it is provided with an easily made sheet copper funnel beneath the vent pipe end.

Hot-water supply pipes to the bath, basin and sink taps are taken from the vent pipe just above the crown of the cylinder. Fig. 91, page 322, shows a typical arrangement. Note carefully the recommended pipe sizes indicated. In all normal circumstances these will ensure a good outflow at all taps, even if they are all in use at the same time. The undersizing of pipework is a common fault which gives rise to many complaints of inefficiency in hot-water systems and inadequate flows at taps. Pipe-sizing by informed technical methods is not difficult, and is infinitely more satisfactory than guesswork. All plumbers should learn how to pipe-size, and this again will be dealt with in a later volume.

Boiler mountings

Tnese are the accessories fitted to, or mounted on, the boiler, which ensure the safe and efficient working of the hot-water system. They may include a thermometer (page 82) which shows the temperature of the hot water; a safety valve (Fig. 92), page 327) which will 'lift' or 'blow off' to relieve any pressure in the system greater than that it is designed to withstand; and a drain-down cock (page 304) which will empty the system when it is to be repaired, the boiler is to be cleaned, or as a common-sense precaution against frost damage where a system is to be left for some time in an unheated or unoccupied house during periods of severe cold weather. See also *Plumbing 2*.

The more expensive modern boilers will have some form of thermostatic control. This is usually arranged to operate under the influence of the different rates of expansion of two strips of different metals. If these are rigidly fixed together throughout their length they will 'bow' because of the greater rate of expansion of one strip than the other. This 'bowing' heat movement of the bi-metal strip moves a plate to or from an air supply hole to the boiler firebred. As the water in the boiler reaches a predetermined temperature, the bi-metal thermostat control device moves to close the air supply and thereby slow down the burning rate of the fire. As the water in the boiler

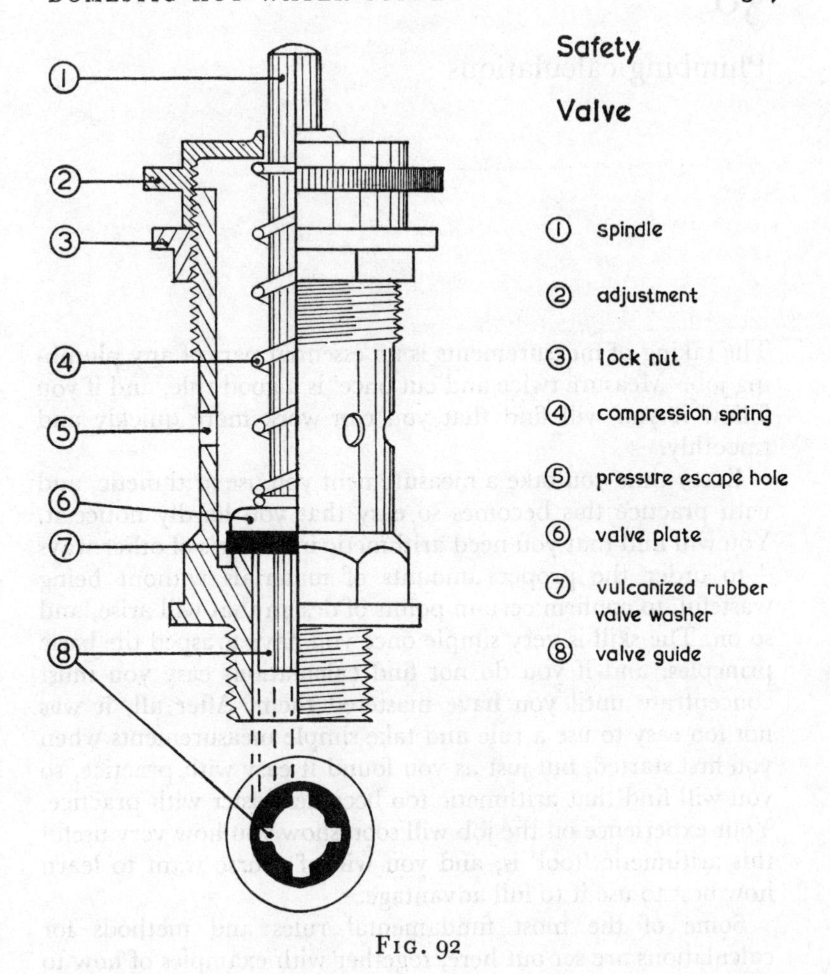

FIG. 92

cools down, the bi-metal strip moves back to open the air supply, causing the fire to burn more strongly and so to heat up the water again to the desired maximum temperature to which the thermostat device is set.

Most solid fuel boilers which have no inbuilt thermostat can be fitted with a thermostat draught regulator at a small cost. This is quickly repaid by the saving in fuel and the improved efficiency of the system.

30
Plumbing calculations

The taking of measurements is an essential part of any plumbing job. 'Measure twice and cut once' is a good rule, and if you follow it you will find that you can work more quickly and smoothly.

Every time you take a measurement you use arithmetic, and with practice this becomes so easy that you hardly notice it. You will find that you need arithmetic in all sorts of other ways —to order the proper amounts of materials without being wasteful, to confirm certain points of design that will arise, and so on. The skill is very simple once you have grasped the basic princples, and if you do not find calculations easy you must concentrate until you have mastered them. After all, it was not too easy to use a rule and take simple measurements when you first started, but just as you found it easy with practice, so you will find that arithmetic too becomes easier with practice. Your experience on the job will soon show you how very useful this arithmetic 'tool' is, and you will of course want to learn how best to use it to full advantage.

Some of the most fundamental rules and methods for calculations are set out here, together with examples of how to use them.

Fractions

Whole numbers will be well known to you. Some examples are: 1, 4, 9, 30, 256, 1002.

A fraction is a part of a whole, for example, four-fifths, one-half, one-quarter, eleven-sixteenths.

When a fraction is written with a *numerator* above and a *denominator* below a dividing line, like this:

$$\frac{\text{numerator}}{\text{denominator}} \text{ or } \frac{3}{8}$$

then it is called a *vulgar fraction.* The denominator shows how many parts the whole is divided into, and the numerator shows how many of these parts are present in this case. For example, $\frac{3}{8}$ indicates that the whole number has been divided into eight parts, and that this fraction contains three of those eighths.

Proper fractions are vulgar fractions with a value of less than 1. You can always recognise them because the denominator is always smaller than the numerator. Examples of proper fractions are: $\frac{1}{2}$, $\frac{3}{4}$, $\frac{1}{4}$, $\frac{3}{16}$, $\frac{27}{50}$, $\frac{236}{3458}$.

Improper fractions are vulgar fractions with a value greater than 1. In this case the numerator will be greater than the denominator: for example,

$$\frac{3}{2} \qquad \frac{24}{4} \qquad \frac{125}{2} \qquad \frac{1023}{1000}$$

these improper fractions can be simplified by dividing the denominator into the numerator: for example, $\frac{3}{2} = 1\frac{1}{2}$, $\frac{25}{4} = 6\frac{1}{4}$, and so on. However, these simplifications are no longer improper fractions, since the numerator is no longer greater than the denominator. The simplification results in *mixed numbers* which are made up of whole numbers and proper fractions: for example, $1\frac{1}{2}$, $6\frac{1}{4}$, $62\frac{1}{2}$, $153\frac{3}{4}$.

Although vulgar fractions are used in plumbing calculations, there is another very useful form of fraction—the *decimal fraction.*

Decimals are thought by some people to be difficult, but they are in fact quite simple You will find that with practice they are easier to use than vulgar fractions. All metric calculations are in decimals, and if any vulgar fractions are involved they will first have to be converted into this form (see page 336).

The word 'decimal' means *in the order of tens.* Look at the following sequence of numbers.

Whole numbers	*decimal point*	*decimal fractions*			
10000 1000 100 10 1		0·1	0·01	0·001	0·0001
		$\frac{1}{10}$	$\frac{1}{100}$	$\frac{1}{1000}$	$\frac{1}{10000}$

Each number in this series is one-tenth of the figure to its left. The *decimal point* separates the whole numbers from the decimal fractions, but you will see that this is also true of the fractions: each one is equal to one-tenth of the figure on its left.

In exactly the same way, if you look at the numbers from right to left, you will see that each number is ten times bigger than the one to its right.

This, then, is the order of tens, or decimal system. The way to use it is shown in the following examples.

Addition and subtraction of decimals

Rule. Set the decimals down just as you would if you were adding and subtracting ordinary figures—that is, units under units, tens under tens, and so on. As a result the decimal points are also one under the other, and this is important.

EXAMPLE 1

Add 1·25, 32·6, 0·02 and 10·736

Solution

```
 1·25
32·6
 0·02
10·376
------
44·606
------
```

EXAMPLE 2

Subtract 17·26 from 29·3

Solution

```
29·30
17·26
-----
12·04
-----
```

Multiplication of decimals

You can multiply in your head by 10, 100, 1000, or any power of ten simply by moving the decimal point as many places to the *right* as there are noughts in the figure by which you are multiplying.

EXAMPLE 3

(i) $3{\cdot}8 \times 10 = 38$
(ii) $0{\cdot}38 \times 100 = 38$
(iii) $0{\cdot}038 \times 1000 = 38$

There are two methods of formal multiplication. The method of working is the same for both, but they are set down differently. Continue to use whichever is best known to you but have a look at the following method since it is commonly used.

EXAMPLE 4

Multiply 22·16 by 2·1.
Rule a. Set the figures down as shown and start to multiply them as though they were whole numbers.

```
 22·16    (multiplicand)
   2·1    (multiplier)
 -----
  2216
 4432
 -----
 46536    (product)
 -----
```

Rule b. Count total number of decimal places (figures after the decimal points) in multiplier and multiplicand (in this example three places).

Rule c. Step off this number of places from the right-hand end of the product:

$$= 46{\cdot}536$$

Plumbing application. An altitude gauge on a boiler reads 3·5 metres. What is the intensity of pressure at the gauge in kN/m^2?

Formula, intensity of Pressure = metres head $\times$ 9·8 kN/m².

Solution 3·5 m $\times$ 9·8 kN/m² =

```
      3·5     (multiplicand)
      9·8     (multiplier)
   -------
     280
    315
   -------
    34·30     (product)
   -------
```

There is a total of 2 decimal places in multiplier and multiplicand. Step off these two places from the right-hand end of the product. The correct answer is then

34·3 kN/m²

Approximations as checks of decimal calculations

It is rather easy to miscount the number of decimal places, especially if you are in a hurry, and for this reason it is always a good idea to check answers *to decimal sums*. You have already seen that decimals are in the order of tens on both sides of the decimal point. If the point were put one place too far to the right then the answer would be ten times too big. And if it were put one place too far to the left it would be one-tenth less than it should be. Both errors could be quite serious; and you can avoid them by working out a rough approximation to the answer.

EXAMPLE 5

Check the answer to Example 4 by a rough approximation: 22·15 $\times$ 2·1 is very nearly 22 $\times$ 2.

The full, correct answer to Example 4 was 46·536. You can work out in your head that 22 $\times$ 2 = 44. This is so nearly the same as 46·536 that it is obvious that the placing of the decimal point must be correct. Suppose, however, that the answer had been incorrectly given as 465·36 or 4·6536. Either of these answers would have been proved wrong by the approximate answer of 44.

Division of decimals

Mental division of 10, 100, 1000 or any other power of 10 is simply done by moving the decimal point as many places to the left as there are noughts in the figure by which you are dividing. You may have to insert some noughts in the answer to fill in any gaps (see Examples (ii) and (iv) below).

EXAMPLE 6

(i) 3·8 ÷ 10 = 0·38
(ii) 0·38 ÷ 10 = 0·038
(iii) 22·4 ÷ 100 = 0·224
(iv) 22·4 ÷ 1000 = 0·0224
(v) 9·5 m ÷ 10 = 0·95 m
(vi) 46 l ÷ 100 = 0·46 l
(vii) 10 kg ÷ 1000 = 0·01 kg
(viii) 35000 mm ÷ 1000 = 35·0 m

The formal method of dividing decimals can be illustrated from the typical plumbing application given here. During the process the rules will be pointed out.

EXAMPLE 7

(This example is included for comparison with *Example 7a* to demonstrate the greater simplicity and ease of metric working.) The pressure on the base of a water column 1 ft square and 1 ft high is 62·5 lb. What pressure acts on each square inch of the base?

Solution: Pressure in lb/in² =

62·5 lb/ft² ÷ 144 (144 in² = 1 ft²)
(dividend) (divisor)

Rule a. This can be set down as shown:

$$\begin{array}{r} \text{quotient} \\ \text{divisor} \overline{)\ \text{dividend}} \end{array}$$

Rule b. If the divisor is not, as it is here, a whole number then it must be made so by moving the decimal place to the right. The decimal place of the dividend would then also have to be adjusted by the same number of places in order to keep its value in proportion to the divisor. As the following examples show:

$62{\cdot}5 \div 2{\cdot}5 = 625 \div 25$ (Divisor)

25)625·

(Both decimal places have been moved one place to the right to make divisor a whole number and keep dividend in proportion).

$62{\cdot}5 \div 2{\cdot}544 = 62500 \div 2554$ (Divisor)

2544·)62500·

(Here the decimal has had to be moved three places. Noughts have been added to the dividend to fill the spaces that occurred.)

$625 \div 2{\cdot}5 = 6250 \div 25$ (Divisor)

25)6250·

(Note that since there is no decimal point in the dividend, only that of the divisor had to be moved one place. A nought has been added to the dividend in order to keep it in proportionate value to the adjusted divisor.)

144)62·5

(i) 0·
144)62·5

(ii) 0·434
144)62·5
(iii) 576
(iv) (v) 490

(i) 144 into 62 will not go. Put a nought above the 2 and a decimal point above that in the dividend.

(ii) 144 into 625 goes 4 times. Put 4 after the decimal point in the quotient

(iii) $4 \times 144 = 576$. Put 576 below 62·5

(iv) subtract 576 from 62·5. This leaves 49·

(v) 144 into 49 will not go. Add a nought to the 49 (this is assumed to have been brought down from the dividend. Noughts added after the decimal point cannot alter the value of the whole.)

(vi) (vii)	432	(vi) 144 into 490 goes 3 times. Put the figure 3 after the 4 in the quotient. (vii) $3 \times 144 = 432$. Put this below the 490 and subtract.
(viii)	580	(viii) the answer to this is 58, into which 144 will not go. Add a nought to the end (as in stage (iii)).
(ix) (x)	576	(ix) 144 into 580 goes 4 times. Put the figure 4 after the 3 in the quotient.
	4	(x) $4 \times 144 = 576$, put this below the 580 and subtract (as in stages (iv) and (vii). Only 4 will be left over.

If a nought were added to the 4, 144 would still not go into it. If you added another nought you would be working to five places of decimals

—that is, to the nearest $\frac{1}{100\,000}$,

and this is quite unnecessary. The working can therefore be stopped at this point, with the quotient as 0·434. The answer to this problem, to three decimal places, is therefore *0·434 lb/in²*

EXAMPLE 7A

The pressure on the base of a water column 1 metre square and 1 metre high is 1000 kilograms. What pressure acts on each square millimetre of the base?

Pressure in kilograms/millimetre

$$= \frac{1000 \text{ kilograms/m}^2}{1\,000\,000 \text{ mm}^2\text{/m}^2}$$

$= 0{\cdot}001$ *kilogram force/mm²*

But 1 kgf = 9·8 newtons

Then 0·001 kgf/mm² = *0·0098N*/mm²

Answers to a given number of decimal places

As has been said, the answer to Example 7 was given to three places of decimals. 0·434 is the same as $\frac{434}{1000}$, and it is not always necessary to work to the nearest $\frac{1}{1000}$ of an inch. It is often enough to work to two places of decimals ($\frac{1}{100}$ in), or even to one decimal place ($\frac{1}{10}$). The best method is to work to one place beyond the number you need, and then to reduce the answer by one place by 'correcting' it. For example:

$$0{\cdot}69 = \frac{69}{100}$$

and this is very nearly equal to 0·70, or $\frac{7}{10}$. Clearly, if you were giving the answer to one decimal place you could correct 0·69 to 0·7 without changing the answer much. To take another example, 0·62 has two places of decimals and represents $\frac{62}{100}$. This is nearer to $\frac{60}{100}$ or $\frac{6}{10}$ than $\frac{70}{100}$ or $\frac{7}{10}$.

Corrected to one place of decimals, the answer to this would therefore be 0·6.

Rule

Count the number of decimal places you need to the right from the decimal point. If the next figure is less than 5, discard it with all other following figures. If it is 5 or over, add 1 to the last figure of the answer and discard the rest. For example:

0·76236 is a number to five places of decimals
corrected to four places of decimals it becomes 0·7624
corrected to three places of decimals it becomes 0·762
corrected to two places of decimals it becomes 0·76
corrected to one place of decimals it becomes 0·8

Conversion to and from decimal fractions

It is useful to know how to convert vulgar fractions to decimal fractions and vice-versa: rules and examples for the calculation are given here.

To convert vulgar fractions into decimal fractions, divide the denominator into the numerator.

EXAMPLE 8A

Convert the vulgar fraction $\frac{1}{2}$ into a decimal fraction.

Solution $\frac{\text{numerator}}{\text{denominator}} = \frac{1}{2} = 2\overline{)\,1{\cdot}0}$ giving 0·5 (see Example 7 for method of division)

```
    0·5
2 ) 1·0
    10
    --
    00
```

Answer: $\frac{1}{2} = 0{\cdot}5$.

EXAMPLE 8B

Convert $\frac{3}{4}$ to a decimal fraction

Solution

$$\begin{array}{r} 0{\cdot}75 \\ 4\overline{)\ 3{\cdot}0} \\ \underline{28} \\ 20 \\ \underline{20} \\ 00 \end{array}$$

Answer: $\frac{3}{4} = 0{\cdot}75$

EXAMPLE 8C

Express $32\frac{1}{4}$ as a decimal.
This is a mixed number, made up of 32 units and 1 quarter of a unit. It can be converted to a decimal form as follows:

Solution 32 +

$$\begin{array}{r} 0{\cdot}25 \\ 4\overline{)\ 1{\cdot}0} \\ \underline{8} \\ 20 \\ \underline{20} \\ 00 \end{array}$$

Answer: $32\frac{1}{4}$ as a decimal $= 32+0{\cdot}25 = 32{\cdot}25$.

To convert decimal fractions into vulgar fractions, make the decimal fraction the numerator, and put a figure 1 under the decimal point, to be the denominator. Add as many noughts to the denominator as there are figures after the decimal place in the numerator.

EXAMPLE 9A

Change 0·75 into a vulgar fraction

Solution

$$\frac{{\cdot}75}{100}$$

Strike out the decimal point

$$\frac{75}{100}$$

Simplify the fraction by cancellation:

Answer: $\frac{3}{4}$

EXAMPLE 9B

Convert 0·02 to a vulgar fraction.

Solution $$\frac{\cdot 02}{100} = \frac{2}{100}$$

Answer: $$\frac{1}{50}$$

EXAMPLE 9C

Change 2·25 to a mixed number.
Solution: Here the number contains 2 whole units plus 0·25 of a unit.

$$2{\cdot}25 = 2 + \frac{25}{100}$$

Answer: $2\frac{1}{4}$

The S.I. metric system

The basic units of measurement are the metre (m) for length, the kilogram (kg) for mass, and the litre (l) for capacity.

Using the following prefixes, multiples and sub-multiples of the above units can be obtained.

Prefix	*Symbol*	
deci	(d)	meaning $\frac{1}{10}$ or 0·1 of a basic unit
centi	(c)	meaning $\frac{1}{100}$ or 0·01 of a basic unit
milli	(m)	meaning $\frac{1}{1000}$ or 0·001 of a basic unit
decca	(da)	10 × one basic unit
hecto	(h)	100 × one basic unit

Prefix	*Symbol*	
kilo	(k)	meaning 1000 × one basic unit
mega	(M)	1 000 000 × one basic unit

For example, 1 m may be subdivided into 100 equal parts. Each part would be 0·01 m or 1 cm (centimetre).

1 cm may be subdivided into 10 equal parts. Each part would then be $\frac{1}{10}$th of $\frac{1}{100}$th of 1 m = $\frac{1}{1000}$ m or 0·001 m and, from the table above, the symbol for 0·001 of a unit is milli (m) hence $\frac{1}{1000}$th of a metre would be called 1 mm (millimetre)

1 kilogram (kg) = 1000 grams.
1 kilometre (km) = 1000 metres

Using the conversion factor of 1 metre = 3·28 ft then 1000 m × 3·28 ft = 3280 ft.

Now, since 1 mile (imperial) = 5280 ft
1 kilometre = 3280/5280 = 0·62 mile
$\simeq \frac{3}{5}$ mile

or 1 mile = 1·6 kilometres

Then miles/hour × 1·6 = kilometres/hour.

For example, 30 mph = 30 × 1·6 km/h = 48 km/h and new motor vehicle speedometers are being calibrated in mph and km/h.

1 litre = 1·76 pints and eventually it will be necessary to order one's favourite beverage in $\frac{1}{2}$ litres rather than pints.

4·5 litres = 1 gallon and petrol and the like will be ordered by the litre. Instead of asking for 2 gallons we shall be ordering 4·5 litres/gal × 2 = 9 litres.

1 kilogram = 2·2 pounds and so our customary weights in pounds will now be roughly half those values in kilograms. To be exact, 1 lb = 0·45 kg.

The following conversion table, used in conjunction with that given in the Chapter entitled 'The International Metric System S.I.' may prove helpful until one gets more used to working in metric units.

CONVERSIONS

British (*imperial*) *to Metric*		*Metric to British* (*imperial*)	
Lengths			
1 inch	= 25·4 mm	1 millimetre	= 0·039 in
1 foot	= 304·8 mm	1 centimetre	= 0·39 in
1 yard	= 0·91 m	1 metre	= 39·37 in = 1·094 yd
1 mile	= 1·62 km	1 kilometre	= 0·62 mile
Mass			
1 ounce	= 28·35 g = 0·02835 kg	1 gram	= 0·035 oz
1 pound	= 0·454 kg	1 kilogram	= 2·2 lb
1 cwt	= 50·8 kg		
1 ton	= 1·016 tonnes	1 tonne	= 0·984 ton
Liquid measure			
1 pint	= 0·568 litres	1 litre	= 1·76 pints
1 gallon	= 4·546 litres		

SOME OTHER USEFUL CONVERSIONS

square inches × 645 = square millimetres (mm^2)
square feet × 0·093 = square metres (m^2)
square yards × 0·836 = square metres (m^2)
cubic feet × 28·3 = litres (l) (1 m^3 = 1000 litres)
lbf/in^2 × 0·7 = head of water in metres
lb/ft^3 × 16·02 = kg/m^3 (kilograms/cubic metre)
lbf/in^2 × 6·894 = kilonewtons/metre² (kN/m^2)
Atmospheric pressure
at 14·7 lb/in^2 = 101·3 kN/m^2
14·5 lb/in^2 = 1 000 000 newtons/m^2 = 100 kN/m^2 = 1 bar
1 in W.G. ≏ 2·5 millibar (m bar or mb)
1 metre head of water ≏ 0·1 bar
10 metre head of water ≏ 1 bar
pressure in bars × 10 ≏ metres head of water.

EXAMPLE 10

Express $\frac{1}{2}$ in $\times$ 7 lb lead pipe in metric terms.

Answer: $\frac{1}{2}$ in $\times$ 7 lb indicates pipe of $\frac{1}{2}$ in nominal bore of such wall thickness that 1 yd run would weigh 7 lb.
7 lb/yd $\times$ 0·454 kg/lb = 3·178 kg/yd
1 metre = 1·094 yard

So 3·178 kg/yd $\times$ 1·094 yd/m = 3·46 kg/m run
and $\frac{1}{2}$ in $\times$ 25·4 mm/in $\simeq$ 13 mm
then $\frac{1}{2}$ in $\times$ 7 lbs/yd = *13 mm $\times$ 3·46 kg/m run*

EXAMPLE 11

Express 130 ft head of water in:

(a) lbf/in^2
(b) kN/m^2
(c) bars (1 bar = 100 000 N/m^2 = 100 kN/m^2)

Answers: (a) 130 ft head $\div$ 2·31 = *56 lbf/in^2*
(b) 130 ft $\times$ 0·3 = 39 m head
and 39 m head $\times$ 9·8 kN/m^2/m head = *382 kN/m^2*
(c) 1 bar = 100 kN/m^2
so 382 kN/m^2 = *3·82 bar*

EXAMPLE 12

Express 130 lbf/in^2 in
(a) ft. head of water } head
(b) m. head of water } head
(c) kN/m^2 } pressure
(d) bars } pressure

Answers: (a) 130 lbf/in^2 $\times$ 2·31 ft/lbf/in^2 = *300 ft head*
(b) 300 ft $\times$ 0·3 m/ft = *90 m head*
(c) 90 m head $\times$ 9·8 kN/m^2/m head = *882 kN/m^2*
(d) 1 bar = 100 kN/m^2
hence 130 lbf/in^2 = 882 kN/m^2 = *8·82 bar*
$\simeq$ 9 bar

EXAMPLE 13

Atmospheric pressure is commonly stated to equal 14·7 lbf/in². What height of water column does this represent (a) in imperial terms and (b) in metric terms?

Answer: (a) lbf/in² × 2·31 = ft head of water.
or lbf/in² ÷ 0·434 = ft head of water
then 14·7 lbf/in² × 2·31 = *34 ft head of water*

Answer: (b) lbf/in² × 6·894 = kN/m²
Then 14·7 lbf/in² × 6·894 = 101·3 kN/m²

$$\text{and metres head} = \frac{\textit{Intensity of press } (\text{kN/m}^2)}{9{\cdot}8}$$

$$= \frac{101{\cdot}3}{9{\cdot}8} = \textit{10·3 m head of water}$$

EXAMPLE 14

Find height of water column producing intensity of pressure equal to atmosphric pressure at 14·5 lbf/in².

Answer: 14·5 lbf/in² × 6·894 = *100 kN/m²*

$$\text{and head water} = \frac{\text{Intensity of pressure (kN/m}^2)}{9{\cdot}8}$$

$$\text{in m} = \frac{100 \text{ kN/m}^2}{9{\cdot}8}$$

$$\simeq \textit{10 metres head of water}$$

Note: 100 000 newtons/metre² = 100 kN/m²
= *1 bar*

EXAMPLE 15

The latent head of solidification (turning to ice) of water is 144 Btu/lb. That is, if 1 lb of water was cooled to 32°F there would be no change in temperature until the 1 lb of water had become 1 lb of ice. Conversely, 1 lb of ice at 32°F would require a heat energy input of 144 Btu before it changed to water during which time there would be no change in temperature. The heat taken from water or added to ice in change of state is not measurable as temperature difference on a thermometer, hence it is said to be 'hidden' or latent heat.

What would be the latent heat of solidification of water in metric terms?

Answer: 1 kg = 2·2 lb

then 144 Btu/lb × 2·2 lb/kg = 316·8 Btu/kg

$$\text{or} \quad \frac{144\ \text{Btu}}{\text{lb}} \times \frac{2{\cdot}2\ \text{lb}}{\text{kg}}$$

the lb symbols cancel out leaving

$$\frac{144\ \text{Btu} \times 2{\cdot}2}{\text{kg}} = 316{\cdot}8\ \text{Btu/kg}$$

now 1 Btu = 1055 joules

so 316·8 Btu/kg × 1055 joules/Btu

$$\text{or} \quad \frac{316{\cdot}8\ \text{Btu}}{\text{kg}} \times \frac{1055\ \text{joules}}{\text{Btu}}$$

again the like Btu symbols cancel out leaving

$$\frac{316{\cdot}8 \times 1055\ \text{joules}}{\text{kg}} = \mathit{334{\cdot}2KJ/kg}$$

and the latent heat of solidfication of water at 0°C or the liquifaction of ice at 0°C is 334 kilojoules/kilogram.

Using a conversion factor applicable to all latent heats, that is 2324 × Btu/lb = J/kg

$$\begin{aligned} 144\ \text{Btu/lb} \times 2324\text{J/kg} &= 334944\text{J/kg} \\ &= 334{\cdot}9\ \text{kJ/kg} \\ &\text{say } \mathit{335\ kJ/kg} \end{aligned}$$

EXAMPLE 16

Using the conversion factor Btu/lb × 2326J/kg find the metric equivalent of the latent heat of steam at 212°F which is known to be 970 Btu/lb at standard atmospheric pressure.

Answer: 970 Btu/lb × 2326 J/kg = 2256220J/kg
= *2256kJ/kg*

Powers of numbers

In the sections on multiplying and dividing decimals (page 329), the 'powers of ten' were referred to. These are:

$10^1 = 10 \times 1$ (the first power of ten)
$10^2 = 10 \times 10$ (the second power of ten)
$10^3 = 10 \times 10 \times 10$ (the third power of ten)
$10^4 = 10 \times 10 \times 10 \times 10$ (the fourth power of ten)
and so on.

The small figure at the top right of the 10's is called an *index* (plural *indices*).

You can see that the index shows how many times the number (any number) has to be multiplied by itself. This multiplication is called *raising the power* of the number.

EXAMPLE 17A

$4^2 = 4 \times 4 = 16$

EXAMPLE 17B

$3^3 = 3 \times 3 \times 3 = 27$

EXAMPLE 17C

$2^5 = 2 \times 2 \times 2 \times 2 \times 2 = 32$

EXAMPLE 17D

$0{\cdot}2^5 = 0{\cdot}2 \times 0{\cdot}2 \times 0{\cdot}2 \times 0{\cdot}2 \times 0{\cdot}2 = 0{\cdot}00032$

Laws of indices

EXAMPLE 18

$2^3 \times 2^5$ could be worked out as follows:

$$
\begin{aligned}
&(2\times2\times2)\times(2\times2\times2\times2\times2)\\
&= 8\times32\\
&= 256
\end{aligned}
$$

You can get the same result by adding the indices together to make a single index and, then raising the number 2 to the power of this combined index:

$$2^3\times2^5 = 2^{3+5} = 2^8 = 256$$

EXAMPLE 19A

$$2^3\times2^2 = 2^5$$

EXAMPLE 19B

$$4^2\times4^2\times4^3 = 4^7$$

EXAMPLE 19C

$$5\times5^2 = 5^3$$

Division is the opposite of multiplication and subtraction the opposite of addition. If numbers can be multiplied by their indices being added together, they can be divided by their indices being subtracted from one another.

EXAMPLE 20

$$
\begin{aligned}
2^5\div2^2 &= (2\times2\times2\times2\times2)\div(2\times2)\\
&= 32\div4\\
&= 8
\end{aligned}
$$

Alternatively by subtraction of indices.

$$2^5\div2^3 = 2^{5-2} = 2^3 = 8$$

EXAMPLE 21

shows how indices are used in problems that involve both multiplication and division:

$$\frac{2^3 \times 2^2 \times 2^2}{2^5}$$

By long method: $= \dfrac{8 \times 4 \times 4}{32}$

$= \dfrac{128}{32}$

Answer: $= 4$.

By adding and subtracting indices:

$\dfrac{2^3 \times 2^2 \times 2^2}{2^5} = 2^{7-5}$

$= 2^2$

Answer: $= 4$

Examples 18 and 19 illustrate the first Law of Indices:

add indices to multiply, while

Examples 20 and 21 illustrate a second law:

subtract indices to divide. There is one other important Law of Indices—the Law of Involution. 'Involution' simply means raising a number to a given power (see Examples 17a, b, and c). With this law, the indices are multiplied together.

EXAMPLE 22

$(2^5)^2$

$(2^5)^2 = 2^{5 \times 2}$ or 2^{10}

$= 1024$

The laws of indices can be expressed as follows:

(1) $a^m \times a^n = a^{m+n}$ (the multiplication law)
(2) $a^m \div a^n = a^{m-n}$ (the division law)
(3) $(a^m)^n = a^{m \times n}$ (the involution law)

where the letter a represents the same number each time, and m is of greater value than n (in Laws 1 and 2).

You have seen how much easier and quicker it is to multiply and divide *multiples of the same number* by these simple laws. It is even more helpful to have a table of values for the various powers of the number you are dealing with. Suppose you take the number 2 as a base, and make a table of its powers. It would look something like this.

Base	*Power of base*	*Number*
2	1	2
2	2	4
2	3	8
2	4	16
2	5	32
2	6	64
2	7	128
2	8	256
2	9	512
2	10	1024
2	11	2048

You can use this table to solve Examples 18 to 21, and any other similar problems in which the base number is 2. For example, you will remember that Example 18 was

$$2^3 \times 2^5$$

You can refer back for the working and the answer, which is 256. If you use the table you can get the answer without any working at all. This is how you do it:

1 Add the indices (in this case $2^{3+5} = 2^8$).
2 Look down column 2 of the table until you find the power 8
3 Run your finger across to the third column to find the equivalent number there. You will find that it is the answer 256.
To take one last example, try solving Example 21 by means of the table.

$$\frac{2^3 \times 2^2 \times 2^2}{2^5} = 2^{7-5}$$

$$= 2^2$$

$$= 4$$

From the table, the answer is also 4.

Logarithms

Logarithms use the same simple principle of the Laws of Indices, and are no more difficult to handle than the table you have just used. A logarithm can be simply defined as *the power to which a given base has to be raised in order to produce a given number.*

If you look at the table of powers above you will see that in order to produce each number in the third column, the base (2) must be raised to the power shown in the second column. For example, to make 256, the base 2 must be raised to its 8th power. Therefore 8 *is the logarithm of 256 to the base* 2. Standard, printed log tables are based on 10 instead of 2—but this is the only real difference. They are, however, just a little more complicated, so that you should spend a little time getting to know your way about a log table. This will help you to do many kinds of boring calculations more easily and quickly: and in fact certain plumbing calculations can only be done with the help of log tables.

Before starting to use a standard log table, look at the table below, which is a simplified version to the base 10.

Base	*Logarithm or power of base*	*Number*
10	0	0
10	1	10
10	2	100
10	3	1000
10	4	10 000
10	5	100 000

Notice that the logarithm is always one less than the number of figures in the given number: for example, the given number 1000 has *four* figures and its logarithm is 3. The given number between 1 and 9 has only one figure, and its log is therefore 0.

EXAMPLE 23

Use the table to evaluate $\dfrac{100 \times 10\,000}{1000}$

Rule: To multiply *add* the logs: to divide them *subtract* the log of the diviser from that of the dividend.

From the simplified table, the log of 100 is 2
the log of 10 000 is 4
the log of 1000 is 3

Then,

	Number	*Log*	
	100	2	
+	10000	4	(add to multiply)
		6	
−	1000	3	(subtract to divide)
		3	

The third power of 10 (from the table) is 1000. The process of converting a logarithm into its equivalent number is of course the exact opposite of finding a logarithm, and it is called 'anti-logging'. Books of log. tables therefore also have separate anti-log tables included in them.

The numbers used in Example 23 were of course all exact powers of ten. What however if the number whose log is required were not an exact power of ten—for example 176? This number is greater than 100 (so that its power must be greater than 2), but smaller than 1000 (so that its power must be less than 3). In other words, its power is 2 plus a fraction. This will be expressed as a decimal. The whole number (2) is called the *characteristic* of the logarithm and the part after the decimal point is called the mantissa. The characteristic of 176 is therefore 2: that of 3206 would be 3, that of 10250 would be 4, and so on.

Rule: The characteristic of any given whole number (the smallest whole number is 1) will be a positive (+) figure, and will be *one less than the number of figures to the left of the decimal point in the given number.*

For example:

Given number	*Number of figures to left of decimal point*	*Characteristic*
1006	4	3
100·6	3	2
10·06	2	1
1·006	1	0

Rule: The characteristic of any given fraction (that is, a number less than 1) will be *negative* (—) and will be *one greater than the number of noughts immediately following the decimal point*. For example:

Given fraction	*Number of noughts*	*Characteristic*
0·0001	3	$\bar{4}$
0·001	2	$\bar{3}$
0·01	1	$\bar{2}$
0·1	0	$\bar{1}$

Note that (1) the minus sign or *bar* shows the characteristic to be that of a fraction, and is called 'bar 4, bar 3', and so on.
(2) One greater than 0 is 1 (see characteristic for the fraction 0·1).

The mantissae are always positive, and are set down in the table so that you can find the exact logarithm of numbers between the powers of ten.

How to use logarithmic tables

You can buy log tables from any large stationer or bookshop. The paperbacks, which list logs and anti-logs only, cost about 4*d*.* Linen-covered versions which include other tables, which you will need later but not at this stage, cost about 1*s*. 10*d*.†

Brief extracts from a typical log table are given below but, of course, you will want to get a complete set as soon as possible.

Suppose you want to find the logarithm of 20. Look down the left-hand column of the table until you find figure 20.
You will find the mantissa in the next column to the right—3010. You will see that the number has two figures, so that its characteristic will be 1. The logarithm would then be written 1·3010.

*Old pence ÷ 2·4 = new pence

†e.g. 1*s*. 10*d*. = 20 old pence and $\frac{20}{2\cdot 4} \simeq 9$ new pence. Nearest new coin 10p.

As you will see, the log tables consist of a series of columns. The one on the extreme left contains the first two *significant* figures of the number whose log you are finding. 'Significant' figures are simply those that matter. In this connection it is

	0	1	2	3	4	5	6	7	8	9	1 2 3	4 5 6	7 8 9
18	·2553	·2577	·2601	·2625	·2648						2 5 7	9 12 14	17 19 21
						·2672	·2695	·2718	·2742	·2765	2 4 7	9 11 14	16 18 21
19	·2788	·2810	·2833	·2856	·2878						2 4 7	9 11 13	16 18 20
						·2900	·2923	·2945	·2967	·2898	2 4 6	8 11 13	15 17 19
20	·3010	·3032	·3054	·3075	·3096	·3118	·3139	·3160	·3181	·3201	2 4 6	8 11 13	15 17 19
21	·3222	·3243	·3263	·3284	·3304	·3345	·3345	·3365	·3385	·3404	2 4 6	8 10 12	14 16 18

useful to consider the value of 0—nought. If this is placed *in front* of a number it means nothing—for example, 035 remains 35 in value. The '0' would not be a significant figure. If it is placed between the figures of a number, or after them, it does mean something: For example, 305, 3005, 3500, and 350 all have different values to 35. If, however, the nought were placed after a decimal point, it would again mean nothing: for example, 35·0 and 35·00 and 35 are all exactly the same in value.

The next nine columns to the right in the log tables are headed 0 to 9, and the following nine columns are headed 1 to 9. These exist so that you can find the logs of three and four figure numbers. You have seen that to find the log of a number with two significant figures you must look in the first column. To find the log of a number with three significant figures, however, you must find the first two figures in the left-hand column, and then the third figure in the next series of columns. In the same way, to find the log of a number with four significant figures, you would get the fourth figure from the last series of columns on the page. For example, suppose you have to find the log of 206. Since the number has three figures you know that the characteristic of the log will be 2. Then, find 20 in the first column of your table. Use a ruler to cast along the same line to find the equivalent figure in column 6 of the next series of columns. You will thus that the mantaissa for 206 is 3139, so that the log of 206 will be 2·3139.

To take another example, suppose you had to find the logarithm of 2064.

1 You know that the characteristic will be 3 since the number consists of 4 figures.
2 Find the first two figures—20—in the left-hand column.
3 Cast along until you reach the mantissa under column 6 of the second series. (In this case it is 3139).
4 The last series of column is sometimes called the *average difference* series. Cast along again until you reach the figure in column 4 of this series (it is 8).
5 Add this figure to the mantissa you found in column 6 of the last series:

$$3139+8 = 3147$$

This is the mantissa for 2064.
6 Since the characteristic is 3, the logarithm will be 3·3147.

Only one point remains to be made clear. This concerns the double row of *differences* for numbers beginning with the figures 1–19.

	0	1	2	3	4	5	6	7	8	9	1	2	3	4	5	6	7	8	9
10	·0000	·0043	·0086	·0128	·0170						5	9	13	17	21	26	30	34	38
						·0212	·0253	·0204	·0334	·0374	4	8	12	16	20	24	28	32	36
11	·0414	·0453	·0492	·0531	·0569						4	8	12	16	20	23	27	31	35
						·0607	·0645	·0682	·0719	·0755	4	7	11	15	18	22	26	29	33
12	·0792	·0828	·0864	·0899	·0934						3	7	11	14	18	21	25	28	32
						·0969	·1004	·1038	·1072	·1106	3	7	10	14	17	20	24	27	31
13	·1139	·1173	·1206	·1239	·1271						3	6	10	13	16	19	23	26	29
						·1303	·1335	·1367	·1399	·1430	3	7	10	13	16	19	22	25	29

You will see that for these figures in the left-hand column, the mantissae are staggered in two lines beyond column 4 of the next series, and that there are two rows of differences—one for each row of mantissae. These are used to give even more accurate logarithms. When you are finding the figure in the appropriate column of the last series, you simply take the one that lines up with the mantissa you have already found in the second series of columns. For example, find the log of 1242. The characteristic will be 3. The mantissa under column 4 of the second series of columns is 0934. This is on the upper of the two staggered lines, so that you will take the upper figure in column 2 of the differences columns. This is 7, so that the logarithm of 1242 will be:

$$0934 + 7 = 0941$$
$$= 3{\cdot}0941$$

With the logarithm of 1252, however, you would take the bottom difference in the appropriate column because this lines up with the mantissa of the third figure. The log of 1252 is 3·0976.

It is equally simple to find the log of 1, 10, or any power of 10. From the table the mantissa of 10 is 0000. The characteristic will be 1 since the number 10 consists of two figures. This also follows from the fact that a log is the power to which the base must be raised to produce a given number, and its characteristic is a kind of index of the power. Therefore, since $10 \times 1 = 10$, or 10 to the power of 1, the log of 10 must be 1·0000.

The log of 1 = 0·0000
The log of 100 = 2·0000
The log of 1000 = 3·0000

and so on. (See also the simplified table on page 348).

Two useful points to remember about the mantissa are:

1 The mantissa part is *always* positive (+) regardless of whether the characteristic is positive (+) or negative (−).
2 For any sequence of figures in a given number, the mantissa found from log tables will be the same for the whole sequence no matter how the numerical value of the number may alter in the power of ten. For example:

Log 20 = 1·3010		The mantissa is the same in each case. The characteristic simply indicates the given number's power of 10.
Log 200 = 2·3010		
Log 0·2 = $\bar{1}$·3010		

and Log 2^2 (that is, 2 squared) $= 0{\cdot}3010 \times 2$

$$\text{Log } \sqrt{2000} = 3{\cdot}3010 \div 2 \text{ or } \frac{3{\cdot}3010}{2}$$

$$\text{Log } \sqrt[5]{20} = 1{\cdot}3010 \div 5 \text{ or } \frac{1{\cdot}3010}{5}$$

$$= 0{\cdot}2620$$

Anti-logarithm tables

Anti-log tables are used in very much the same way as log tables, to find the actual value of a number from its log. For example, suppose you had to multiply two simple numbers such as 48 and 20. You can see straight away that the answer to this will be 960. By logs, however, you would work this out as follows:

Number	*Log*
48	1·6812
20	1·3010 (add to multiply)
	2·9822

To evaluate the log 2·9822, you must use *anti-log tables*, of which the following is an extract:

	0	1	2	3	4	5	6	7	8	9	1	2	3	4	5	6	7	8	9
·95	8913	8933	8954	8974	8995	9016	9036	9057	9078	9099	4	4	6	8	10	12	15	17	19
·96	9120	9141	9162	9183	9204	9226	9247	9268	8290	9311	2	4	6	8	11	13	15	17	19
·97	9333	9354	9376	9397	9419	9441	9462	9484	9506	9528	2	4	7	9	11	13	15	17	20
·98	9550	9572	9594	9616	9638	9661	9683	9705	9727	9750	2	4	7	9	11	13	16	18	20
·99	9772	9795	9817	9840	9863	9886	9908	9931	9954	9977	2	5	7	9	11	14	16	18	20

1 In the left-hand column of a standard anti-log table you will find the first two figures of the mantissa—in this case, ·98.
2 Cast along this line until you reach the second column in the next series. You will find the figure 9594.
3 Cast further along the line until you reach column 2 in the last series of columns. Here you will find the figure 4.
4 Add 4 to the figure you found before

$$9594 + 4 = 9598$$

5 Since the characteristic of the log is 2, the answer will be 959·8. This is so close to the answer found by ordinary multiplication (960), that it can be regarded as correct.

Now that you know how logs work and can use the log tables, you can begin to profit from this valuable aid to calculations.

EXAMPLE 24

Do you remember how long and boring was the working out of Example 7? Now use logs to find the solution to the same problem which, briefly, was 62·5 ÷ 144.

Rule: To divide, subtract the log of the divisor (144) from that of the dividend (62·5). From the tables the log of 62·5 is 1·7959 and that of 144, 2·1584.

Number	*Log*
62·5	1·7959
−144	2·1584 (subtract to divide)
	$\bar{1}$·6375

As you can see, it is perfectly straightforward to subtract one mantissa from the other, but the characteristics are a different matter. How do you take 2 from +1? Suppose you have £1 to your credit in the Bank and then you withdraw £2. You now owe the Bank £1 and your 'credit' would be −£1. All that happens is that you are left with a minus quantity: −1. The logarithm of the answer to the problem is therefore minus 1·6375, or $\bar{1}$·6375 ('bar one' ·6375).

Such characteristics are called *negative characteristics*, and they always occur with logarithms of numbers less than 1. Remember that a *negative characteristic* is marked by a bar, and will always be *one greater than the number of noughts following the decimal point*. In other words the negative characteristic is an index to the number of noughts after the decimal point in a number less than 1. For example the characteristic of 0·6 would be $\bar{1}$; the characteristic of 0·06 would be $\bar{2}$; the characteristic of 0·006 would be $\bar{3}$; and so on.

Example 24 can therefore be finished, and the result will be $\bar{1}$·6375. From the anti-log tables, the final answer can be found to be 0·4340, or 0·434. Notice that the position of the decimal point has been decided by the characteristic, $\bar{1}$, which indicates that there are no noughts after the decimal point: if it had been $\bar{2}$, for example, the answer would have been 0·0434.

The nought on the end of the answer, on the other hand, is not a significant figure, and can therefore be left off.

This example may seem to you to take longer than ordinary long division: but you will find that with practice it becomes so easy and natural that it is much faster. Here is another example, and if you work it by both long division and logs, you will find that the log way is the quicker way.

EXAMPLE 25

Divide 48·02 by 323·2.

Number	*Log*	
48·02	1·6814	
323·2	2·5095	(subtract to divide)
	$\bar{1}$·1719	

Anti-log $\bar{1}$·1719 gives Answer: 0·1486.

If you do the sum by long division you will find that the answer is 0·1516. This variation occurs because log tables cannot be perfectly accurate, and this difference of $\frac{3}{1000}$ is so small that it is insignificant.

It is also useful to be able to deal with *powers of numbers* by logs. All you have to do is to multiply the log of the number by the power to which it has to be raised. For example 32^5 could be worked out as

$$32 \times 32 \times 32 \times 32 \times 32 = 3355439$$

However, the answer 33550000 can be far more quickly and easily obtained from log tables.

Number	*Log*
32	1·5051

multiply by 5 for 5th power = 7·5255. Anti-log = 33550000

EXAMPLE 26

The volume of a cylinder can be found using the formula:

$$\text{Vol.} = \pi r^2 h \text{ where } \pi = 3 \cdot 142$$
$$r = \text{radius of base of cyl.}$$
$$h = \text{height of cyl.}$$

or: $\text{Volume} = \dfrac{\pi D^2 h}{4}$ where D = diameter of base

h = height of cylinder

$$\frac{\pi}{4} = 0 \cdot 7854$$

Find the capacity in litres of a hot store cylinder which is 600 mm in diameter and 1650 mm in height

N.B. 1 litre = 1000 cm^3 600 mm = 0·6 m
1 litre = 0·001 m^3 1650 mm = 1·65 m
or 1000 litre = 1 m^3

Solution by logs

Number	*Log*
0·7854	$\bar{1}$·8951
+0·6 m	$\bar{1}$·7782
+0·6 m	$\bar{1}$·7782
+ 1·65 m	0·2175
	$\bar{1}$·6690

Anti-log $\bar{1}$·6690 gives answer 0·4667 m^3 capacity of vessel. Note that the negative (1) characteristic indicates no noughts following the decimal point.
Now, since there are 1000 litres in 1 m, the cylinder in question will hold 0·4667 $m^3 \times$ 1000 = 466·7 litres (say *467* litres).

Note: the inherent simplicity of metric calculation renders the use of logs unnecessary in an example like this, as the following illustrates:

converting linear dimensions to metres

$$0 \cdot 7854 \times 0 \cdot 6 \text{ m} \times 0 \cdot 6 \text{ m} \times 1 \cdot 65 \text{ m} \times 1000 \text{ litres/m}^3$$
$$= 467 \text{ litres}$$

EXAMPLE 27

The working capacity of a lift and raise pump can be found by means of the formula:

$$\text{litres pumped/h} = 0{\cdot}0008 \times D^2 \text{ (cm)} \times \text{length of stroke (cm)} \times \text{No. of strokes/min} \times 60 \text{ min/h}$$

Find how many litres are discharged per hour from a pump 100 mm diameter, with a 305 mm stroke working at 30 strokes per minute.

	Number	*Log*	
	10·0 cm	1·000	(multiply by 2 to raise
		2	to second power)
		2·0000	
+Constant	0·0008	$\bar{4}{\cdot}9031$	
+length of stroke	30·5 cm	1·4833	
+30 strokes/min	30	1·4771	
+60 min/h	60	1·7782	
		3·6417	

Anti-log = 4382 litres

Again, this could be quite simply worked without using logarithms which have been employed in this example just for practice in their use.

Yet another form of calculation you will meet which is much more easily and quickly done with logs is *evolution*: that is, square roots, cube roots, and so on. The process of extracting the root of a given number is the exact opposite of raising its power. For example, when you are extracting the root,

$$\sqrt[3]{27} = 3$$

whereas when you are raising the power,

$$3^3 = 27$$

Rule: To extract the root of a number by logs, divide the log of the number by the root. For example, to take a simple number, the square root of 16 is 4. This would be found by logs as follows:

Number	*Log*	
16	$\frac{1 \cdot 2041}{2}$	(divide by 2 to find square root)

Anti-log 0·6020 gives result 3999.
The characteristic shows that this should be 3·999, and this, by correction to one place of decimals, is 4.

Answer: 4.

EXAMPLE 28

Use logs to find the following values:

(a) square root of 144 ($\sqrt{144}$)
Log $144 \div 2 = 2 \cdot 1584 = 1 \cdot 0792$
Anti-log of 1·0792 gives 12·00

Answer: 12

(b) cube root of 27 ($\sqrt[3]{27}$)

$$\text{Log } 27 \div 3 = \frac{1 \cdot 4314}{3} = 0 \cdot 4771$$

Anti-log of 0·4771 gives 3·000

Answer: 3.

(c) fifth root of 32 ($\sqrt[5]{32}$)

$$\text{Log } 32 - 5 = \frac{1 \cdot 5051}{5} = 0 \cdot 3010$$

Anti-log of 0·3010 gives 2·000

Answer: $\sqrt[5]{32} = 2$

When finding roots of numbers having a log with a negative characteristic, the following rule has to be applied:

Make the negative characteristic exactly divisable by the root index.

A few examples should make this clear.

(d) Find the square root of 0·5

$$= \frac{\text{Log of } 0{\cdot}5}{2}$$

$$= \frac{\bar{1}{\cdot}6990}{2}$$

Now, the root index is $\sqrt[2]{0{\cdot}5}$ although the index 2 is usually omitted when square roots are involved and it would appear as $\sqrt{0{\cdot}5}$. However, the index 2 will not divide exactly into $\bar{1}$, but it would divide exactly into $\bar{2}$.

Now, $\bar{2}+1$ is the same as $\bar{1}$ and so, to make the negative part of $\bar{1}{\cdot}6990$ exactly divisible by 2 we write $\bar{2}+1{\cdot}6990$, which is the same as $\bar{1}{\cdot}6990$ and

$$2\,\underline{|\bar{2}+1{\cdot}6990}\qquad \bar{1}+0{\cdot}8495 = \bar{1}{\cdot}8495$$

Anti-log $\bar{1}{\cdot}8495 = 1{\cdot}7071$

Answer: $\sqrt{0{\cdot}5} = 0{\cdot}707$

(e) Find the cube root of 0·27, i.e. $\sqrt[3]{0{\cdot}27}$

Solution:

$$\frac{\text{Log } 0{\cdot}27}{3} = \frac{\bar{1}{\cdot}4314}{3}$$

$$= \frac{\bar{3}+2{\cdot}4314}{3}$$

$$= \bar{1}+0{\cdot}8105$$

Anti-log $= 0{\cdot}6465$

Answer: $\sqrt[3]{0{\cdot}27} = 0{\cdot}6465$

Proof: $0{\cdot}6465^3$ (0·6465 cubed) $= \text{Log } 0{\cdot}6465 \times 3$
$= \bar{1}{\cdot}8105 \times 3$
$= \bar{1}{\cdot}4315$
Anti-log $= 0{\cdot}27$

Note: $3 \times 8 = 24$. Put down the 4 and carry 2 positive units. Then $3 \times \bar{1} = \bar{3}$ to which we add $+2$ and $\bar{3}+2 = \bar{1}$.

(f) Find fifth root of 0·027

Solution: $\sqrt[5]{0\cdot027} = \frac{\log 0\cdot027}{5} = \frac{\bar{2}\cdot4314}{5}$

$= \frac{\bar{5}+3\cdot4314}{5}$

$= \bar{1}+0\cdot6863$

and antilog $\bar{1}\cdot6863 = 0\cdot4856$

$\therefore \sqrt[5]{0\cdot027} = 0\cdot4856$

The next example requires you to use all the rules for logs which you have dealt with so far.

EXAMPLE 29

Find the value of $\sqrt[5]{\frac{12^2 \times 32}{40}}$

Number	Log	
12	1·0792	(multiply by 2 to raise
	2	to second power)
	2·1584	
+ 32	1·5051	(add logs to multiply)
	3·6635	
− 40	1·6021	(subtract logs to divide)
	2·0614	

Divide by 5 for fifth root:

$$\frac{2\cdot0614}{5} = 0\cdot4123$$

Anti-log of 0·4123 gives 2·584

Answer: 2·584

EXAMPLE 30

How many litres of water would one metre run of 32 millimetre bore pipe hold?

Working this in imperial values from first principles and remembering:

Volume of pipe (ft³) = cross sectional area (ft²) × length (ft)

$$= \frac{\pi D^2}{4} \text{ (ft)} \times \text{L (ft)}$$

or $\quad 0{\cdot}7854 \times \text{D}^2 \text{ (ft)} \times \text{L (ft)}$

but 1 ft³ water = 6·25 gallons

so capacity in gallons = $0{\cdot}7854 \times D^2 \text{ (ft)} \times L \text{ (ft)} \times 6{\cdot}25$

If length $L = 1$ ft

then capacity in gallons =

$$0{\cdot}7854 \times \frac{1{\cdot}25 \text{ in}}{12 \text{ in/ft}} \times \frac{1{\cdot}25 \text{ in}}{12 \text{ in/ft}} \times 1 \text{ ft} \times 6{\cdot}25$$

now, since $\dfrac{0{\cdot}7845 \times 6{\cdot}25}{12 \times 12}$ will be constant in all such problems, this can be worked out to become a one figure constant.

and = 0·034

The above formula can then be simplified to:

gallons capacity = D^2 (inches) × L (feet run) × 0·034

and since 32 mm = $1\frac{1}{4}$ in

1 m = 3·28 ft

Answer: (imperial) gallons = 1·25 in × 1·25 in × 3·28 ft × 0·034

= *0·17 gallons*

Note: 1 litre = 1000 millilitres³ (ml)³ or 1000 cm³

= 1·76 pint

= 0·22 gallon

and 0·17 gallons ÷ 0·22 = *0·77 litres* say *0·8 litres*

Working the example in metric values we can still use a simplified formula but shall have to find a new constant. The reasoning is as follows:

1 litre = $\frac{1}{1000}$th m³ = 1000 ml³ = 1000 cm³

putting area and length in cm, we have:

$$\text{capacity in litres} = \frac{0{\cdot}7854 \times D^2(\text{cm}) \times \text{Length (cm)}}{1000 \text{ cm}^3/\text{litre}}$$

now $\frac{0{\cdot}7854}{1000}$ will be constant in all such problems.
It can be simplified to $= 0{\cdot}00078$
say $= 0{\cdot}0008$

and the simplified formula:

$$\text{litres capacity} = D^2 \text{ (cm)} \times L \text{ (cm)} \times 0{\cdot}0008$$

and the *Answer*

litres capacity $= 3{\cdot}2 \text{ cm} \times 3{\cdot}2 \text{ cm} \times 100 \text{ cm} \times 0{\cdot}0008$ of pipe
$=$ *0·8 litres* which agrees with metric conversion of the imperial worked example.

So remember this useful simplified formula:

litres capacity of pipe or cylindrical vessel $= D^2 \text{ (cm)} \times L \text{ (cm)} \times 0{\cdot}0008$

where D = diameter of pipe or cylinder.
L = length of pipe or cylinder.
$0{\cdot}0008$ = constant derived as shown above.

EXAMPLE 31

This simplified formula with its 0·0008 approximate constant (0·0007854 to be exact), could be used to solve *Example 26* as follows:

Capacity in litres $= D^2 \text{ (cm)} \times L \text{ (cm)} \times 0{\cdot}0008$
$= 60 \text{ cm} \times 60 \text{ cm} \times 165 \text{ cm} \times 0{\cdot}0008$
$=$ *475 litres* which, allowing for the approximation of the constant, as mentioned above, gives virtually the same answer.

It is interesting to note the water holding capacities of commonly used pipes as under:

Nominal bore inches	millimetres	Gallons per foot run	Litres per metre run
$\frac{1}{2}$	13	0·008	0·134
$\frac{3}{4}$	19	0·019	0·28
1	25	0·034	0·5
$1\frac{1}{4}$	32	0·053	0·8
$1\frac{1}{2}$	38	0·076	1·12
2	50	0·136	2·0
4	100	0·544	8·0

Note: 1 gallon = 4·5 litres
1 metre = 3·28 feet

so gallons/ft run × 4·5 litres/gallon × 3·28 ft/m = litres/metre run.

Thus: 32 mm pipe holds 0·053 gal/ft × 4·5 × 3·28 = 0·78
say *0·8 litres*

as found in previously worked examples.

Or, from the table above, 1 m run of 32 mm bore pipe holds 0·8 litres and

litres		metres		litres/metre run
(water content of pipeline)	=	(run of pipe)	×	(from table)

Now try the following examples for yourself:

Use logs to solve
(a) 32×20
(b) 183×4
(c) $3{\cdot}142 \times 6$
(d) $32 \div 20$
(e) $\sqrt{64}$
(f) $\dfrac{1{\cdot}462 \times 72{\cdot}4}{11{\cdot}37}$
(g) The log of 36 is 1·5563: without referring to the tables, write down the logs of:
36^3; 360; 3·6; $\sqrt{36}$; $\sqrt[5]{3600}$